Australian vegetation

EDITED BY R.H. GROVES

Australian vegetation

The editor is at the CSIRO Division of Plant Industry, Canberra, Australia

CAMBRIDGE UNIVERSITY PRESS

Cambridge

London New York New Rochelle

Melbourne Sydney

Published by the Press Syndicate of the University of Cambridge
The Pitt Building, Trumpington Street, Cambridge CB2 1RP
32 East 57th Street, New York, NY 10022, USA
296 Beaconsfield Parade, Middle Park, Melbourne 3206, Australia

First published 1981

Filmset in Hong Kong by
Asco Trade Typesetting Limited
Printed in Hong Kong by
South China Printing Co.

British Library Cataloguing in Publication Data

Australian vegetation.

1. Botany – Australia
I. Groves, R.H.
581.9′94   QK431   80-40421

ISBN 0 521 23436 0 hard covers
ISBN 0 521 29950 0 paperback (Australia only)

to JEAN

CONTENTS

CONTRIBUTORS

D.H. Ashton, Department of Botany, University of Melbourne, Parkville, Victoria, 3052

S.V. Briggs, NSW National Parks and Wildlife Service, P.O. Box 84, Lyneham, ACT, 2602

W.H. Burrows, Queensland Department of Primary Industries, P.O. Box 689, Rockhampton, Queensland, 4700

A.B. Costin, CRES, Australian National University, P.O. Box 4, Canberra, ACT, 2600

G.M. Cunningham, Soil Conservation Service of New South Wales, Goulburn, New South Wales, 2580

H. Doing, Department of Plant Ecology, Agricultural University, Transitorium, De Dreijen 11, Wageningen, The Netherlands.

A.M. Gill, CSIRO, Division of Plant Industry, P.O. Box 1600, Canberra City, ACT, 2601

A.N. Gillison, CSIRO, Division of Land Use Research, P.O. Box 1666, Canberra City, ACT, 2601

R.H. Groves, CSIRO, Division of Plant Industry, P.O. Box 1600, Canberra City, ACT, 2601

T.M. Howard, 5 Days Rd, Maroota South, New South Wales, 2756

R.W. Johnson, Queensland Department of Primary Industries, Queensland Herbarium, Meiers Rd, Indooroopilly, Queensland, 4068

J.H. Leigh, CSIRO, Division of Plant Industry, P.O. Box 1600, Canberra City, ACT, 2601

L.D. Love, 46 Vancouver Ave, Toongabbie, New South Wales, 2146

P.W. Michael, Department of Agronomy and Horticultural Science, University of Sydney, Sydney, New South Wales, 2006

R.F. Parsons, Department of Botany, La Trobe University, Bundoora, Victoria, 3083

G. Singh, Department of Biogeography and Geomorphology, Research School of Pacific Studies, Australian National University, P.O. Box 4, Canberra, ACT, 2600

R.L. Specht, Department of Botany, University of Queensland, St Lucia, Queensland, 4067

J.G. Tracey, CSIRO, Division of Plant Industry, Private Bag No. 3, P.O., Indooroopilly, Queensland, 4068
D. Walker, Department of Biogeography and Geomorphology, Research School of Pacific Studies, Australian National University, P.O. Box 4, Canberra, ACT, 2600
J. Walker, CSIRO, Division of Land Use Research, P.O. Box 1666, Canberra City, ACT, 2601
L.J. Webb, CSIRO, Division of Plant Industry, Private Bag No. 3, P.O., Indooroopilly, Queensland, 4068
O.B. Williams, CSIRO, Division of Land Use Research, P.O. Box 1666, Canberra City, ACT, 2601

PREFACE

Australian vegetation has interested both botanists and naturalists since Europeans first encountered Australia and its plant life. The ubiquity yet distinctiveness of species of *Eucalyptus*, the colour and variety of the vegetation around the early settlement at Sydney, the sometimes-bizarre plant forms of southwestern Australia, and the monotony of the inland mallee areas; all these features and others evoked different responses in different people. Banks and Solander in 1770 had their interest quickened and their botanical horizons widened. The very naming of Botany Bay bears witness to this. Charles Darwin in 1836, on the other hand, found the vegetation around Albany, Western Australia, dreary in the extreme.

The vegetation of Australia still evokes an ambivalent response from more general observers, both those from within the country and those from other continents. A *Nothofagus* or a *Eucalyptus regnans* forest experienced on a misty morning in spring in southeastern Australia can evoke awe in some, but the omnipresent eucalypts of coastal Australia dull the senses and seem monotonous to those with a more 'Europocentric' vision.

No matter what the general response to Australian vegetation, modern botanists are usually intrigued and stimulated by it and want to know more about it. Almost 200 years after European settlement we know quite a lot about Australian vegetation, as the following chapters and their reference lists show. The effects of fire on vegetation and individual species, the role of soil nutrients in delimiting vegetation distribution, the selective grazing of animals are just three of the subjects which Australian ecologists have contributed to international botanical science. Other subjects, such as pollination biology, are greatly in need of more research.

This book is aimed at stimulating the interest of graduate students and fellow botanists around the world, especially those fortunate enough to see for themselves the unique Australian vegetation and flora. To professional botanists resident in Australia both audiences can bring new insights and enthusiasms to the study of Australian vegetation. If reading the chapters that follow provides the catalyst for further research on the ecology of Australian vegetation then the efforts of the individual authors will have been worthwhile.

Some of the chapters include distribution maps of vegetation types. A recent map of all major vegetation types, using the most recent terminology, appears on pp. xiv–xv.

Readers unfamiliar with Australian geography can find place names in a recent atlas such as *The Reader's Digest Atlas of Australia*. Places mentioned in the text but not listed in that atlas have their latitude and longitude cited after their first mention.

The chapters were completed between September 1978 and August 1979; later research may not have been included.

ACKNOWLEDGMENTS

The production of a volume such as this one depends on the willing help of a number of people, some of whom I mention specifically:

Dr L.T. Evans, friend, colleague and fellow-enthusiast for research on the ecology of Australian plants and vegetation, for his counsel at all stages of this volume;

the authors of the chapters, for their forbearance with a sometimes-harassed editor;

Professor N.C.W. Beadle, for generously sharing the contents, prior to its publication, of his own excellent and longer book on the same subject;

Mrs K. Cawsey, for help with references;

Mr M. Gray and Mr A. Chapman, for help with checking my taxonomy in the preparation of the Index of Scientific Names;

Ms P. Kaye, for assistance with reading drafts and proofs;

Mr C.J. Totterdell, for expert help with choosing photographs;

and finally, Jean, Josephine and Jeremy, to whom I owe an unrepayable debt of time.

To all these people and to others un-named who have given me of their time and their counsel, I extend my thanks.

R.H. Groves
Canberra, 1979

1 Closed-forests
2 Open-forests
3 Woodlands
4 Open-woodlands
5 Scrubs and heaths
6 Shrublands
7 Open-shrublands
8 Herblands

▨ with hummock grass understorey

Characteristic genus or family

a *Astrebla*
c *Casuarina*
d *Dichanthium*
e *Eucalyptus*
k *Chenopodiaceae*
m *Melaleuca*
n *Nothofagus*
p *Callitris*
s *Cyperaceae*
w *Acacia*
x Mixed or other

The natural vegetation of Australia. A map prepared by Dr J.A. Carnahan and based on the one in *Australia: a Geography*, ed. D.N. Jeans, Sydney University Press, 1977.

INTRODUCTION

1

Phytogeography of the Australian floristic kingdom

H. DOING

The structure and composition of vegetation in any one place is determined by both present and past environments as well as by the flora available. Even at the level of major vegetation formations, there are essential differences in vegetation structure, depending on the phytogeographical situation; this is evident in Australia more than anywhere else. The 'hummock grassland', 'mallee scrubland', 'wet sclerophyll forest' and much of the 'dry sclerophyll forest' of Australia (Beadle & Costin, 1952) cannot be fitted into global classifications based on the vegetation of other continents (see e.g. Doing, 1971; Schmithüsen, 1976). Although such differences always have an eco-physiological background, the relative importance of morphological characters, e.g. scleromorphy, woodiness, deciduousness, succulence etc., is partly dependent on the taxa present in a certain area for historical reasons. For floristic composition of the vegetation, this relationship is self-evident. A phytogeographical survey is therefore essential for a proper understanding of Australian vegetation.

This contribution (including the map) is mainly a revision of a paper published earlier (Doing, 1970). It is based on available literature, personal communications and field reconnaissances from 1963–8 (while the author was a member of the CSIRO Division of Plant Industry at Canberra) and in 1971.

General characteristics of the area and definitions
Australia is, from a phytogeographical point of view, a fascinating continent. It is the most homogeneous and isolated of the plant kingdoms on the earth (Good, 1974). Its coastline is relatively short when compared with its surface. If we calculate for each continent the

relationship between the length of its actual coastline and the circumference of a circle having the same surface as that continent (Table 1.1), this becomes even clearer.

Australia is divided into two very unequal parts by the Great Divide, which is a watershed rather than a mountain range effective as a barrier for plant distribution. The highest ranges, from northern Queensland to southern Victoria, are in most parts not higher than 1500 to 1800 m, and are interrupted by several wide gaps, e.g. near Townsville in the tropical, near Rockhampton in the sub-tropical, and near Newcastle in the temperate zone. Apart from some small, isolated alpine areas in the southeast, large seasonal fluctuations in temperature in general, and low temperatures in particular, are absent. This is caused by the geographical situation of the continent, viz. its position between 10° and 44°S latitude. The only large and important barrier is the complex of dry lowlands north of the Great Australian Bight (Nullarbor Plain and Great Victoria Desert) which separates the great southern humid or sub-humid areas of southeastern and southwestern Australia. There is no barrier of comparable plant geographical importance in the tropical areas of northern Australia (Blake, 1953).

Geologically, Australia is old. There have been only comparatively minor morphological changes since the Tertiary and glaciation in the Pleistocene has covered relatively small surfaces. In most parts, erosion cycles have reached an advanced stage and there are relatively homogeneous soil landscapes (Stephens, 1961). Apart from the east coast region, climatic changes are rarely abrupt, although local differences in rainfall (the most important single factor nearly everywhere in Australia) may be considerable during shorter periods. Australia's geographic isolation

Table 1.1. *Length of coast, surface and relationship of coast length to surface (Rel. 1) and to circumference of a circle having the same surface (Rel. 2) for the five continents (without islands)*

	Length of coast ($\times 10^3$ km)	Surface ($\times 10^6$ km^2)	Rel. 1 (km : km^2)	Rel. 2
Africa	30	30	1 : 1000	1.5 : 1
America				
(north + central + south)	104	36	1 : 340	5.0 : 1
Asia	71	34	1 : 470	3.5 : 1
Australia	19	8	1 : 400	1.9 : 1
Europe	87	10	1 : 110	7.6 : 1

from other continents by the Indian, Southern and Pacific Oceans generally is considered as old, but that from Asia (New Guinea) as decreasing in relatively recent geological times (Beard, 1977; and the next chapter, Walker & Singh, 1981).

The only major latitudinal belts at sea-level in Australia are the tropical, the sub-tropical and the warm-temperate belt. Climatic definitions of latitudinal belts, as adopted in this paper, are presented in Table 1.2. In the southern highlands the lower alpine altitudinal belt (Lang, 1970) is well developed locally, but it is never extensive. Areas covered by permanent snow are absent. The mountains in the tropical and sub-tropical belts do not even reach a natural tree-line (cf. Webb, 1964). Transitional zones between latitudinal belts, especially in the east, are wide, and altitudinal belts up to timber-line are not clearly marked because nearly all trees are evergreen, and conifers in the relevant areas are inconspicuous or rarely form pure stands. Although arid and semi-arid areas (for definitions of the various degrees of aridity, see Table 1.3) are predominant, real deserts (e.g. rocks and sand dunes devoid of vegetation) are mostly recent and man-made.

If all these factors are considered, it is not surprising that Australia is homogeneous phytogeographically. This applies to genera in particular: a large proportion of the more important genera (Burbidge, 1963) has a wide distribution in either an east–west, a north–south or in both directions. Average species areas are comparatively small and the percentages of genera (*c.* 30%) and species (nearly 90%) endemic to

Table 1.2. *Thermal definitions of latitudinal zones of the earth*

Thermal zone	Mean annual temperature (°C)	Mean temperature warmest month (°C)	Mean temperature coldest month (°C)
Equatorial	>25	—	>25
Tropical	>22	—	18–25
Sub-tropical + mediterranean	16–22	—	10–18
Warm-temperate + sub-mediterranean	12–16	—	5–10
Cool-temperate	5–12	—	> −5
Boreal	<5	>10	< −5
Sub-antarctic	<5	5–10	> −5
Arctic + antarctic	—	0–10	< −5
Polar	—	<0	—

Table 1.3. *Degrees of humidity and aridity. After Holdridge*
et al., *1971*

Ratio of potential evapotranspiration/precipitation	
> 0.25	super-humid
0.25–0.50	per-humid
0.50–1.00	humid
1.00–2.00	sub-humid
2.00–4.00	sub-arid
4.00–8.00	arid
8.00–16.00	per-arid
16.00–32.00	super-arid
> 32.00	waterless (seared)

Australia are large. The total number of species, estimated at *c.* 15 000, is not outstandingly high, compared with other parts of the world. There are, however, some areas which may be called floristically rich even by the highest standards (Keast, 1959*a*; Good, 1974). There are not many, and no large endemic families, but there are a few more (notably the Epacridaceae, Goodeniaceae and Restionaceae) that have their major centre of development in Australia.

These facts were used (Keast, 1959*a*; Good, 1974) as a basis to support the theory that most Australian families have been present for a long time but may not have originated in Australia. This would mean that Australia's isolation and special ecological conditions have led to series of 'species explosions' by adaptive radiation (Carlquist, 1974), but not to a high degree of endemism of the flora at the higher taxonomic levels. In general, Australian phytogeographers tend to show a peculiar reluctance to admit a high degree of originality of the flora of their country (Gardner, 1944; Beadle, 1966). The problem of the origin of the Australian flora can only be treated effectively in a much wider context and will not be discussed here any further. Many facts from the literature on this subject, at least in the English language, were collected and discussed by Darlington (1968), but new evidence was presented more recently by Raven & Axelrod (1972) and Beard (1977). The experimental results of van Zanten (1978), related to dispersal of moss spores, suggest that explanations of distribution patterns may vary with the groups of organisms involved.

There is no doubt of the striking originality of the plant cover of Australia, i.e. of its vegetation formations. This is caused in the first place by the dominance of a limited group of taxa (Carnahan, 1976). A very large proportion of Australia is covered by plant communities in which one or more of the following taxa are dominant (i.e. within a vegetation layer covering a higher percentage of the total surface than all other taxa together): *Acacia* sect. Phyllodineae, *Astrebla, Atriplex, Banksia, Callitris, Danthonia, Dichanthium,* Epacridaceae, *Eremophila, Eucalyptus, Hakea, Heteropogon, Leptospermum, Maireana, Melaleuca, Myoporum, Nothofagus, Plectrachne, Stipa, Themeda* and *Triodia.* Apparently, the vegetation of Australia is more original than its flora, and more original in some areas than in others.

Existing literature and approach adopted in this chapter
In this chapter the botanical geography of Australia will be discussed mainly in the form of a proposal for a partition of the continent (including Tasmania and smaller continental islands) into regions, provinces etc., according to principles as applied by Meusel (1965, 1978). This includes adoption of the following hierarchy of phytogeographical 'territoria': Kingdom – Region – Province – Sector – District, with subdivisions according to convenience. No division below the level of province (occasionally subprovince) is attempted here. The number of regions (7) and provinces (25) is much larger than in previous publications on Australia. This is in agreement with the distinctive character of their floras (e.g. degree of endemism) and with trends in modern textbooks (e.g. Walter & Straka, 1970).

Although it is not my aim to give a review of all Australian phytogeographical literature, a short survey of the existing situation is needed as a basis for the subsequent discussion.

Botanical exploration of Australia started in 1699, when William Dampier collected 40 species on the coast of tropical Western Australia, more than 90 years after the discovery of Australia by Willem Jansz and nearly 90 years before the first European settlers arrived at Botany Bay on the east coast. The first major work of phytogeographical importance was that by Hooker (1860). It took more than two centuries after Dampier's arrival before botanical geography was firmly established in Australia with the publication of Diels' book (1906) on Western Australia. In many respects this book is still unique in Australian botanical literature and it is a gold-mine of interesting observations. Diels' map of botanical 'districts' in temperate Western Australia is still used, but his vegetation map of Australia has been much improved since. On the same subject, Gardner (1944) was one of the first authors to claim that the 'Australian' flora element is derived from the palaeo-tropic and the antarctic element, and Beadle (1966), in his most stimulating comparison of the Australian 'sclerophyll' and 'rainforest' floras, seems to go even further by suggesting that there is only one Australian flora with one common (tropical) origin.

Other contributions to the botanical geography and chorology of Australia are found in the new *Flora of New South Wales* (Contributions of the New South Wales National Herbarium, Sydney, 1961–present) and some studies on eucalypts, e.g. Blake (1953) and Jackson (1965), the latter being one of the very few Australian publications showing detailed species distribution maps. In addition, a large proportion of the Australian ecological maps (e.g. Beadle, 1948; Moore, 1953; Costin, 1954) are botanical–geographical rather than vegetation maps. The units shown on these maps – mostly called 'alliances' and mainly based on combinations of species dominant in the upper vegetation layer – are climax communities. Seral communities (e.g. pasture and weed communities) and finely patterned mosaics are not mapped. This is partly due to the scale of the maps and partly to the after-effects of Clementsian ecology. The species lists describing the mapping units are mixtures of native and introduced species (if present), and generally no plot analyses are published. For these reasons the mapping units are geographical areas, based on mixed criteria of vegetation and flora.

The vegetation units of most Australian surveys carried out by the former CSIRO Division of Land Research & Regional Survey (e.g. their Land Research Series Numbers 2, 4, 6, 7, 11), although their floristic analysis is fragmentary, are essentially similar in character. In this way botanical geographical information is available for various parts of Australia, although a synthesis of this material has not been attempted yet, except on very small-scale maps (Beard, 1969; Carnahan, 1976).

Until relatively recently, chorology – especially comparative chorology – was virtually non-existent in Australia. Much of the available information on the distribution of genera was compiled by Burbidge (1960). In agreement with the chorological approach (which was not stated explicitly) some general 'zones', 'focal areas', transitional areas and an 'overlap' were indicated in Burbidge's paper, but no botanical geographical division of Australia was attempted. The situation of the Australian 'region' in the southern hemisphere was further elucidated by Wace (1965) and by Darlington (1968), with an emphasis on southern affinities. Northern and eastern affinities were treated by van Steenis (1963) and van Steenis & van Balgooy (1966). The consequences of recent views on plate tectonics have been discussed by only a few authors, however (Raven & Axelrod, 1972; Beard, 1977).

It will be clear that the map of the Australian Kingdom and its Subkingdoms, Regions and Provinces presented in this chapter (Figure 1.1) is tentative and preliminary. As long as not much more information is available on the chorology of Australian plants and on the boundaries of vegetation formations and smaller vegetation units (on a floristic basis), including the remote parts of the continent, the present map may, however, be useful as a general guide to the main subdivisions of Australia as a phytogeographic Kingdom, especially for non-Australian botanists. It may also serve as a schematic basis for comparison with similar maps in other fields, e.g. zoogeographical maps (Keast, 1959*b*; Whitley, 1959).

As mentioned before, the map is based on mixed criteria of flora and vegetation. This procedure seems justified in general because phytogeographic areas may be based not only on coincidence of species areas in the first place, but also on the abundance of species, attributing a greater weight to common than to rare species and also taking into account the density of species within various parts of their areas. If one accepts this, it is obviously permissible also to use the distribution areas of plant communities and vegetation formations to delimit phytogeographic areas. This conforms to the principles set out by Meusel (1965). Formation boundaries are at least partially based on the boundaries of distribution or of occurrences of high densities of dominant species, and mostly the distribution of numerous other species also tends to coincide with these formation boundaries. This applies especially to a country like Australia, where in many areas the only maps available are formation maps. Thus, the map published here could not have been conceived without the use of Williams' (1955) map of 'vegetation regions' of Australia and Carnahan's (1976) map of 'natural vegetation'. For distribution data on Australian plants it is based mainly on recent Australian floras (including Blakely, 1955 and Beard, 1965) and on Burbidge (1960).

Description of the areas distinguished on the map
In the following paragraphs the units of the map will be discussed briefly and illustrated by some examples of the distribution of trees or shrubs, and also occasionally of herbaceous species. Woody species have been chosen, firstly because only for this group (mainly *Eucalyptus*) were a reasonable number of distribution maps and data available, and secondly, because general ecological experience in Australia as a rule confirms that there is good agreement between the distribution of dominant trees on the one hand, and that of herb and shrub species on the other, at least in forested areas.

Figure 1.1. Botanical geographical map of the Australian plant kingdom.

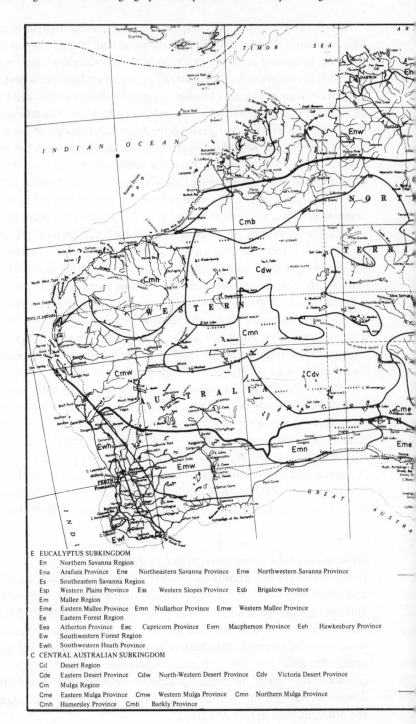

E EUCALYPTUS SUBKINGDOM
En   Northern Savanna Region
Ena   Arafura Province   Ene   Northeastern Savanna Province   Enw   Northwestern Savanna Province
Es   Southeastern Savanna Region
Esp   Western Plains Province   Ess   Western Slopes Province   Esb   Brigalow Province
Em   Mallee Region
Eme   Eastern Mallee Province   Emn   Nullarbor Province   Emw   Western Mallee Province
Ee   Eastern Forest Region
Eea   Atherton Province   Eec   Capricorn Province   Eem   Macpherson Province   Eeh   Hawkesbury Province
Ew   Southwestern Forest Region
Ewh   Southwestern Heath Province
C CENTRAL AUSTRALIAN SUBKINGDOM
Cd   Desert Region
Cde   Eastern Desert Province   Cdw   North-Western Desert Province   Cdv   Victoria Desert Province
Cm   Mulga Region
Cme   Eastern Mulga Province   Cmw   Western Mulga Province   Cmn   Northern Mulga Province
Cmh   Hamersley Province   Cmb   Barkly Province

LEGEND

Railways	—o——o—
Main Roads	------------
State Boundary	—·—·—·—
International Boundary	▬▬▬▬

If not only the areas of taxa, but also dominance and abundance in plant communities of endemic taxa are taken into consideration, the Australian Kingdom is perhaps the best defined of all plant kingdoms on the earth. From this point of view it does not include New Zealand or New Guinea, in contrast to the usual zoogeographical 'Realms' (see, e.g., *The Atlas of the Earth*, 1972). It is interesting that this agrees with Good's (1974) delimitation, for example, which is based on purely floristic criteria more than most other textbooks of phytogeography. The general character of the Australian flora has received due attention in the literature, and therefore is not repeated here in any detail.

The two subkingdoms

The initial division of the Australian Kingdom into a '*Eucalyptus*' (E) and a 'Central Australian' Subkingdom (C) is new in its present form, although the Central Australian Subkingdom is more or less identical with the 'Eremaean' region of most Australian authors. It is the only grouping of phytogeographic regions sufficiently reflecting the concentric arrangement of climatic regions (dominated by the aridity factor) and stressing the fact that the southwestern Australian flora is more closely related to the eastern than the central Australian flora, the southwestern flora being incompletely isolated from the eastern. Yearly rainfall at the limit between E and C is approximately 600 mm in the north and 250 mm in the south. The boundary between both subkingdoms is obscured by the phenomenon that certain members of the *Eucalyptus* flora (e.g. *E. camaldulensis* and *E. terminalis*) penetrate into the Central Australian Subkingdom along many watercourses and waterholes, and there are even endemic *Eucalyptus* in such environments. On the other hand, some species of the Central Australian flora (e.g. *Acacia aneura*) penetrate into the driest parts of the *Eucalyptus* Subkingdom in some areas, and others are found in the latter in littoral habitats, such as salt marshes.

The relict occurrence of *Livistona* in the Macdonnell Ranges (Cmn) is only one example out of many which points to the former predominance of the *Eucalyptus* flora over the entire continent. For various reasons it seems justified to regard the Central Australian flora partly as a relatively recent expansion of coastal floras (Diels, 1906; Burbidge, 1960) and partly as an impoverished *Eucalyptus* (including rainforest) flora (Beadle, 1966). In this context it is not surprising that a rainforest flora was much more widespread than the present one during the Tertiary (Duigan, 1951; Specht, 1958*b*; Beard, 1977).

The list of taxa (including many families and genera) having a wide distribution throughout the *Eucalyptus* Subkingdom and missing or rare in the Central Australian Subkingdom includes: *Actinotus, Banksia, Callistemon, Calytrix, Cladium, Conospermum, Daviesia, Hovea, Leptorhynchos, Leucopogon, Lomandra, Mitrasacme, Monotoca, Oxylobium,* Philydraceae, *Pultenaea, Restio, Ricinocarpos, Schoenus, Stackhousia, Stylidium* and many others. The Central Australian Subkingdom has no families which are missing from the *Eucalyptus* Subkingdom, but there are about 40 endemic genera (e.g. *Macgregoria, Newcastelia, Rutidosis* and *Threlkeldia*) and the flora is almost entirely endemic at the specific level. Burbidge (1960), Good (1974) and the various Australian floras provide more complete lists of taxa, endemic or characteristic of the Australian Kingdom as a whole or of the Subkingdoms.

The difference between both Subkingdoms is very spectacular. Natural vegetation in E is mostly dominated by species of *Eucalyptus* (most with relatively small areas), by other Myrtaceae or by rainforest trees, whilst the rest of the flora in most areas is notably rich in species of Amaryllidaceae, Cyperaceae, Dilleniaceae (*Hibbertia*), Droseraceae, Epacridaceae, Euphorbiaceae, Goodeniaceae, Liliaceae (and related families), Loganiaceae, Myrtaceae, *Olearia*, Orchidaceae, Papilionaceae, Pittosporaceae, Proteaceae, Rhamnaceae, Rutaceae (*Boronia*), Stylidiaceae, etc. In C the dominance of *Eucalyptus* (the few species of this genus present in the restricted areas where water accumulates have mostly large areas) is partly replaced by that of *Acacia* (sect. Phyllodineae). The total number of taxa per surface unit is mostly much smaller. Prominent are Amaranthaceae (*Trichinium*), *Callitris*, Chenopodiaceae (*Maireana*), Compositae (*Helipterum*), *Dodonaea, Eriostemon*, Gramineae (e.g. *Triodia*), Myoporaceae (*Eremophila*), *Zygophyllum*, etc. Most of these groups are, however, also well represented in E.

It is also possible to distinguish the northern from the southern part of the Flora-Kingdom on the basis of 'tropical' (e.g. Acanthaceae, Asclepiadaceae, *Bauhinia*, Bombacaceae, *Cassia*, Combretaceae (*Terminalia*), *Erythrina, Ipomoea*, Meliaceae, *Pandanus, Rhizophora*) versus 'temperate' groups (e.g. *Baeckea, Epilobium, Euphrasia, Leucopogon*, Ranunculaceae, Restionaceae, Umbelliferae, *Wahlenbergia*). Among the former are the pantropical taxa. Most authors have not considered this difference as a reason for a major subdivision of the Australian territory.

The Central Australian Subkingdom is subdivided into two regions: a (predominantly sandy) 'Desert Region' and a (mostly rocky) 'Mulga Region'.

The *Eucalyptus* Subkingdom is divided into five regions: two eucalypt savanna regions, one region dominated by eucalypt shrub ('mallee') or treeless 'saltbush', depending on soil conditions and two forest regions (eucalypt or rainforests). All seven regions will now be described briefly.

The desert region. Full climatic deserts are virtually absent from Australia. The 'Desert Region' (Cd) can be described as a sparsely vegetated, partly treeless area with interior drainage in the heart of Australia, mainly consisting of sandhills, sandplains, salt lakes and 'gibberplains' (stony deserts). Climate is tropical to sub-tropical and per-arid, locally even super-arid (cf. Tables 1.2, 1.3). This region falls into three more-or-less isolated parts: an eastern province (Cde), including the Simpson and Sturt Deserts (28°00′S; 141°15′E) and the area of the great salt lakes (Lake Eyre, Lake Frome, etc.) which has the lowest rainfall of Australia (<150 mm per year); a northwestern province (Cdw), including the Great and Little Sandy and Balwina (Beard, 1969) Deserts (<350 mm rainfall, combined with extremely high temperatures); and finally the Victoria Desert province (Cdv) (<200 mm rainfall). Even here there are some characteristic woody species, e.g. *Acacia dictyophleba*, *A. strongylophylla* (Cd); *A. peuce* (Cde); *A. basedowii*, *Casuarina decaisneana*, *Eucalyptus concinna* (Cdw); *E. gongylocarpa* (Cdv). However, grasses such as *Plectrachne* and *Triodia* species are most prominent in the landscape.

The Mulga region. Situated around these central provinces more-or-less concentrically are huge areas which in the central parts of the region are scarcely less arid, but with rocky (e.g. in the Musgrave, Petermann and Macdonnell Ranges) or stony soils (Gibson Desert), many (periodically dry) watercourses and depressions with heavy soils. Vegetation is therefore more varied in structure and much richer in species. Tree and shrub deserts, tussock and hummock grasslands and saltbush are the main constituents of the landscape. This has been indicated as the 'Mulga Region' (Cm).

It is divided into three sub-tropical and two tropical (northern) provinces. Phyllodineous *Acacia* species are prominent in the former, scattered eucalypts and open grasslands in the latter part of the region. It is necessary to distinguish three proper 'Mulga provinces': a western (Cmw, largely coinciding with Gardner's Austin and Coolgardie districts), an eastern (Cme, from South Australia to southwestern Queensland) and a northern mulga province (Cmn, extending along the tropic

of Capricorn over 2500 km from west to east). The latter can be subdivided into at least three subprovinces: a western (west of the
Warburton Range (26°06′S; 141°15′E) with some Western Australian
flora elements), a central (from the Warburton Range in Western
Australia to the Burke and Eyre Ranges in western Queensland) and an
eastern subprovince. Both tropical provinces have a slightly higher
(summer) rainfall (up to 650 mm at their northern limit) and are
ecologically comparable to the 'Sahel' in Africa. There is a western
'Hamersley Province' (Cmh) and an eastern 'Barkly Province' (Cmb).
The latter can be regarded as transitional between the Central Australian
provinces and the Northwestern Savanna Province (Enw). Apart from
the Eastern Forest Region and part of the Savanna Regions (provinces
Enb and Ess) Cm is the only Australian region with considerable areas
above 500 m altitude. Some characteristic woody species are: *Acacia
aneura, A. brachystachya, A. kempeana, A. tetragonophylla, Eucalyptus
microtheca* (Cm); *A. georginae, A. maitlandii, E. argillacea, E. gamophylla* (Cmn); *A. burkittii, A. cana, E. gillii* (Cme); *A. doratoxylon* (also
in Esp), *A. farnesiana, E. pyrophora* (Cmn + Cme); *A. adsurgens, A.
chisholmii, A. eremaea, A. exocarpoides, A. goochii, A. grasbyi, A. kochii,
A. palustris, A. tysonii, A. xerophila, E. ebbanoensis* (Cmw); *A. glabri-
flora, A. pyrifolia, A. sclerophylla, E. kingsmillii* (Cmh); *A. luerssenii,
E. odontocarpa, E. pallidifolia* (Cmb). The 'sclerophyll hummock grassland', mostly dominated by *Triodia* species (e.g. *T. basedowii, T. longi-
ceps, T. pungens, T. wiseana*) is characteristic of Cmh + Cmb + Cmn,
and with partly different physiognomy and floristic composition (e.g.
with *Plectrachne schinzii*), of the sand dune areas of Cd. Cmw contains a
number of taxa related to the southwest Australian flora, which is
optimally developed in Ewh and Ewf.

The northern savanna region. This region has a tropical, strongly seasonal
(with summer rains), semi-humid climate. Some important woody taxa
in this area (most of them endemic) are: various Bombacaceae, *Brachychiton diversifolium, Callitris intratropica, Eucalyptus alba* (also in Timor
and New Guinea), *E. grandifolia, E. microtheca, E. miniata, E. phoenicea,
E. setosa* and *E. tetrodonta*. Some are deciduous (*Eucalyptus, Cochlospermum* sp.); these are mainly concentrated in special habitats (e.g.
among boulders). Together with a narrow coastal belt further southwards, this region and Eea roughly coincide with the part of Australia
where no frosts occur.

The northwestern part of Cape York Peninsula, the area around
Darwin (including Bathurst and Melville Islands) and, to a lesser degree,

the northwestern parts of the Kimberley area (Western Australia) receive a relatively high rainfall (>1000 mm) and consequently have a number of 'rainforest' species which are missing from drier areas. These areas can be distinguished as a far northern province, Ena, which I have called the Arafura Province. There are some endemic tree species, e.g. *Livistona benthamii* (Specht, 1958a). Some of the tropical taxa, present here, but missing in Enw and Ena, are found among the Apocynaceae, *Croton* sp., *Dioscorea*, Ebenaceae, *Ficus* sp., *Nypa*, *Pimelea* sp., Rubiaceae, Sapotaceae, *Tacca*, *Terminalia* sp., Tiliaceae, Vitaceae.

The Northwestern Savanna Province (Enw) is more distinct in charac- ter than the Northeastern Province, and is separated, e.g., by the endemic species *Eucalyptus clavigera*, *E. foelscheana*, *E. latifolia*, *E. ptychocarpa* and *E. tectifica*. The Northeastern Province has some minor endemic species, e.g. *E. brownii* and *E. microneura*. Some trees are restricted to Ene and Esb, e.g. *E. cambageana* and *E. peltata*. The importance of the Great Artesian Basin as a barrier between both areas was discussed by Perry & Lazarides (1964).

The southeastern savanna region. This region (Es) has a sub-tropical (tropical in the northern part of Esb) to temperate climate with, in the temperate parts of the region, a predominance of sub-tropical years, alternating with 'mediterranean' years (Emberger, 1959). Conversely, a certain amount of winter rain can be counted on in the northern part (Esb), in contrast to the tropical savanna provinces.

The Western Plains Province (Esp, for a more detailed division see *Flora of New South Wales*) is semi-arid, sub-tropical to mediterranean and has as endemic eucalypts, *E. largiflorens*, *E. microcarpa* and *E. woollsiana*.

The Western Slopes Province (Ess) is semi-humid, warm to cool temperate (or sub-mediterranean in some years) and is characterised by *Eucalyptus albens*, *E. blakelyi*, *E. bridgesiana*, *E. macrorhyncha*, *E. melliodora* (in some directions extending into neighbouring provinces) and *E. rossii*. The drier areas of Tasmania, with *E. linearis*, *E. pauciflora*, *E. rubida*, *E. tasmanica* (only the first and the last are endemic in Tasmania) can be best regarded as a rather sharply distinguished subprovince of Ess.

The Brigalow Province (Esb) is characterised by various formations dominated by brigalow (*Acacia harpophylla*) and by *Brachychiton rup- estris* and *Eucalyptus melanophloia*. Grasses (e.g. *Paspalidium* spp.) are often dominant in the herb layer. Important accompanying species are

Casuarina cristata, Geijera parviflora, E. populnea or *E. trachyphloia*, and they link this province with the rest of Es. There are also a number of rainforest species (e.g. lianes) scattered throughout this province and some treeless areas, e.g. with 'Mitchell' grass (*Astrebla* spp.) or *Triodia mitchellii*. Heavy, relatively fertile soils are widespread. The province largely coincides with the part of eastern Australia where the Great Divide is distant from the coast and therefore the distinction between coastal and inland climate and vegetation is less sharp than elsewhere. The whole region has an elevation of >200 m, partly >500 m. It has many species in common with Eec (e.g. *Eucalyptus crebra, E. tessellaris*), and density of natural vegetation is often remarkably high, but floristically it is much more closely linked to Esp.

The Mallee region. This region has a semi-arid mediterranean climate (di Castri & Mooney, 1973). It provides a certain link between the Eastern and Western Forest Regions. Dominant vegetation formations are mallee (a tall eucalypt shrub) and saltbush (a form of semi-desert, mainly of succulent chenopodiaceous dwarf shrubs, in Australia often indicated as 'shrub steppe'). Several vegetation layers with relatively low density in the drier parts of Australia may occur independently, e.g. as 'mallee', 'mulga', 'bluebush' (*Maireana* spp. dominant), 'saltbush' or 'spinifex', or in various combinations, e.g. 'mallee-bluebush' or 'mulga-spinifex': It seems clear that floristic criteria should be applied for a coherent classification of vegetation as well as for delimitation of phytogeographical areas.

The woody species, common to the Eastern (Eme) and the Western Provinces (Emw) are the 'mallee' eucalypts *E. angulosa, E. calycogona, E. dumosa, E. flocktoniae, E. foecunda, E. gracilis, E. oleosa* and *E. pileata*. Other species are restricted either to the Eastern Province (*Acacia continua, A. rupicola, Eucalyptus behriana, E. diversifolia, E. fruticetorum, E. gracilis* and *E. odorata*) or to the Western Province (*Acacia erinacea* and many other acacias, *Eucalyptus brockwayi, E. corrugata, E. crucis, E. cylindrocarpa, E. dielsii, E. diptera, E. ewartiana, E. salmonophloia, E. salubris* and *E. stricklandi*). Several of the species areas are much smaller than the provinces.

The Nullarbor Province (Emn) is floristically poor, and probably has no endemic trees or shrubs. Since the difference between mallee and saltbush vegetation is more closely related to soil conditions than to climate (in Eme there were also originally large areas of saltbush), and the species of the Nullarbor Province generally are the same as those of similar vegetation types in Emw (or Eme), the province has been included in the *Eucalyptus* Subkingdom rather than in the Central Australian Subkingdom. However, its characteristics are mainly negative. As far as aridity of the climate is concerned, low rainfall is partly compensated for here by relatively low temperatures.

The eastern forest region. The region (Ee) is the richest and most complicated of all regions. Climate varies from tropical to cool temperate at sea-level, and in all provinces a range of altitudinal belts is present, locally (Eeb, Eet) including an alpine belt. The only common trait of these climates is the absence or infrequency of arid periods, especially in summer. Yearly precipitation is > 1500 mm in the northern and > 650 mm in the southern part of the region. Apart from the higher areas, in which frosts are frequent, the region may be defined ecologically as the area where the growth of rainforests is possible, provided topographic and soil conditions are favourable (Webb, 1959, 1965). Taxa present throughout this region, but restricted to it in their Australian distribution, are mainly at the family and genus level, with *c.* 150 endemic genera in total. Some characteristic groups are *Archontophoenix, Bauera, Blandfordia, Centella, Cladium,* Cyatheaceae, *Doryphora,* Gesneriaceae, Hymenophyllaceae, *Lomatia, Orites, Sarcochilus, Telopea, Tieghemopanax,* Winteraceae, *Zieria.* Although there are not many trees ranging from north to south through the region, e.g. *Acmena smithii* (Myrtaceae) and *Eucalyptus tereticornis,* neither of them endemic to the region, the changes are much more gradual in this direction than those from the coast towards the inland regions. There are southern and northern area limits of species all along the coast from Cape York to Tasmania and to distinguish provinces is not easy. On the other hand, many more data are available, especially on the southern half of the region, than on most other parts of Australia, and finally some areas, e.g. Gympie area and Bunya Mountains (Queensland), Twofold Bay (New South Wales) and Flinders Range (South Australia) are mentioned so frequently as area limits that there is little doubt of their importance for the delimitation of provinces.

The Atherton Province (Eea) is tropical with high rainfall and a relatively short dry period in winter. It is close to New Guinea and somewhat isolated from the other Ee provinces. Its rainforests might be called 'sub-equatorial'. Floristically the province is imperfectly explored, but many endemic species are known, e.g. *Agathis microstachya* and *Ag. palmerstonii* (Araucariaceae), *Balanops australiana* (Balanopaceae), *Eucalyptus leptophleba, Podocarpus amarus.* In addition, many (mainly tropical) taxa of this province are found nowhere else in Australia (Burbidge, 1960), e.g. *Agathis robusta*, Balanopaceae, Balanophoraceae, Calycanthaceae, Dichapetalaceae, Elaeagnaceae, Musaceae, Nepenthaceae, Ochnaceae, Podostemonaceae, Theaceae and there are even two endemic families. The relationship with other Ee provinces is demonstrated by the occurrence of common endemics. The following cases may illustrate this. *Castanospermum australe* (Papilionaceae) and *E. grandis* are only known from Eea + Eec + Eem, *Agathis robusta* and *E. pellita* only from Eea + Eeh, *Araucaria bidwillii* from Eea + Eem and, finally, *Brachychiton acerifolium*, *Casuarina torulosa*, *Laportea photinophylla* (Urticaceae), *Podocarpus elatus*, *Syncarpia glomulifera* (Myrtaceae) and *Toona australis* (Meliaceae) only from Eea + Eec + Eem + Eeh. On the other hand, the province has many taxa in common with the Arafura Province (Ena), e.g. *Nepenthes*.

The Capricorn Province (Eec) is (in the lowlands) tropical with drier and cooler winters than Eea. It is not as mountainous as the other Ee provinces and therefore its western limit is not as well defined. Its northern and southern connections are also numerous and in this sense it is in many respects a transitional area. It is relatively poor in endemic taxa. Species in common with Eem are *Araucaria cunninghamii* (also in New Guinea), *E. triantha* and *Flindersia australis* (Rutaceae); species in common with Eem + Eeh are *E. propinqua*, *E. robusta* and *Tristania conferta* (Myrtaceae).

The Macpherson Province (Eem) is sub-tropical in the lowland parts. In addition to its character of being an 'overlap' between the eastern Australian tropical and temperate floras (Burbidge, 1960) it has many endemic taxa, viz. Akaniaceae, Petermanniaceae, 34 genera (Burbidge, 1960) and numerous species, e.g. *Acacia bakeri*, *Eucalyptus grandis*, *E. intermedia*, *Grevillea robusta* (Proteaceae), *Sloanea woollsii* and many other rainforest trees (Francis, 1951). Important species in common with Eeh are *Callicoma serratifolia* and *Ceratopetalum apetalum* (Cunoniaceae), *Doryphora sassafras* (Monimiaceae), *Eucalyptus globoidea*, *E. gummifera*, *E. maculata*, *E. micrantha*, *E. microcorys*, *E. paniculata*, *E. pilularis*, *E. punctata*, *E. racemosa*, *E. resinifera*, *E. robusta*, *E. saligna*, *E. siderophloia*, *Ficus rubiginosa* (Moraceae) and *Livistona australis* (Palmae).

The Hawkesbury Province (Eeh) is rich in floristic, climatological and pedological variation. The general trend in the Eastern Forest Region, that highland species in one area are only found on lower altitudes in a more southern area, is most conspicuous in this province. The climate varies from cool temperate in the western part of the province (Northern and Central Tablelands of New South Wales) to sub-tropical or nearly so in coastal areas. There are rainforests (e.g. on the slopes of the Dorrigo Plateau and north of Nowra), eucalypt savannas (e.g. on the volcanic soils around Armidale and Braidwood) and eucalypt forests in various forms. The Hawkesbury Sandstone (around Sydney) carries one of the richest and most characteristic floras of Australia (Beadle, Evans, Carolin & Tindale, 1972). Endemic tree species of the province are *Acacia parramattensis, Callitris muelleri, Eucalyptus capitellata, E. fastigata, E. haemastoma, E. piperita* and *Podocarpus spinulosus*. The province naturally falls into a northern and a southern part (subprovince), imperfectly divided from each other by the Hunter River valley. Between the subprovinces there are some important differences, e.g. the presence of *Nothofagus moorei* (Fagaceae) in the north and of *Eucalyptus botryoides* and *E. longifolia* in the south. Another interesting group forms the species common to Eem + Eeh + Eeb + Eet, e.g. *Acacia melanoxylon* (also in Ess) and the tree-ferns *Cyathea australis, Dicksonia antarctica* and *Todea barbara*.

The Bassian Province (Eeb) derives its name from Bass Strait (between Victoria and Tasmania), and was introduced in 1896 as a zoogeographic territory by Spencer (quoted by Keast, 1959*b*). Its climatic variation is even greater than that of the Hawkesbury Province, varying from alpine to warm-temperate. A floristic description of altitudinal belts in this province is given by Lang (1970). The mosaic of plant communities is also richly varied, ranging from 'fjaeldmark' vegetation (McVean, 1969) to mangrove and from temperate rainforest to treeless temperate grassland. Endemic tree species are *Cyathea marcescens, Eucalyptus bicostata, E. delegatensis, E. obliqua* (extending into Eeh and Eem), *E. regnans* and *E. sieberi*. In addition, there are some endemic species in common with Eet: *Cyathea cunninghamii, E. ovata, Nothofagus cunninghamii* and *Podocarpus lawrencei* (a shrub). As in Eeh, two subprovinces can be distinguished, viz. the Australian and the Tasmanian Subprovince. Missing in Tasmania are, for example, *Avicennia marina* (Verbenaceae) (the only mangrove species of the southern coast of Australia), *Eucalyptus baxteri, E. cypellocarpa* (both endemic in the Australian Subprovince) and *E. dives*. On the other hand, the Tasmanian Subprovince also has some endemics, viz. *Eucalyptus coccifera, E. globulus* (with two minor occurrences in southern Victoria), *E. gunnii* and *E. subcrenulata*. Many herbaceous species of higher altitudes are restricted to this province.

Floristic kingdom phytogeography 21

The West-Tasmanian Province is the area where most of the Tasmanian endemics are concentrated, e.g. *Anodopetalum biglandulosum* (Cunoniaceae), *Athrotaxis* (Taxodiaceae, three spp., endemic genus), *Dacrydium franklinii, Microcachrys* (monospecific genus) and *Phyllocladus aspleniifolius* (all Podocarpaceae), *Eucalyptus simmondsii, E. vernicosa* and *Nothofagus gunnii*. The climate is sub-alpine to cool-temperate and humid to per-humid.

The southwestern forest region. This region is the main area of the famous Western Australian flora, extremely rich in endemics and bizarre sclerophyll plant forms. Although there are many parallels with the Eastern Forest Region, the differences are even greater than could be expected in connection with the longitudinal distance. Plant families are largely the same as in Ee in similar environments, but nearly 25% of the genera and a very large percentage of the species are endemic in Ew. Examples among the *c.* 125 endemic genera are: *Anigozanthos, Beaufortia, Calothamnus, Cephalotus, Chamaelaucium, Conostylis, Dasypogon, Dryandra, Hydatella, Kingia, Macropidia* and *Tremandra*, some of them with many species. If species dominance and abundance in the major vegetation types is used as a criterion, the usual situation is that genera in the majority of cases are the same in Ee and Ew but species are nearly always different. This means that differences are of the same order as those between Europe and North America if similar climatic areas are compared. Geographic isolation and historical factors are, as usual, by no means the only bases for explanation of flora and vegetation differences. The southwestern Australian climate is, in contrast to that in southeastern Australia, mediterranean, and geological and soil conditions (predominance of laterites, loose sandy soils or limestone in the west) are also different in most areas. Some woody species, endemic in Ew, are *Acacia cyanophylla, A. extensa, A. pulchella, Banksia prionotes* (Proteaceae), *Eucalyptus falcata, E. megacarpa, E. rudis, Melaleuca parviflora* (Myrtaceae), *Macrozamia riedlei* (Cycadaceae) and *Nuytsia floribunda* (Loranthaceae).

The Southwestern Heath Province (Ewh) is sub-humid (annual rainfall *c.* 500–750 mm). Sclerophyll shrub and dwarf shrub vegetation ('heath'), comparable to the South African 'fynbos', is conspicuous on poor sandy soils, in particular in the southeastern and northwestern parts of the province. In the Stirling Range (northeast of Albany) are the only mountain tops in the region above 1000 m. They are famous for their micro-endemism of *Darwinia* species (Carlquist, 1974). In the central part (e.g. the Northam area) eucalypt woodland, somewhat similar to the 'savanna woodlands' of Ess, is found. The mediterranean climate and the relative scarcity of grasses in southwestern Australia (Gardner, 1944) mean that this similarity remains rather superficial. On the other hand, the province has some *Eucalyptus* species in common with the Eastern Mallee Province (Eme), e.g. *E. angulosa, E. calycogona, E. conglobata* and *E. incrassata*.

Diels (1906) proposed a subdivision of temperate Western Australia into a number of phytogeographical districts (perhaps better to be evaluated as 'sectors' in the hierarchy used here), which is still used in a revised and extended form by modern authors (e.g. Beard, 1965). This is a somewhat more detailed division than that into subprovinces in the present classification. The Stirling and Irwin Districts appear not to be as well characterised as the Avon and Eyre Districts, and in view of this a Northern (Irwin + Avon Districts) and a Southern (Stirling + Eyre Districts) Subprovince may be distinguished, both with many endemic eucalypts and acacias. Some of the most important endemics of the Northern Subprovince are *Eucalyptus astringens, E. loxophleba, E. rhodantha* and *E. wandoo*; some of the Southern Subprovince are *E. doratoxylon, E. platypus, E. preissiana* and *E. tetragona*.

The Southwestern Forest Province (Ewf) is humid (rainfall > 750 mm per year), and forests are dominant here as climax communities. Some endemic tree species of the province are *Agonis flexuosa* (Myrtaceae), *Banksia grandis, B. verticillata, Casuarina fraserana, Eucalyptus calophylla, E. decipiens, E. marginata* and *E. patens*. This province is connected with the Bassian Province of the Eastern Forest Region (Eeb) by a number of species with discontinuous distribution in both provinces, e.g. *Dillwynia cinerascens* (Papilionaceae) (Green, 1964). In the southern part of the province forests are much more luxurious than elsewhere in Western Australia, which also indicates an ecological relationship to Eeb. A number of important woody species are restricted to this area, e.g. *Acacia pentadenia, Casuarina decussata, Eucalyptus cornuta, E. diversicolor* (the giant 'karri'), *E. guilfoylei, E. haematoxylon, E. jacksonii* and *Podocarpus drouynianus* (a shrubby species). Species restricted to the Northern Subprovince have mostly a limited distribution in the limestone plains between the coast and the escarpment (the Darling Range), e.g. *E. decipiens, E. foecunda, E. gomphocephala* and *E. lane-poolei*.

Concluding remark

To conclude this chapter, it must be stressed that destruction of native vegetation by man is so fast in some Australian districts, e.g. in coastal heath north of Newcastle (Eeh), in the Brigalow forests (Esb) and in Western Australia (Ewh) that it may be assumed that species and perhaps even genera of higher plants are disappearing before they are discovered and described. This is one reason why the publication of taxonomic and phytogeographic data is urgent, even if the resulting papers are as incomplete and imperfect as the present chapter.

References
Beadle, N.C.W. (1948). *The Vegetation and Pastures of Western New South Wales*. Sydney: Government Printer.
Beadle, N.C.W. (1966). Soil phosphate and its role in molding segments of the Australian flora and vegetation. *Ecology*, *47*, 992–1007.
Beadle, N.C.W. & Costin, A.B. (1952). Ecological classification and nomenclature. *Proceedings of the Linnean Society of New South Wales*, *77*, 61–82.
Beadle, N.C.W., Evans, O.D., Carolin, R.C. & Tindale, M.D. (1972). *Flora of the Sydney Region*. Sydney: A.H. & A.W. Reed.
Beard, J.S. (1965). *Descriptive Catalogue of West Australian Plants*. Chipping Norton, NSW: Surrey Beatty & Sons.
Beard, J.S. (1969). The natural regions of the deserts of Western Australia. *Journal of Ecology*, *57*, 677–711.
Beard, J.S. (1977). Tertiary evolution of the Australian flora in the light of latitudinal movements of the continent. *Journal of Biogeography*, *4*, 111–18.
Blake, S.T. (1953). Botanical contribution of the northern Australia regional survey. I. Studies on northern Australian species of *Eucalyptus*. *Australian Journal of Botany*, *1*, 185–352.
Blakely, W.F. (1955). *A Key to the Eucalypts*, 2nd edn. Canberra: Forestry & Timber Bureau.
Burbidge, N.T. (1960). The phytogeography of the Australian Region. *Australian Journal of Botany*, *8*, 75–212.
Burbidge, N.T. (1963). *Dictionary of Australian Plant Genera*. Sydney: Angus & Robertson.
Carlquist, S. (1974). *Island Biology*. New York & London: Columbia University Press.
Carnahan, J.A. (1976). Natural vegetation. In *Atlas of Australian Resources*, 2nd Series. Canberra: Department of National Resources.
Castri, F. di & Mooney, H.A. (ed.) (1973). *Mediterranean Type Ecosystems. Origin and Structure*. Ecological Studies, 7. Berlin: Springer.
Costin, A.B. (1954). *A Study of the Ecosystems of the Monaro Region of New South Wales*. Sydney: Government Printer.
Darlington, P.J. (1968). *Biogeography of the Southern End of the World*. New York: McGraw-Hill.
Diels, L. (1906). *Die Pflanzenwelt von West-Australien südlich des Wendekreises. Die Vegetation der Erde*, ed. A. Engler & O. Drude, vol. VII. Leipzig: Engelmann. (Reprint (1976), Vaduz: A.R. Gantner.)
Doing, H. (1970). Botanical geography and chorology in Australia. *Landbouwhogeschool Wageningen, Miscellaneous Papers*, *6*, 81–98.
Doing, H. (1971). Vegetation formations in Australia. *Acta Botanica Neerlandica*, *20*, 258–9.
Duigan, S.L. (1951). A catalogue of the Australian Tertiary flora. *Proceedings of the Royal Society of Victoria*, *63*, 41–56.

Emberger, L. (1959). La place de l'Australie méditerranéene dans l'ensemble des pays méditerranéens du Vieux Monde. In *Biogeography and Ecology in Australia* (Monographiae Biologicae VIII), ed. A. Keast, R.L. Crocker & C.S. Christian, pp. 259–73. The Hague: Junk.

Francis, W.D. (1951). *Australian Rain-forest Trees*, 2nd edn. Sydney: Halstead Press.

Gardner, C.A. (1944). The vegetation of Western Australia. *Journal of the Royal Society of Western Australia*, *28*, xi–1xxxvii.

Good, R. (1974). *The Geography of the Flowering Plants*, 4th edn. London: Longman Green & Co.

Green, J.W. (1964). Discontinuous and presumed vicarious plant species in southern Australia. *Journal of the Royal Society of Western Australia*, *47*, 25–32.

Holdridge, L.R., Grenke, W.C., Hatheway, W.H., Liang, T. & Tosi, J.A. (1971). *Forest Environments in Tropical Life Zones*. Oxford: Pergamon Press.

Hooker, J.D. (1860). *The Botany (of) the Antarctic Voyage*, part 111. *Flora Tasmaniae*, vol. 1. London: Lovell Reeve.

Jackson, W.D. (1965). Vegetation. In *Atlas of Tasmania*, ed. J.L. Davies, pp. 31–35. Hobart: Lands & Survey Department.

Keast, A. (1959*a*). The Australian environment. In *Biogeography and Ecology in Australia* (Monographiae Biologicae VIII), ed. A. Keast, R.L. Crocker & C.S. Christian, pp. 15–35. The Hague: Junk.

Keast, A. (1959*b*). The reptiles of Australia. In *Biogeography and Ecology in Australia* (Monographiae Biologicae VIII), ed. A. Keast, R.L. Crocker & C.S. Christian, pp. 115–35. The Hague: Junk.

Lang, G. (1970). *Die Vegetation der Brindabella Range bei Canberra*. Abh. Math.-Naturwiss. Kl. 1970, 1. Mainz: Akademie der Wissenschaften und der Literatur.

McVean, D.N. (1969). Alpine vegetation of the central Snowy Mountains of New South Wales. *Journal of Ecology*, *57*, 67–86.

Meusel, H. (ed.) (1965, 1978). *Vergleichende Chorologie der Zentral-Europäischen Flora*. Jena: G. Fischer.

Moore, C.W.E. (1953). The vegetation of the south-eastern Riverina. *Australian Journal of Botany*, *1*, 485–567.

Perry, R.A. & Lazarides, M. (1964). Vegetation of the Leichardt-Gilbert area. *CSIRO Australia Land Research Series*, *11*, 152–91.

Raven, P.H. & Axelrod, D.I. (1972). Plate tectonics and Australasian paleobiogeography. *Science*, *176*, 1379–86.

Schmithüsen, J. (ed.) (1976). *Atlas zur Biogeographie* (*Meyers grosser physischer Weltatlas*, Bd 3). Mannheim, Wien, Zürich: Bibliographisches Institut.

Specht, R.L. (1958). *a*. The Gymnospermae and Angiospermae collected on the Arnhem Land expedition. *b*. The geographical relationships of the flora of Arnhem Land. In *Records of the American-Australian Expedition to Arnhem Land*, ed. R.L. Specht & C.P. Mountford, vol. 3, pp. 185–317, 415–78. Melbourne: University Press.

Steenis, C.G.G.J. van (ed.) (1963). *Pacific Plant Areas*, vol. 1. Manila: Bureau of Printing.

Steenis, C.G.G.J. van & Balgooy, M.M.J. van (ed.) (1966). Pacific plant areas, vol. 2. *Blumea Supplement*, vol. 5, pp. 1–312.

Stephens, C.G. (1961). The soil landscapes of Australia. CSIRO Australia Soil Publication Number 18.

The Atlas of the Earth (1972). London: Mitchell Beazley, George Philip.

Wace, N.M. (1965). Vascular plants. In *Biogeography and Ecology in Antarctica* (Monographiae Biologicae XV), ed. J. van Mieghem & P. van Oye, pp. 201–66. The Hague: Junk.

Walker, D. & Singh, G. (1981). Vegetation history. In *Australian Vegetation*, ed. R.H. Groves, pp. 26–43. Cambridge University Press.

Walter, H. & Straka, H. (1970). *Arealkunde*, 2. Aufl. Stuttgart: E. Ulmer.

Webb, L.J. (1959). A physiognomic classification of Australian rainforests. *Journal of Ecology*, *47*, 551–70.

Webb, L.J. (1964). An historical interpretation of the grass balds of the Bunya Mountains, south Queensland. *Ecology*, *45*, 159–62.

Webb, L.J. (1965). The influence of soil parent materials on the nature and distribution of rain forests in south Queensland. In *Proceedings of the Symposium on Ecological Research in Humid Tropics Vegetation, 1963*, pp. 3–14. Kuching, Sarawak: UNESCO.

Whitley, G.P. (1959). The freshwater fishes of Australia. In *Biogeography and Ecology in Australia* (Monographiae Biologicae VIII), ed. A. Keast, R.L. Crocker & C.S. Christian, pp. 136–49. The Hague: Junk.

Williams, R.J. (1955). Vegetation regions. In *Atlas of Australian Resources* 1st Series. Canberra: Department of National Development.

Zanten, B.O. van (1978). Experimental studies on transoceanic long-range dispersal of moss spores in the Southern Hemisphere. *Journal of the Hattori Botanical Laboratory*, *44*, 455–82.

2

Vegetation history

D. WALKER & G. SINGH

The foundations

There is compelling geophysical evidence that the crustal plates which carry Australia and Antarctica remained at least partially joined in high latitudes until about fifty five million years ago (Late Palaeocene). Their separation at about that time was the final major event in the fragmentation of Gondwanaland, South America and Greater India having drifted off much earlier, whilst New Zealand had left Australia's eastern flank about twenty five million years previously (Coleman & Packham, 1976; Johnson, Powell & Veevers, 1976; Veevers & McElhinny, 1976; Veevers & Cotterill, 1978). Interpretations of the fossil record in relation to these movements and the landform evolution of the continent (Ollier, 1977) by Kemp (1978), Martin (1978), Galloway & Kemp (1980) and Walker (1981) provide the basis of the following account.

In the latest phases of their contiguity, as the Cretaceous passed into the Palaeocene about sixty five million years ago, the flora of Australia and adjacent Antarctica developed some, at least, of the characteristics of a cool-temperate, non-sclerophyllous rainforest including araucarians, podocarps, *Phyllocladus*, *Dacrydium*, Proteaceae (*Proteacidites*) Winteraceae and some *Nothofagus*. It is uncertain how far such vegetation extended toward the centre of the continent which, by this time, had emerged from the sea. Somewhat later, as Australia and Antarctica separated, there was restricted marine incursion particularly on parts of the present southern coast (Eucla and Murray basins) and fluviatile and lacustrine deposition was restricted to a few inland basins of limited extent (e.g. Lake Eyre basin). Fragments of continental crust became detached from Australia's east coast and were carried eastward by sea-floor spreading until they came under the influence of the Pacific crustal plate and were sheared northward and westward with respect to the Australian continent. In this way, Fiji, the New Hebrides and central New Guinea and the Papuan peninsula, amongst other islands or their parts, had originated and begun to move toward their modern positions fifty million years ago. By about this time the fossil record of the rainforest demonstrates a greater variety of taxa, some of them perhaps indicative of slightly warmer conditions, e.g. *Cupanieidites* (Sapindaceae), *Anacolosidites* (Olacaceae), *Beaupreadites* (Proteaceae), Myrtaceae. Although the fossil sites are unevenly distributed (Figure 2.1) and doubtless exaggerate the importance of wet-adapted plants, it is reasonable to conclude that, as Australia began its drift to lower latitudes, it carried with it a Gondwanic endowment mainly forming a diversifying rainforest over the greater part of the continent and the fragments breaking from it.

Figure 2.1. Locations of Tertiary pollen analytical sites in Australia (after Kemp, 1978, and Martin, 1978).

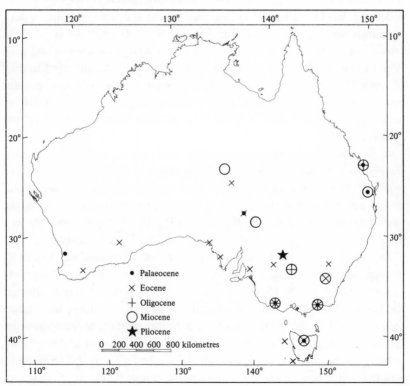

Martin (1978) singles out the mid-Eocene (*c.* forty five million years ago) as a time of significant change in the Australian flora. Not surprisingly, since by then Australia was quite isolated from other continents, the changes were quantitative rather than qualitative, *Nothofagus* attaining great abundance at the apparent expense of the Proteaceae in the southern half of the continent. The uplift of the eastern highlands had begun and alluviation of the continental basins was becoming more extensive. In the Late Eocene marine transgression was more widespread than before. Events such as these, the northward drift into lower latitudes and the generally falling temperatures evidenced from deep sea cores (Shackleton & Kennett, 1975), must have had profound effects on atmospheric circulation and Australia's climate probably became more diversified than it had been formerly.

The Oligocene is the 'Dark Age' of Australian vegetation history and all that can usefully be said is that it may have seen the extinction of some 'tropical' taxa (e.g. *Anacolosidites*) from the south and the beginnings of the emergence of eucalypts and acacias as significant elements in the vegetation. The sea retreated episodically, much of its former bed becoming basins of terrestrial sedimentation; the uplift of the eastern highlands was more-or-less completed and the northernmost edge of the crustal plate, now welded to fragments migrated from the south and east, began to rise and form New Guinea. By the end of the middle Miocene (*c.* ten million years ago) the Australian plate was approaching its present position with respect to southeast Asia and a major new, if extremely unstable, interface between floras long isolated from one another was established.

According to oxygen isotope determinations on marine sediments, temperatures which had remained fairly stable in the long term from the mid-Oligocene, resumed their downward drift and began to show fluctuations which became characteristic in the Quaternary, as early as the mid-Miocene (Shackleton & Kennett, 1975). By that time, too, the last major marine transgression of the continent had already ended.

At about this time the indigenous flora of Australia was evidently undergoing considerable differentiation. *Nothofagus*, although recurring in some places in the Pliocene, quickly loses its pre-eminence amongst the fossils. There are indications that, particularly in areas least likely then to have been assured of high, year-round rainfall, more open vegetation, including grassland, was appearing. It is perhaps significant that, although marsupials must have been in Australia from the Cretaceous onward, their fossils are most commonly associated with deposits of

Miocene and younger ages. Their browsing and grazing habits must have had profound effects on the emerging and spreading vegetation types which in turn perhaps elicited reciprocal development amongst the animals (cf. Keast, 1972).

During the past ten million years the disposition of the Australian land mass in relation to Antarctica, Asia and New Zealand has been much as it is today. New Guinea, which has continued to rise and accumulate sedimentary masses around its skirts, and Tasmania have been periodically joined with and separated from the continent by oscillations of sea level. Some volcanic activity, which had occurred episodically and discontinuously along the whole length of the eastern highlands throughout the Tertiary, almost certainly continued here and there into the Quaternary. Periodically, the highest mountains developed ice caps, particularly in Tasmania, the Snowy Mountains of New South Wales and New Guinea where there is still perennial snow above about 4550 m above sea-level in Irian Jaya. Over the greater part of the Australian continent, however, the processes of geomorphic development which had dominated the whole of the Tertiary period continued, modified by the changing temperature, circulation and precipitation patterns which probably reached their greatest mutability during the Quaternary.

Although the size of the continent, the chronological uncertainties and the general lack of fossil evidence from Early and Middle Pleistocene render generalisation difficult, it seems that the main expansion of open woodland and grassland vegetation probably took place some time during the Late Pliocene to Early Quaternary period (three to one million years ago). It is arguable, however, that at least the final episodes, which may have been associated with the considerable floristic changes (Figure 2.2), were very much more recent and continue at an exaggerated rate at the present day (Singh, Kershaw & Clark, 1980). Certainly the arrival of man on the continent more than 40 000 years ago and his use of fire as a management tool (Jones, 1975) must have had potent effects on the vegetation and the fauna. When he reached New Guinea is even less certainly known, but possibly by 9000 and certainly by 6000 years ago he was actively gardening in the island's highlands (Golson, 1977). The flow of Malesian plants into the continent, and more particularly into New Guinea, facilitated by the juxtaposition of Australia to Southeast Asia from the mid-Miocene, was augmented by the carriage of useful plants by early visitors from that region. And in due course, but only during the past 200 years or so, European settlers wittingly and otherwise have introduced a great variety of aliens from many parts of the world (see Chapter 3, Michael, 1981).

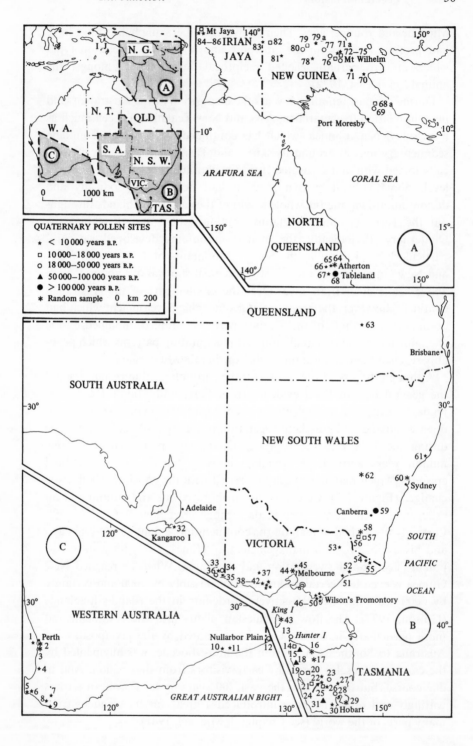

QUATERNARY POLLEN SITES

★ < 10 000 years B.P.
□ 10 000–18 000 years B.P.
○ 18 000–50 000 years B.P.
▲ 50 000–100 000 years B.P.
● > 100 000 years B.P.
✳ Random sample

Important phenomena

The fate of the rainforest
In Australia, the term 'rainforest' is usually reserved for floristically and structurally complex, predominantly broad-leaved, non-sclerophyllous forest. Many of its families and some of its genera have geographical ranges which are predominantly tropical outside Australia. It often contains gymnosperms (e.g. *Agathis, Podocarpus*) which in some, particularly cool-temperate, regions may predominate (e.g. *Athrotaxis* and *Dacrydium* in Tasmania). Rainforest grows in some of the wettest parts of Australia and over the great part of New Guinea. However, under

Figure 2.2. Locations of Quaternary pollen analytical sites in Australia and New Guinea. 1. Rottnest, W. Aust.; 2. Narrows Bore, W. Aust.; 3. Fremantle Bore, W. Aust.; 4. Myalup Swamp, W. Aust.; 5. Flinders Bay Swamp, W. Aust.; 6. Scott River Swamp, W. Aust.; 7. West Lake Muir, W. Aust.; 8. Weld Swamp, W. Aust.; 9. Boggy Lake, W. Aust.; 10. Norina Rock Shelter, W. Aust.; 11. Madura Cave, W. Aust.; 12. N145, W. Aust.; 13. Cave Bay Cave, Hunter Island, Tas.; 14. Broadmeadows, Tas.; 15. Pulbeena Swamp, Tas.; 16. Rocky Cape, Tas.; 17. Lemonthyme Creek, Tas.; 18. Pieman Damsite, Tas.; 19. Henty Bridge, Tas.; 20. Lake Margaret, Tas.; 21. Lake Vera, Frenchman's Cap, Tas.; 22. Brown Marsh, Tas.; 23. Monpeelyata Canal, Tas.; 24. Tarn Shelf, Tas.; 25. Eagle Tarn, Mt Field National Park, Tas.; 26. Beatties Tarn, Mt Field National Park, Tas.; 27. Lake Tiberias, Tas.; 28. Pipe Clay, Tas.; 29. Remarkable Cave, Tas.; 30. Adamson's Peak (unnamed cirque), Tas.; 31. Blakes Opening, Tas.; 32. Lashmars Lagoon, Kangaroo Island, S. Aust.; 33. Marshes Swamp, S. Aust.; 34. Lake Leake, S. Aust.; 35. Browns Lake, S. Aust.; 36. Wyrie Swamp, S. Aust.; 37. Lake Keilambete, Vict.; 38. Lake Mumblin, Vict.; 39. Lake Gnotuk, Vict.; 40. Sadlers Swamp, Vict.; 41. Lake Bullenmerri, Vict.; 42. Cobrico Swamp, Vict.; 43. City of Melbourne Bay, Vict.; 44. Melbourne Sediments, Vict.; 45. Lancefield Swamp, Vict.; 46. Five Mile Beach Swamp, Vict.; 47. Mt La Trobe, Vict.; 48. Tidal River, Vict.; 49. Darby Beach, Vict.; 50. Cotters Lake, Vict.; 51. Loch Sport, Vict.; 52. Lake Curlip, Vict.; 53. Bunyip Bog, Vict.; 54. Delegate River Swamp, Vict.; 55. Rooty Breaks Swamp, Vict.; 56. Twynam Cirque, N.S.W.; 57. Blue Lake, N.S.W.; 58. Toolong Range Block Stream, N.S.W.; 59. Lake George, N.S.W.; 60. Long Reef, N.S.W.; 61. Fingal Bay Swamp, N.S.W.; 62. Cowra, N.S.W.; 63. Gilgai Soil, N.S.W.; 64. Butcher's Creek Coal Seam, Qld.; 65. Lake Euramoo, Qld.; 66. Quincan Crater, Qld.; 67. Bromfield Swamp, Qld.; 68. Lynch's Crater, Qld.; 68a. Laravita, P.N.G.; 69. Kosipe, P.N.G.; 70. Lake Wanum, P.N.G.; 71. Lake Yanamugi, P.N.G.; 71a. Many short cores from Mt Wilhelm, P.N.G.; 72. Summit Bog, Mt Wilhelm, P.N.G.; 73. Brass Tarn, Mt Wilhelm, P.N.G.; 74. Imbuka, Mt Wilhelm, P.N.G.; 75. Kamanimambuno, Mt Wilhelm, P.N.G.; 76. Manton's, Mt Hagen, P.N.G.; 77. Draepi, Mt Hagen, P.N.G.; 78. Oldfield's site, P.N.G.; 79. Inim, P.N.G.; 79a. Birip, P.N.G.; 80. Sirunki, P.N.G.; 81. Tari P.N.G.; 82. Telefomin, P.N.G.; 83. Yakas Tarn, P.N.G.; 84. Ijomba, Mt Jaya, Irian Jaya; 85. Yellow Valley, Mt Jaya, Irian Jaya; 86. Ertzberg, Mt Jaya, Irian Jaya.

comparably wet climates, soil conditions and perhaps disturbance history determine its replacement by wet sclerophyll forest dominated by *Eucalyptus* spp. and otherwise with a flora of strongly autochthonous Australian affinity.

Already in the Upper Cretaceous the gymnosperm component of the rainforest, *Nothofagus* '*brassii*' and *Proteacidites* (Proteaceae) were established where southern Australia and New Zealand abutted on the Antarctic continent. The Palaeocene saw the presence of additional angiosperms, amongst them *Ilex*, Euphorbiaceae, Winteraceae, *Casuarina*, Myrtaceae and *Nothofagus* '*fusca*'. By the Eocene the families Olacaceae (*Anacolosidites*), Sapindaceae (*Cupanieidites*), Bombacaceae (*Bombacacidites*) and Santalaceae (*Santalumidites*) are represented in the fossil floras. Although the investigated sites are few and unevenly distributed (Figure 2.1), it seems that forests with some of the floristic characteristics of modern rainforests grew across the southern half of Australia by about sixty million years ago. Because Australia was still remote from Asia it is most reasonable to suppose that the flora of which these forests were composed was Gondwanic in origin. The modern extra-Australian ranges of some of the taxa must therefore be the results of later migration from Australia or ancient endowments from other parts of Gondwanaland or, in the case of the gymnosperms, from earlier land masses. Martin (1977) has shown how the present world range of *Ilex*, for example, must have been established through more than one ancient dispersal route, which included Australia, and subsequent expansions and extinctions. There is some variety in the fossil assemblages of broadly Eocene age which suggests that there was probably regional differentiation within the rainforest and perhaps areas of quite different vegetation only dimly reflected by fossils.

It seems that rainforests in which *Nothofagus* '*brassii*' and gymnosperms played important roles continued to flourish, at least in the southeastern quarter of Australia, and probably more widely, through the Oligocene and into the Miocene. There was some geographical diversity, as for example, the prevalence of *Nothofagus* '*fusca*' and *Dacrydium franklinii* in the Yallourn (Victoria) deposits, and even the suggestion that, towards the southern centre, forest might have been restricted to galleries along rivers in Middle to Late Miocene time (Callen & Tedford, 1976). Probably the closest living approximations to these forests are to be found in the middle altitudes of New Guinea (cf. Duigan, 1966) where, however, the flora is much more diverse, and in the cool-temperate forests of Tasmania.

Throughout this period, Australia had drifted into lower latitudes. Atmospheric circulation had probably intensified and climatic differentiation and uncertainty increased as a result. The orogeny of the eastern highlands, including Tasmania, must have added to the climatic diversity and ensured a relatively wet eastern seaboard. New Guinea began to rise, constantly providing new habitats (including cool montane ones) close to the equator, in which the Gondwanic inheritance from Australian continent fragments, overland invaders from the south and overseas immigrants from the north and west were able to establish and speciate. It is impossible to trace the effects of these environmental changes on the vegetation but it is reasonable to suppose that they contributed to changes which first became obvious in some fossil records in the Late Miocene. From them it seems possible that rainforests had already become less general in inland Australia, their place having been taken in part by sclerophyll forests and perhaps also by more open savannas and shrublands. By the beginning of the Quaternary, about two million years ago, this tendency was more pronounced and it seems likely that the rainforests had become limited to the eastern fringe of the continent, Tasmania and New Guinea. Although there is no relevant fossil evidence, we may suppose that there was considerable differentiation within these forests. Cool Tasmania, distant from new sources of appropriate taxa, retained more of the character of the old Gondwanic, southern Australian vegetation than did Queensland and New Guinea which were climatically different and open to enrichment from southeast Asia.

The Quaternary period, with its pronounced climatic fluctuations, has probably witnessed the episodic contraction and expansion of patches throughout the rainforest's range. Its composition has also changed. Evidence for these assertions comes from a set of pollen diagrams from the Atherton Tableland in northeastern Queensland (Kershaw, 1970, 1971, 1975, 1978) and from numerous pollen diagrams from the highlands of New Guinea (Williams, McDougall & Powell, 1972; Powell *et al.*, 1975; Hope, 1976; Walker & Flenley, 1979). All the Atherton Tableland sites are from within the present natural limit of rainforest. Together they show that from *c.* 120 000 to *c.* 80 000 years ago (using a chronology extrapolated from later radiocarbon dates) rainforest comparable with some types now living, rich in angiosperms, but containing the locally-extinct *Dacrydium*, occupied the region. This period was followed, until *c.* 30 000 years ago, by a sequence of forest types in all of which *Araucaria*, *Podocarpus* and *Dacrydium* were more important than they are anywhere in the region today, whilst *Casuarina* and eucalypts began to be important somewhere in the region. From *c*. 30 000 to *c*. 10 000 years ago rainforest seems to have vacated the region, replaced by sclerophyll woodland. It then returned but with gymnosperms in much less significant quantities. During each period of its dominance, and not least in the last 10 000 years, the composition of the rainforest has evidently changed from time to time; certainly what we see now is not a facsimile of the Tertiary ancestor. The resilience of some components of the rainforest in southeastern Victoria is exemplified by the history of repeated reduction and re-expansion of the area of *Nothofagus cunninghamii* at Wilson's Promontory, during the last 13 000 years (Howard & Hope, 1970; Ladd, 1979). In the New Guinea highlands the record covers only the last 30 000 years and is most notable for its evidence of a 1500 m rise in tree-line since the height of the last glaciation about 18 000 years ago. As important as this gross altitudinal shift is, however, the evidence is that it, and the corresponding changes which had attended the oncoming cold, were not simply expansions and contractions of otherwise stable vegetation zones but involved important changes in the composition of the vegetation units which together make up the rainforest.

In summary, it seems that rainforests of Gondwanic origin expanded and diversified throughout Australia until at least twenty million years ago. Thereafter they have suffered overall contraction and dissection of their area, doubtless punctuated by expansion from time to time and place to place, coupled with floristic enrichment particularly in the north. In New Guinea there are still great areas of natural and near-natural rainforest left. In Australia this formerly great and extensive vegetation type is reduced to disturbed fragments many of which continue to be exploited for timber.

The effects of fires during the Quaternary
Much of Australia is occupied by plants adapted to living with fire of various intensities and frequencies. They accommodate to, and have even come to rely on, this extreme environmental phenomenon in a variety of anatomical and physiological ways, e.g. underground lignotubers, thick bark, heat-stimulated flowering, serotinous seeds. There can be no doubt that these responses to fire have evolved over a very long period of time and that the capacity to produce them was crucial in determining which of the higher taxa in the continent's floristic endowment should have provided the denizens of the increasing number of fire-prone localities since the later part of the Tertiary. But, despite a few scattered geological

transformations attributed to very ancient fires (Kemp, 1980), it is only for the Quaternary that continuous records of changing fire impact are available.

The fossil record may provide information about the history of fire in the landscape by implication from the known biology of the plants recorded in it, and more directly from the amounts of charcoal, often in microscopic particles, that deposits contain. Churchill (1968) was the first to remark on the significance of charcoal in pollen-analysed deposits but was unable to attribute the changes in abundance of the three eucalypt species he studied to changes in the impact of fire, although he pointed out that it must have been an important determinant of the vegetation in southwest Western Australia for more than 5000 years. More recently three Quaternary sites on the Australian continent have provided both pollen and charcoal diagrams (Singh *et al.*, 1980).

At Lake George, on the Southern Tablelands of New South Wales, four major vegetation fluctuations have been identified during the last 350 000 years and interpreted mainly as the effects of 'glacial–interglacial' climatic cycles. In general, fire activity was evidenced during the 'interglacial' periods of sclerophyll forest or woodland vegetation but was nearly absent from the intervening colder herbfield and grassland periods. Before about 120 000 years ago, the forest periods were characterised by fire-sensitive taxa (*Casuarina, Podocarpus, Astelia, Drimys, Cyathea, Lycopodium*) and fire was occasional, produced only very small amounts of charcoal in the deposits and may readily be conceived of as natural in origin. During the interval thought to be the chronological equivalent of the last interglacial, however, these fire-sensitive plants evidently declined in the face of a major expansion of *Eucalyptus* and myrtaceous shrubs. At the same time fire activity, as measured by the amount of fine charcoal in the sediments, increased. The trend of replacement of fire-sensitive by fire-tolerant vegetation persisted despite minor fluctuations, as when, during a cooler 'interstadial' interval, the fire-sensitive plants temporarily recovered, until during the last 10 000 years *Casuarina* is the only member of the fire-sensitive suite substantially represented in the fossil record. On the whole, however, the warmer episodes of the last 120 000 years (last 'interglacial' and Holocene) seem to have had open eucalypt woodlands with myrtaceous shrub understorey and a herbaceous, grassy, ground layer. Climatically, the last 'interglacial' and the Holocene are not thought to have been sufficiently different from preceding warm periods for this to be an adequate explanation of the major change of ecological mode which occurred

120 000 years ago. There is no doubt that increased fire frequency or intensity (or both) was an important part of the mechanism of the change. Although the earliest known remains of human settlement in Australia are only 36 000 years old (Bowler, 1976), it is possible that the much earlier increase in the importance of fire in determining the vegetation around Lake George was associated with man's arrival there (Singh *et al.*, 1980).

On the Atherton Tableland of northeastern Queensland, charcoal particle counts indicate that fire was associated with the expansion of sclerophyll vegetation at the expense of rainforest about 75 000 years ago. Thereafter until about 40 000 years ago, however, the amount of charcoal entering the sediments studied was modest, suggesting fire incidence consistent with the persistence of quantities of *Araucaria* and other gymnosperms in the vegetation (p. 32). The decline and the virtual disappearance of this element and the assertion of sclerophyll woodland dominance from about 40 000 years ago was contemporaneous with the greatest fire activity, once more attributed to human actions (Singh *et al.*, 1980). It is perhaps significant, however, that rainforest returned to the region, and fires were fewer in it, during the climatic changes which took place, world-wide, at the beginning of the Holocene (p. 37).

Even in the highlands of New Guinea there is evidence that fire has been a force in vegetation change, although a specifically fire-adapted suite of species has not emerged as a result. Data from a number of sites there strongly suggest that fire accompanied or followed replacement of forest by grassland and was strongly correlated with human occupation (G.S. Hope, personal communication). On Mt Wilhelm, for instance, the forest limit has been depressed and the sub-alpine grasslands extended by increased burning on the mountain from about 250 years ago (Corlett, 1979).

The evidence so far available indicates that fire has had potent effects on the vegetation of Australia in the past and is probably responsible for many of the vegetation patterns which now exist. The timing of its greatest impact, and the manner of its expression, has varied from place to place, and has been greatest where a long-adapted flora was available to respond positively to it. Although permissive climates and adequate fuel are essential pre-requisites for increases in the importance of fire in the landscape, human activity may prove to have been the deciding factor and so to have affected significantly a large proportion of Australia's vegetation cover.

Events at the Pleistocene-Holocene transition
The main vegetational, geomorphic and climatic changes during the Late Quaternary in Australia have been reviewed by Bowler *et al.* (1976). In Tasmania, the Late Pleistocene forest limit was close to sea-level, particularly in the west; the greater part of what was then the southern-most projection of the Australian continent was covered by grasslands, herbfields, heaths, or barren. Between about 11 500 and 10 000 years ago shrubberies and scrubs, in which *Microcachrys tetragona, Nothofagus gunnii, N. cunninghamii*, and *Podocarpus alpina* played parts, ascended the mountains whilst *Eucalyptus*, later with *Phyllocladus*, occupied the lower slopes in southern Tasmania. The expansion of eucalypt wood-lands into the midlands of the region came somewhat later (Macphail, 1975, 1976). In the northwest, eucalypt forest began to replace grassland about 11 000 years ago (E. Colhoun, personal communication); on Hunter Island (Hope, 1978) a similar process was not complete until after 8000 years ago. Although there were regional differences in the course of vegetation development, closed-forests of *Nothofagus cunninghamii* came to occupy the west and the slopes of isolated mountains in the centre and far south, whilst eucalypt forest covered much of eastern Tasmania (Macphail, 1975, 1976).

In the mountains of Victoria and southern New South Wales, herb-fields and grasslands of the Late Pleistocene were invaded by open eucalypt woodlands (presumably *Eucalyptus pauciflora* and its shrub associates) from lower altitudes. On the Mt Buffalo Plateau (Victoria) (36°42′S; 146°47′E), the tree-line passed 1330 m about 11 000 years ago (Binder, 1978; Binder & Kershaw, 1978). On Mt Kosciusko (New South Wales) evidence from sites at about 1900 m above sea-level, beyond the present extent of forest, indicates that the vegetation of formerly more-or-less barren mountain tops and plateaux began about 17 000 years ago. The process brought alpine herbfields and fjaeldmark by about 13 000 years ago but the most pronounced change came about 8700 years ago when the tree-line rose to its present altitude of about 1750 m above sea-level, if not somewhat higher (Raine, 1974).

In New Guinea, at the opposite end of the continent of the time, the forest limit began to rise from about 2300 m above sea-level about 16 000 years ago and reached slightly above its present elevation of 3800 m above sea-level about 7000 years later, most of the movement occurring between 11 000 and 9000 years ago. Grassland, herbfield and shrubberies were restricted in the process which was certainly more complicated than the uphill march of pre-existing forest zones (Walker & Flenley, 1979; Walker & Hope, 1981).

In the lower parts of the continent, where effective precipitation might be expected to have been a more important proximate cause of vegetation change than temperature, evidence for its occurrence is no less positive. On the Atherton Tableland (northern Queensland) rainforest expanded at the cost of sclerophyll woodland between 10 000 and 7000 years ago, depending on location (Kershaw, 1970, 1971, 1975). In the south, near the borders of Victoria and South Australia, significant changes clearly took place about 10 000 years ago. In the region of Lake Leake and Wyrie Swamp (South Australia) (37°39′S; 140°20′E) open eucalypt (perhaps *E. pauciflora*) woodlands were invaded by *Casuarina stricta* (Dodson, 1974*a*, 1975*a*, 1977), whilst near Lake Keilambete (Victoria) *Melaleuca* and *Leptospermum*, formerly hardly significant amongst grasses and herbs (including Chenopodiaceae), became more important (Dodson, 1974*b*). At Wilson's Promontory, on the coast of Victoria, a eucalypt forest gave way to *Melaleuca–Leptospermum* heath about 11 000 years ago and by 2000 years later had further deteriorated towards a bushy swamp, presumably as a result of climatically-induced waterlogging (Ladd, 1979).

There is no doubt that Australian vegetation reacted to the world-wide climatic changes which marked the transition from the Late Pleistocene to the Holocene. But there is no clearly identifiable date at which these reactions started. Once begun, their repercussions were felt in some places for a few thousand years. In some cases, shifts of the bounds between major vegetation formations were involved whilst in others changes in the importance of existing components of the local vegetation were the results.

Seral vegetation changes

Although there have been many accounts of seral changes inferred from present vegetation patterns or from experimental studies (e.g. Specht, Rayson & Jackman, 1958; Barrow, Costin & Lake, 1968; Webb, Tracey & Williams, 1972; Howard, 1973; Ashton, 1976), the fossil record has rarely been resolved in sufficient detail to allow unequivocal statements about the courses and rates of changes which go beyond the time-span of experiments. Nevertheless, it provides a few important hints to condition ecological thinking.

Before fire had totally altered the composition of the vegetation around Lake George (p. 35), the forests of the 'interglacial' periods were not unchanging through long stretches of time. In particular the abundances and occurrences of the less common taxa (e.g. *Drimys, Eucalyptus*) were not constant, suggesting continuous or episodic fluxes in the composition of the forests in which fire-sensitive, shade-tolerant, as well as fire-tolerant, light-demanding, taxa took part. The major climatically-prompted transitions from herbaceous to forest vegetation were also staged; herbfield and grassland was replaced by ferny shrubbery of cool-temperate taxa which later gave way to *Casuarina* followed by a range of other plants (e.g. *Podocarpus, Drimys*) which together synthesised the forest. Although at Lake George and on the Atherton Tableland fire may have had a role in the temporary perturbation of closed-forests, seral change and pattern maintenance were very much more complex during periods in which fire was a major and persistent ecological force. It is likely that seral processes in much of Australia's vegetation now bear little resemblance to their counterparts in the millennia before burning became so important.

Amongst the sand dunes near Port Stephens (New South Wales) the elimination of *Eucalyptus–Angophora* dune forest, between 6000 and 3000 years ago, was an epidsodic process punctuated by temporary recoveries. Even during periods of relative dune stability and reforestation, the record of water plants indicates fluctuations in the water levels in the dune slack, and therefore the dune soil (Macphail, 1974). At Wilson's Promontory (Victoria), marine incursion converted a fresh-water swamp to salt marsh about 7000 years ago which, after some sand accumulation, changed to the *Melaleuca ericifolia* heath growing there now (Ladd, 1979). Marshes Swamp in southeastern South Australia become progressively drier between 8000 and 5000 years ago, its vegetation changing from *Myriophyllum–Triglochin* communities to a sedgy myrtaceous scrub (Dodson & Wilson, 1975).

Despite the difficulties of sufficiently refined identification of fossil pollen grains, there is here a field of pollen analytical application now ripe for exploitation in Australia.

The effects of European settlement
It seems almost certain that Aboriginal activities, particularly burning, had strongly affected the pattern of Australian vegetation during thousands of years before European settlement. The early European pastoralists, however, evidently cleared forests and used fire in a different, perhaps more permanently destructive, way than had the Aborigines. Rather frequent burning before massive litter had accumulated was perhaps replaced by rarer but hotter conflagrations.

The spread of European settlement with its great variety of land uses including grazing, agriculture, forestry, mining and urbanisation has wrought enormous changes in Australian vegetation. Almost 60% of the continent has been grazed and large areas of forest cleared to facilitate this. Agricultural practices usually favour exotic plants at the expense of the native flora. These events of the last two centuries are too recent to have left more than traces in the fossil record. Where they are found, however, the appearance of exotic plants (e.g. *Pinus*, *Plantago lanceolata*, Cynareae) usually also marks decreases in native trees and shrubs (Dodson, 1974*a*, 1975*b*; Scarlett, Hope & Calder, 1974; Hope, 1978).

The process continues to the present day. At no time in its long history has Australian vegetation and its dependent animals been subject to such rapid and destructive change as during the last 200 years.

We are grateful to Mr R.T. Corlett, Dr M. Macphail and Dr I.J. Raine for permission to refer to their Ph.D. theses and to Mr L. Pancino for drawing the maps.

References

Ashton, D.H. (1976). The development of even-aged stands of *Eucalyptus regnans* F. Muell. in central Victoria. *Australian Journal of Botany, 24*, 397–414.

Barrow, M.D, Costin, A.B. & Lake, P. (1968). Cyclical changes in an Australian fjaeldmark community. *Journal of Ecology, 56*, 89–96.

Binder, R.M. (1978). Stratigraphy and pollen analysis of a peat deposit, Bunyip Bog, Mt Buffalo, Victoria. *Monash Publications in Geography Number 19*.

Binder, R.M. & Kershaw, A.P. (1978). A late-Quaternary pollen diagram from the southeastern highlands of Australia. *Search, 9*, 44–5.

Bowler, J.M. (1976). Recent developments in reconstructing late Quaternary environments in Australia. In *The Origin of the Australians*, ed. R.L. Kirk & A.G. Thorne, pp. 55–77. Canberra: Australian Institute of Aboriginal Studies.

Bowler, J.M., Hope, G.S., Jennings, J.N., Singh, G. & Walker, D. (1976). Late Quaternary climates of Australia and New Guinea. *Quaternary Research, 6*, 359–94.

Callen, R.A. & Tedford, R.H. (1976). New late Cenozoic rock units and depositional environments, Lake Frome area, South Australia. *Transactions of the Royal Society of South Australia, 100*, 125–67.

Churchill, D.M. (1968). The distribution and prehistory of *Eucalyptus diversicolor* F. Muell., *E. marginata* Donn ex Sm., and *E. calophylla* R.Br. in relation to rainfall. *Australian Journal of Botany, 16*, 125–51.

Coleman, P.J. & Packham, G.H. (1976). The Melanesian borderlands and India-Pacific plates' boundary. *Earth Science Reviews, 12*, 197–233.

Corlett, R.T. (1979). Human impact on the subalpine vegetation of Mt Wilhelm, Papua New Guinea. Ph.D. thesis, Australian National University.

Dodson, J.R. (1974a). Vegetation history and water fluctuations at Lake Leake, south-eastern South Australia. I. 10 000 B.P. to present. *Australian Journal of Botany*, *22*, 719–41.

Dodson, J.R. (1974b). Vegetation and climatic history near Lake Keilambete, Western Victoria. *Australian Journal of Botany*, *22*, 709–17.

Dodson, J.R. (1975a). Vegetation history and water fluctuations at Lake Leake, south-eastern South Australia. II. 50 000 B.P. to 10 000 B.P. *Australian Journal of Botany*, *23*, 815–31.

Dodson, J.R. (1975b). The pre-settlement vegetation of the Mt Gambier area, South Australia. *Transactions of the Royal Society of South Australia*, *99*, 89–92.

Dodson, J.R. (1977). Late Quaternary palaeoecology of Wyrie Swamp, southeastern South Australia. *Quaternary Research*, *8*, 97–114.

Dodson, J.R. & Wilson, I.B. (1975). Past and present vegetation of Marshes Swamp in south-eastern South Australia. *Australian Journal of Botany*, *23*, 123–50.

Duigan, S.L. (1966). The nature and relationships of the Tertiary brown coal flora of the Yallourn area in Victoria, Australia. *Palaeobotanist*, *14*, 191–201.

Galloway, R.W. & Kemp, E.M. (1980). Late Cainozoic environments in Australia. In *Ecological Biogeography in Australia*, ed. A. Keast, pp. 51–80. The Hague: Junk.

Golson, J. (1977). No room at the top: agricultural intensification in the New Guinea Highlands. In *Sunda and Sahul, Prehistoric studies in South-east Asia, Melanesia and Australia*, ed. J. Allen, J. Golson & R. Jones, pp. 601–38. London: Academic Press.

Hope, G.S. (1976). The vegetational history of Mt Wilhelm, Papua New Guinea. *Journal of Ecology*, *64*, 627–64.

Hope, G.S. (1978). The late Pleistocene and Holocene vegetational history of Hunter Island, north-western Tasmania. *Australian Journal of Botany*, *26*, 493–514.

Howard, T.M. (1973). Studies in the ecology of *Nothofagus cunninghamii* Oerst. I. Natural regeneration on the Mt Donna Buang massif. *Australian Journal of Botany*, *21*, 67–78.

Howard, T.M. & Hope, G.S. (1970). The present and past occurrence of beech (*Nothofagus cunninghamii* Oerst.) at Wilson's Promontory, Victoria, Australia. *Proceedings of the Royal Society of Victoria*, *83*, 199–210.

Johnson, B.D., Powell, C. McA. & Veevers, J.J. (1976). Spreading history of the eastern Indian Ocean and Greater India's northward flight from Antarctica and Australia. *Geological Society of America Bulletin*, *87*, 1560–6.

Jones, R. (1975). The Neolithic, Palaeolithic and the hunting gardeners: man & land in the Antipodes. In *Quaternary Studies*, ed. R.P. Suggate & M.M. Cresswell, pp. 21–34. Wellington: The Royal Society of New Zealand.

Keast, A. (1972). Australian mammals: zoogeography and evolution. In *Evolution, Mammals and Southern Continents*, ed. A. Keast, F.C. Erk & B. Glass, pp. 195–246. Albany, N.Y.: State University of New York.

Kemp, E.M. (1978). Tertiary climatic evolution and vegetation history in the southeast Indian Ocean region. *Palaeogeography, Palaeoclimatology, Palaeoecology*, *24*, 169–208.

Kemp, E.M. (1980). Pre-Quaternary fire in Australia. In *Fire and the Australian Biota*, ed. A.M. Gill, R.H. Groves & I.R. Noble, pp. 3–21. Canberra: Australian Academy of Science.

Kershaw, A.P. (1970). A pollen diagram from Lake Euramoo, north-east Queensland, Australia. *New Phytologist*, *69*, 785–805.

Kershaw, A.P. (1971). A pollen diagram from Quincan Crater, north-east Queensland, Australia. *New Phytologist*, *70*, 669–81.

Kershaw, A.P. (1975). Stratigraphy and pollen analysis of Bromfield Swamp, north-eastern Queensland, Australia. *New Phytologist*, *75*, 173–91.

Kershaw, A.P. (1978). Record of last interglacial–glacial cycle from northeastern Queensland. *Nature*, *272*, 159–61.

Ladd, P.G. (1979). A Holocene vegetation record from the eastern side of Wilson's Promontory, Victoria. *New Phytologist*, *82*, 265–76.

Macphail, M. (1974). Pollen analysis of a buried organic deposit on the backshore at Fingel Bay, Port Stephens, New South Wales. *Proceedings of the Linnean Society of New South Wales*, *98*, 222–33.

Macphail, M. (1975). Late Pleistocene environments in Tasmania. *Search*, *6*, 295–300.

Macphail, M. (1976). The history of vegetation and climate in southern Tasmania since the late Pleistocene. Ph.D. thesis, University of Tasmania.

Martin, H.A. (1977). The history of *Ilex* (Aquifoliaceae) with special reference to Australia: evidence from pollen. *Australian Journal of Botany*, *25*, 655–73.

Martin, H.A. (1978). Evolution of the Australian flora and vegetation through the Tertiary: evidence from pollen. *Alcheringa*, *11*, 181–202.

Michael, P.W. (1981) Alien plants. In *Australian Vegetation*, ed. R.H. Groves, pp. 44–64. Cambridge University Press.

Ollier, C.D. (1977). Early landform evolution. In *Australia: a Geography*, ed. D.N. Jeans, pp. 85–98. Sydney: University Press.

Powell, J.M., Kulunga, A., Moge, R., Pono, C., Zimike, F. & Golson, J. (1975). Agricultural traditions of the Mt Hagen area. *Department of Geography, University of Papua New Guinea, Occasional Paper*, *12*.

Raine, I.J. (1974). Pollen sedimentation in relation to the Quaternary vegetation history of the Snowy Mountains of New South Wales. Ph.D. thesis, Australian National University.

Scarlett, N.H., Hope, G.S. & Calder, D.M. (1974). Natural history of the Hogan Group. 3. Floristics and plant communities. *Papers & Proceedings of the Royal Society of Tasmania*, *107*, 83–98.

Shackleton, N.J. & Kennett, J.P. (1975). Palaeotemperature history of the Cenozoic and the initiation of the Antarctic glaciation: oxygen isotope analyses in DSDP sites 277, 279, 281. In *Initial Reports of the Deep Sea Drilling Project 29*, pp. 743–55. Washington: United States Government Printing Office.

Singh, G., Kershaw, A.P. & Clark, R. (1980). Quaternary vegetation and fire history of Australia. In *Fire and the Australian Biota*, ed. A.M. Gill, R.H. Groves & I.R. Noble, pp. 23–54. Canberra: Australian Academy of Science.

Specht, R.L., Rayson, P. & Jackman, M.E. (1958). Dark Island heath (Ninety-mile Plain, South Australia). VI. Pyric succession: changes in composition, coverage, dry weight and mineral nutrient status. *Australian Journal of Botany*, *6*, 59–88.

Veevers, J.J. & Cotterill, D. (1978). Western margin of Australia: evolution of a rifted arch system. *Geological Society of America, Bulletin*, *89*, 337–55.

Veevers, J.J. & McElhinny, M.W. (1976). The separation of Australia from other continents. *Earth Science Reviews*, *12*, 139–59.

Walker, D. (1981). Speculations on the origins and evolution of Sunda-Sahul rainforests. In *Proceedings of the ATB Conference in Venezuela*, ed. G. Prance, (in press). New York: Columbia University Press.

Walker, D. & Flenley, J.R. (1979). Late Quaternary vegetation history of the Enga Province of upland Papua New Guinea. *Philosophical Transactions of the Royal Society of London* Series B, *286*, 265–344.

Walker, D. & Hope, G.S. (1981). Vegetation history. In *Biogeography and Ecology in New Guinea*, ed. J.L. Gressitt, (in press). The Hague: Junk.

Webb, L.J., Tracey, J.G. & Williams, W.T. (1972). Regeneration and pattern in the subtropical rainforest. *Journal of Ecology*, *60*, 675–95.

Williams, P.W., McDougall, I. & Powell, J.M. (1972). Aspects of the Quaternary geology of the Tari-Koroba area, Papua. *Journal of the Geological Society of Australia*, *18*, 333–47.

3

Alien plants
P.W. MICHAEL

In the early times of a colony, there is comparatively little
difficulty in distinguishing the colonists from the native species;
but as the surface of the land becomes artificially disturbed, the
habits of all its plants are influenced, – the endemic plants are
driven from their native places, and take refuge in hedgerows,
ditches and planted copses, and from there associating with the
introduced plants, are apt to be classed in the same category
with them; whilst the introduced wander from the cultivated
spots and eject the native, or taking their places by them, appear
like them to be truly indigenous. (Hooker, 1860).

Alien plants, in the sense used by Watson (1847), are plants 'now more
or less established but either presumed or certainly known to have been
originally introduced from other countries'. They are taken here to
include those plants from other countries which are sufficiently well
established to be considered as true constituents of the Australian flora.
They include by far the greater number of plants naturalised in
Australia. In an unpublished list of some 800 weeds and naturalised
plants of New South Wales prepared for a Pan-Pacific Science Congress
by Blakely (1923), about 85% of the species included are aliens.

The most satisfying definition of naturalised plants is that given by
Thellung (1912), largely following de Candolle (1855), who presented the
first comprehensive and still useful treatment of naturalised plants in
various parts of the world. Thellung's definition is quoted here in full.

Nous appelons complètement naturalisée et, par abréviation,
naturalisée, une espèce qui, n'existant pas dans un pays avant sa
période historique (au point de vue de l'exploration botanique!),
venant à y être transportée par l'action volontaire ou incon-

sciente de l'homme ou par une cause inconnue, s'y trouve en-
suite avec tous les caractères des plantes spontanées indigènes,
c'est-à-dire, croissant et se multipliant par ses moyens naturels
de propagation (graines, tubercules, bulbilles, drageons, fragments de tiges ou de rhizomes, etc., suivant l'espèce), sans le
secours direct de l'homme, se manifestant avec plus ou moins
d'abondance et de régularité dans les stations qui lui convien-
nent, et ayant traversé des séries d'années pendant lesquelles le
climat a offert des circonstances exceptionelles.

A species is called completely naturalised or, in short, naturalised, if, not having been present in a country or region before
its historic period (from the point of view of botanical exploration) it has been transported there by the intentional or
accidental action of man or by a cause unknown and is now
found there with all the characteristics of spontaneous indigenous plants, that is to say, increasing and multiplying by natural
means (seeds, tubers, bulbs, suckers, fragments of stems or
rhizomes etc., according to the species) without the direct help
of man, occurring regularly, in greater or less abundance, in
situations which suit it, and having gone through series of years
during which there have been extreme climatic conditions. (My
translation.)

If Thellung's 'pays' is taken to mean both 'country' and 'region', as in
my translation, Australian naturalised natives are included. Australian
naturalised native species were, before European exploration and settlement, confined to more limited areas. These plants include those which
have escaped following their cultivation as ornamentals, for example,
Acacia baileyana, occurring originally in a limited area of New South
Wales, and the Western Australian *Albizia lophantha*, and those which
have greatly increased the areas they originally occupied in response to
clearing and/or grazing of native communities, for example, *Bassia
birchii* and *Chloris truncata*.

In this chapter, attention is given initially to the difficulties experienced
in determining whether Australian plants are alien or not. A brief
account of the extent of the alien flora is then presented with lists of the
most commonly represented families and selected species characteristic
of broad climatic regions. The origins, history and spread of certain
species are discussed and comparisons and contrasts are made between
the distribution of aliens in Australia and in their areas of origin. Brief
mention is made of the variation within alien species in Australia. No

account is given of phytocoenological relationships, a subject which has been given only scant attention in Australia.

Many of the observations made in this chapter are based on my own, unpublished work.

Natives or aliens

In Australia, as in other countries, a high proportion of the alien species are weeds of cultivation, pastures, roadsides and waste places. These weedy aliens may be called pioneer species because of their ability to colonise disturbed or denuded land. During the history of land development in Australia relatively few native species have behaved in this way. This is clearly indicated in the list of Blakely (1923).

There are a number of tropical or sub-tropical species, questionably native to Australia, which have moved southwards into more temperate areas. Some of these may be pan-tropical or pan-sub-tropical species, for example, *Sida rhombifolia* and *Bidens pilosa*. Certain species may have been introduced to northern Australia by visitors from southeast Asia before European settlement. Macknight (1976) presented evidence that *Tamarindus indicus* was introduced by Macassans in the eighteenth century, for example.

It is, indeed, often difficult to establish whether a particular species is native or alien. Plants which were described or renamed early from material collected in Australia must not be assumed, for that reason alone, to be native, for they could have been introduced along with or very soon after European settlement. Robert Brown, in 1802–4, noted a number of alien plants near Sydney, including *Urtica urens, Silene anglica, Anagallis arvensis, Euphorbia peplus* and *Poa annua* (Britten, 1906; Maiden, 1916). He also collected the American species, *Gnaphalium pensylvanicum*. The common weed, *Lepidium hyssopifolium*, long thought to be native in Australia, has recently been reported as being indistinguishable from the South African species, *L. africanum* (Ryves, 1977). *Senecio lautus*, in the broad sense, includes a number of native forms described by Ali (1969). Amongst these presumed native forms is one common as a weed in pastures in the central and northern coastal regions of New South Wales. O.M. Hilliard (personal communication) has, however, identified the common weed as *S. madagascariensis*, a native of South Africa and neighbouring regions (Hilliard, 1977). *Apium leptophyllum* was recorded by Bentham (1863–78) as occurring, presumably native, in both Australia and South America, but it seems certain that it is native to sub-tropical America whence it has spread throughout the sub-tropical areas of southeast Asia, Australia and elsewhere. On the other hand, Australian native material named *Eryngium rostratum* by Bentham (1863–78) is different from the Chilean plants originally described as such and should be referred to *Eryngium ovinum*.

Some species are represented by both alien and native forms. A good example is *Oxalis corniculata*, in the broad sense, which includes four or five native forms which warrant specific status, for example, *O. exilis*, as well as perhaps two or three alien forms which are common weeds in gardens and glasshouses. It is likely too that *Cynodon dactylon*, known to have been introduced from India, includes native forms.

Extent of the alien flora

Australia, with an area of 7.7×10^6 km^2, extends latitudinally from about 10° to 44°S and shows great diversity in climates and soils. Of an estimated total number of species of vascular plants somewhere between 15 000 and 20 000, about 10% are aliens. Accurate estimates of numbers are difficult because of problems in determining taxonomic status and insufficient knowledge, especially of the flora of the 3.0×10^6 km^2 of land north of the Tropic of Capricorn. Alien plants are often poorly collected and the native or alien status of some species is, as has been indicated already, difficult to determine.

The following treatment is based on a thorough examination especially of the existing Australian floras, and of the works of Burbidge (1963), Chippendale (1971), Clifford & Ludlow (1972) and Jones & Clemesha (1976).

Ferns are represented by five families, five genera and six species, all of which were introduced as garden plants. The most significant are two species of *Pityrogramma*, widespread in northern Queensland, *Salvinia molesta*, a floating fern common in dams and streams especially along the coast from central New South Wales to northern Queensland (Figure 3.1), *Selaginella kraussiana* in southeastern Australia and *Cyrtomium falcatum* which has colonised cliff faces along the coast in Sydney.

Gymnosperms are represented by only one genus, *Pinus*, with a number of species, the most common being *P. radiata*, escaped from plantations in southeastern Australia.

Angiosperms are represented by about 110 families and more than 700 genera. About 30 families are represented by one genus only and many genera by only one species. Families represented each by four or more genera, and comprising some 600 genera in all, are listed below accord-

Figure 3.1. *Salvinia molesta*, native of Brazil, covering a lake in the Botanic Gardens, Townsville, Queensland. Photo: P.W. Michael.

ing to the classification of Cronquist (1968), with the numbers of genera reported in each family in brackets. About two-thirds of the genera are alien. Those families preceded by a dagger (†) are found in a wide range of naturalised floras throughout the world, for example in Albany and Bathurst, eastern Cape Province, South Africa (Salisbury, 1919; Martin & Noel, 1960); Argentina (Hauman, 1928); British Isles (Clapham, Tutin & Warburg, 1952); California (Robbins, 1940); Japan (Hisauchi, 1940); Montpellier, southern France (Thellung, 1912); New Zealand (Allan, 1940; Healy, 1944); Quebec (Rousseau, 1968) and Texas (Texas plants checklist, 1962).

† Ranunculaceae (4); Papaveraceae (6); Urticaceae (4); Cactaceae (7); Aizoaceae (10); † Caryophyllaceae (21); Portulacaceae (4); † Chenopodiaceae (5); † Amaranthaceae (7): † Polygonaceae (4); † Malvaceae (12); Cucurbitaceae (5); † Cruciferae (36); † Primulaceae (4); † Rosaceae (17); † Leguminosae (63), including Mimosaceae (8), Caesalpiniaceae (7) and Papilionaceae (48); Onagraceae (5); † Euphorbiaceae (9); † Umbelliferae (23); Gentianaceae (4); † Apocynaceae (4); Asclepiadaceae (6); † Solanaceae (13); † Convolvulaceae (5); † Boraginaceae (13); Verbenaceae (6); † Labiatae (22); † Scrophulariaceae (23); Bignoniaceae (4); Acanthaceae (4); Pedaliaceae, including Martyniaceae (4); † Rubiaceae (9); † Compositae (106); Cyperaceae (5); † Gramineae (90); † Liliaceae, including Alliaceae, Amaryllidaceae and Alstroemeriaceae (15) and Iridaceae (23).

This list includes 25 of the 29 families recorded by Hooker (1860) in the earliest substantial, though not exhaustive, list of species, 139 in all, naturalised in Australia. The four other families included by Hooker were Fumariaceae (now represented in Australia by 2 genera), Oxalidaceae (2), Geraniaceae (3) and Plantaginaceae (1).

To indicate briefly the wide range of alien species in the Australian flora some widespread species belonging to the families Caryophyllaceae and Leguminosae, the order Lamiales, and the families Compositae and Gramineae are given here. They are listed according to whether they occur essentially in southern Australia in areas with mediterranean or more temperate climates or in northern Australia in areas with sub-tropical or tropical climates.

Southern Australia. *Stellaria media, Trifolium subterraneum, Ulex europaeus, Amsinckia* spp., *Echium plantagineum, Cirsium vulgare, Hypochoeris radicata, Phalaris aquatica, Lolium perenne.*

Northern Australia. *Drymaria cordata, Stylosanthes humilis, Mimosa pigra, Stachytarpheta* spp., *Hyptis suaveolens, Tithonia diversifolia, Tridax procumbens, Panicum maximum, Axonopus compressus.*

The area occupied by each of the large number of alien species in Australia varies greatly from the vast areas of southern Australia colonised by *Arctotheca calendula* and *Hordeum* spp. to the scattered occurrences of *Chrysanthemum leucanthemum* in the wetter areas of southeastern Australia; from the large tree, *Cinnamomum camphora,* widespread in the far north coast of New South Wales and the isolated occurrences throughout southern New South Wales of *Ailanthus altissima,* a species which extends its distribution locally by suckering, to the occasional infestations of *Alternanthera philoxeroides* in watercourses in eastern New South Wales; from the dense stands of *Ligustrum* spp. in the Sydney area to the restricted open areas colonised by *Hebenstretia dentata* in and around Newcastle; from the wide areas of *Homeria* spp. in Western Australia to the occasional garden escape, *Freesia refracta* in wetter parts of southern Australia; from the thousands of hectares of *Oxalis pes-caprae* in South Australia and the abundant weeds, *O. corymbosa* and *O. latifolia* in Sydney gardens to the single weedy occurrence of *O. tetraphylla* in an old garden near Sydney, *O. perdicaria* (Figure 3.2) known only to occupy a few hectares near Grenfell in New South Wales and the tiny isolated occurrences of *O. brasiliensis* along the Hunter River.

In an overall view of Australian vegetation we are concerned essentially only with those alien plants which are now widespread and occupy vast areas, or which are prominent features of the vegetation of particular regions or which repeatedly occur in similar situations, for example, *Euphorbia peplus* in gardens and *Polypogon monspeliensis* in damp places.

Classification of the alien flora

The great diversity in alien species and in the areas and situations they occupy lead to serious difficulties in attempts to classify them according to the complicated system outlined in detail by Thellung (1912) and further elaborated in cumbersome fashion by Ponert (1977). Even the much simpler classification of naturalised plants devised by Hauman (1928) for Argentina is scarcely workable because the different groups and categories may overlap. He divided these plants into three groups. 1. Anthropophilic weeds occurring in situations modified by man (anthropophilic plants are those whose presence is closely associated with man's activities (Thellung, 1912)); 2. species introduced unintentionally which colonise natural situations; and 3. plants escaped from cultivation. Group 1 is further subdivided into weeds of cultivation (including a special category for weeds known only in gardens) and weeds of non-cultivated areas, for example, roadsides, paths and rubbish heaps. Group 3 is further subdivided into plants of modified and of natural situations.

There are obvious difficulties in distinguishing between modified and natural situations even in Australia where European settlement began just less than 200 years ago. There are few aliens which have invaded apparently undisturbed native vegetation in Australia. *Baccharis halimifolia* in uncleared swamps of *Melaleuca quinquenervia* is one such example (Westman, Panetta & Stanley, 1975).

Figure 3.2. *Oxalis perdicaria*, native of South America, grown from bulbs collected from the site of its only known Australian occurrence near Grenfell, New South Wales. Photo: P.W. Michael.

Origins, history and spread

In ascertaining the origin of Australian alien plants one has to rely mostly upon the scant information in existing floras and standard references like Willis (1973) and Good (1974). Anderson (1939) listed the countries or regions of origin of 390 alien species, excluding grasses, in New South Wales. He recorded 93 (24%) from Europe, Asia and Africa; 65 (17%) from Europe and Asia; 44 (11%) from Europe; 31 (8%) from South America; 30 (8%) from Mediterranean regions; 24 (6%) from North America; 18 (5%) from the Americas; 12 (3%) from tropical America; 11 (3%) from Central America and the West Indies and the remaining 39 (10%) from other areas.

Everist (1960) grouped 578 plants naturalised in Queensland in 1959 as follows: 234 (40%) from Europe, Western Asia and North Africa; 114 (20%) from New World Tropics; 75 (13%) from Old World Tropics; 51 (9%) from extra-tropical South America; 48 (8%) from extra-tropical North America and the remaining 56 (10%) from Eastern Asia; South Africa; Cosmopolitan Tropics and from unknown or uncertain origins. In comparing these numbers with those naturalised plants recorded by Bailey (1883, 1890, 1899–1902, 1913), Everist (1960) indicated that there was little difference in the proportions originating from the various regions despite the increase in numbers from 132 in 1883 to 341 in 1913 and 578 in 1959.

Ewart & Tovey (1909) noted that less than 10% of the 364 aliens they recorded in Victoria were of American origin. Contrast this with the 17%, excluding grasses, recorded by Anderson (1939) and about 17% of grasses recorded by Cross & Vickery (1950) for New South Wales and with the 37% recorded by Everist (1960) for Queensland. As we move from southern Australia into sub-tropical and tropical regions of northern Australia the proportion of aliens from sub-tropical and tropical America (including the West Indies) increases markedly. Because of the history of European settlement the proportion of aliens from Europe and the Mediterranean regions is high even in Queensland.

Examples of aliens originating from some of the broad regions are given below.

Europe

> *Malva parviflora, Verbascum virgatum, Carduus nutans, Cirsium vulgare, Dactylis glomerata, Poa annua;*

Mediterranean regions

> *Ranunculus muricatus, Trifolium subterraneum, Parentucellia latifolia, Cynara cardunculus, Koeleria phleoides, Phalaris minor;*

North America
>   *Solanum elaeagnifolium, Amsinckia* spp., *Ambrosia arte-
>   misiifolia, Solidago canadensis, Andropogon virginicus, Panicum
>   capillare;*

South America
>   *Nicandra physalodes, Verbena bonariensis, Conyza bonariensis,
>   Xanthium spinosum, Paspalum dilatatum, Nassella trichotoma;*

South Africa
>   *Emex australis, Oxalis pes-caprae, Arctotheca calendula,
>   Rhynchelytrum repens, Sporobolus africanus, Homeria* spp.

It is important to try and relate the origins of the ancient weeds of cultivation which form a considerable proportion of the alien plants of Australia to the chief primary (essentially neolithic) regions of agriculture as outlined and illustrated by Darlington (1963) following Vavilov (1926, 1935) and Kuptsov (1955), namely the Indian and Indonesian regions expanding to cover southeast Asian and Pacific regions; Ethiopian, near Asiatic, Mediterranean and central Asian regions expanding to cover Europe and Western Asia; Nigerian region expanding to cover a great part of Africa north and south of the tropics; Mexican and Peruvian regions expanding to cover a great part of the Americas and the West Indies and the Chinese region expanding to cover eastern Asia.

Better still, the origin of Australian alien plants may largely be traced to the regions of origin of crop plants as outlined by Darlington (1963) following many authors. These regions include southwest Asia, the Mediterranean, Europe, Abyssinia, Central Africa, Central Asia, Indo–Burma, southeast Asia, China, Mexico, United States of America, Peru, Chile and Brazil–Paraguay.

In a consideration of the history of some Australian alien species, attention will now be given, where possible, to their places of origin. The early history of many of our alien plants has been documented by Maiden (1920), McBarron (1955), Everist (1960), Mann (1970), Michael (1972), Auld (1977) and Mitchell (1978), and the reader is referred to these papers for additional information on some of the species mentioned.

Xanthium spinosum, whose world history is admirably documented by Widder (1923), originated in Chile whence it was transported to Australia on animals or in grain in the first half of the nineteenth century, just as it had been transported to Europe much earlier. It may also have been introduced to Australia as an ornamental. *Xanthium occidentale*, originating in North America and the West Indies, was apparently introduced to Australia with cotton seed. *Xanthium italicum*, native to the United States of America and Mexico, was probably introduced in the same way. A form of *X. orientale* appears to have been introduced to Australia from California. This variable species, which I believe is native to the west coast of the United States of America, extending into Mexico, was probably introduced to Europe by the Spanish quite early; and from southern Europe, where it is now quite common, it appears to have been re-introduced to America to areas around the St. Lawrence River. *Xanthium cavanillesii*, known only from two general localities in New South Wales, both associated with very old settlements, must have been introduced soon after European settlement from South America. The two occurrences are of different forms of the species, indicating different times or ways of introduction.

The four most significant alien species of *Echinochloa* present in Australia can be traced to two original centres, the most important being the Indo–Burma region, the source of different forms of *E. colona* and *E. crus-galli* and the obligate weed of rice, *E. oryzoides*. *Echinochloa microstachya* originated in North America. A long-awned form of *E. crus-galli* and *E. oryzoides* appear to have been introduced, along with *E. microstachya* to Australia in rice seed from California. These two former occur in rice-fields in southern Europe and Iraq also.

Hordeum leporinum and *H. glaucum*, both native to the Mediterranean region, must have been introduced quite early (Cocks, Boyce & Kloot, 1976) and undoubtedly many times from various places in Europe or on the way to Australia. In their discussion of the origin of the *H. murinum* complex in Australia, Cocks *et al.* (1976) neglect to mention the probable importance of South America, especially in view of the early trade between Portugal, Spain and the American colonies and the prominence of both species in the Mediterranean region.

Some of the thistles, notably *Cirsium vulgare* and *Silybum marianum*, natives of Europe and/or the Mediterranean regions, were introduced to Australia in the very early days of settlement, *Cirsium vulgare* as a contaminant of pasture seed and *Silybum marianum*, both as a medicinal plant and as Scotch thistle. Many Australian alien plants were introduced originally as ornamentals and in a study of catalogues of Australian Botanic Gardens and nurseries, as referred to by Michael (1972), numerous species will be found. A number of weeds of great economic importance are included, for example *Eupatorium adeno-*

phorum, *Homeria* spp., *Lantana camara*, *Oxalis pes-caprae*, *Eriocereus martinii*, and *Opuntia* spp. *Amaranthus powellii*, a common weed of cultivation throughout southern Australia, may well have been introduced with *A. hypochondriacus* which was grown as an ornamental in the very early days of the settlement of New South Wales. *Amaranthus powellii* belongs to a group of species native to the Americas, including also *A. hybridus*, *A. quitensis* and *A. retroflexus*, all pioneers especially occurring along riverbanks or in other open areas in their places of origin (Sauer, 1967).

The spread of aliens after introduction to Australia has, as elsewhere, essentially followed settlement. Plants not transported by man himself with his animals, crop or pasture seeds, have been spread by wind, water or birds and have colonised disturbed situations. Ornamental plants grown in more and more home gardens have led to more and more foci from which the plants have been able to move into previously un-colonised areas. Intentionally-sown pasture species, sometimes, like *Trifolium subterraneum* originally appearing here and there of their own accord, have escaped from cultivation.

In South Australia, where *Oxalis pes-caprae* was introduced before 1850, we may imagine bulbs being grown in all home gardens in the closely settled plains north of Adelaide. With occasional removal and discarding of excess growth from gardens, cultivation and floods along the close streams rising in the Mount Lofty Ranges and traversing the plains, considerable spread was accomplished in the second half of last century. Such favourable combinations of factors do not seem to have occurred over large areas in other southern states.

In contrast, the recent explosion of the American *Parthenium hysterophorus* in Queensland (Haseler, 1976; Armstrong, 1978) was apparently initiated from the sowing in 1960 of a single large area of impure grass seed imported from Texas. This plant was first collected in the Upper Brisbane Valley in 1955, where it still occurs, but it required a massive inoculum to stimulate large-scale spread. The plant now occupies many thousands of square kilometres.

The rapid spread of this plant is exceptional in the history of aliens in Australia. Doley (1977) and Williams & Groves (1980) have predicted a much wider potential distribution. Doing, Biddiscombe & Knedlhans (1969) and Medd & Smith (1978) have predicted further increases in the area occupied by *Carduus nutans* ssp. *nutans* which was first collected in Australia in 1950, almost certainly introduced in seed of *Lolium perenne* from New Zealand.

There is little concrete information on the rate of spread of aliens in Australia. McVean (1966) has calculated that *Chondrilla juncea* spread throughout part of southern Australia at an average rate of about 24 km per year for 40 years after its initial introduction to Australia. In a period of ten years, rapid increases in the area occupied by *Andropogon virginicus* in the neighbourhood of Sydney have been observed. *Ambrosia tenuifolia* from South America, long established around Newcastle, has greatly extended its range between Newcastle and Sydney. The South American *Facelis retusa* originally collected near Newcastle is now extending beyond Sydney. Macarthur (Select Committee, 1852) first noted *Xanthium spinosum* along the Nepean River in the late 1830s. By the early 1850s it had become a weed of public concern. For many years and probably since the early days of Sydney *Parietaria judaica* has been firmly established in a relatively small area near the harbour, but in the last few years, with the tremendous disturbance associated with new building and freeway operations, there has been a sudden increase in area occupied by the species.

Moore (1959) has given an account of changes in vegetation in southeastern Australia following clearing of woodlands, grazing and use of superphosphate and the sowing of *Trifolium subterraneum*. These management factors, leading to increased soil fertility, have encouraged the spread of many weedy aliens, including plants well-known as nitrophilous species in Europe, for example, *Cirsium vulgare* and *Onopordum acanthium*, confined in the initial stages of settlement to stock camps and rabbit warrens.

Distribution of aliens in Australia and areas of origin
This subject has been reviewed briefly in relation to important Australian pasture weeds by Michael (1970), who differentiated between those aliens originating in Europe or the Mediterranean region, areas with a long history of disturbance due to man's activities, and those originating from the Cape Province of South Africa, a region subjected to a much shorter period of disturbance by man. The former species may have developed much greater adaptability to different environments than the latter, and, accordingly, when introduced to Australia have colonised areas not exactly comparable with the areas of optimum range in Europe.

Echium plantagineum occurs in lower latitudes in Australia than in the Northern Hemisphere (Moore, 1967). *Carduus nutans* ssp. *nutans*, a biennial with a qualitative vernalisation requirement (Medd, 1977), occurs in areas in Australia with higher winter temperatures and lower relative humidities than in Europe (Doing, Biddiscombe & Knedlhans, 1969). It is difficult to relate the distribution of *Avena fatua* and *A. ludoviciana* in Australia to their distribution in Europe and the Mediterranean region. Although *A. ludoviciana* is common in the Mediterranean region it is of little importance in areas of Australia with mediterranean-type climate, where *A. fatua*, characteristic of the more northern parts of Europe, prevails. In Australia, *A. ludoviciana* is most abundant in the wheat belt of northern New South Wales and southern Queensland where there is appreciable summer rainfall.

On the other hand, species from the Cape Province, for example, *Oxalis pes-caprae*, *Arctotheca calendula* and *Homeria* spp., occupy essentially similar climatic zones in Australia and South Africa. It appears from the data presented by Gray (1976) that *Chrysanthemoides monilifera* ssp. *monilifera* and ssp. *rotundata*, from the southwestern Cape Province and the southeastern coastal areas of South Africa respectively, each occupy corresponding climatic zones in Australia.

Habitat preferences of plants in Europe may be reflected in their distribution in Australia. A good example is *Trifolium subterraneum* ssp. *yanninicum*, essentially confined in Europe to waterlogged soils in the Balkan peninsula and naturalised only in similar situations in southern Australia (Morley & Katznelson, 1965).

Species originating in the Americas may sometimes show clear distributional relationships, but, more often, the relationships are confused. *Eremocarpus setigerus*, native to California with mediterranean-type climate, is naturalised in Australia only in Western Australia and South Australia in areas with similar climate. *Conyza canadensis* var. *canadensis*, abundant especially in the eastern United States of America and in Canada, occurs only sporadically in coastal and tableland regions of eastern Australia, whilst *C. parva*, widespread in the eastern United States of America, the West Indies, Mexico, Central America and northern South America, occurs only in the coastal, mostly sub-tropical, parts of eastern Australia.

McMillan (1975) related occurrences of four *Xanthium* spp. in Australia to particular areas in the United States and in South America from which he believes them to have been introduced, essentially on the basis of photoperiodic requirements. All the Australian collections of *X. occidentale* (referred to as *X. chinense*), examined by McMillan represented a type photoperiodically adapted to the southeastern United States and similar to populations occurring on the Mississippi Delta south of New Orleans, Louisiana, situated at about 30°N. In Australia, however, *X. occidentale* extends in great abundance from latitudes below 20° to 33°S. *Xanthium cavanillesii* from near Sydney was shown to be nearly identical with the same species from Buenos Aires, Argentina, at about the same latitude (34°S).

There are many examples of pairs of closely related species which show overlapping distributions in Australia. *Digitaria ciliaris* of sub-tropical or tropical origin and *D. sanguinalis* of more temperate origin often occur together in central New South Wales but, at lower latitudes in Queensland, *D. ciliaris* occurs to the virtual exclusion of *D. sanguinalis*. The reverse situation applies at higher latitudes in Victoria. The American species *Gnaphalium americanum*, of sub-tropical and tropical origin, and *G. spicatum*, of more temperate origin, show essentially the same pattern.

Historical factors which have greatly influenced the occurrence of aliens in Australia may often obscure ecological relationships. *Echium vulgare*, from Europe, was apparently introduced to Australia much earlier than *E. plantagineum* (Piggin, 1977) and is now confined to a few of the older settled areas in southeastern Australia, where it is often accompanied by the generally much more successful *E. plantagineum*. The introduction and subsequent establishment of *Oxalis perdicaria* in a single small area in New South Wales referred to earlier could be termed an historical accident. The abundance of *Amaranthus quitensis* in the Hunter Valley, New South Wales and its virtual absence elsewhere in Australia (Michael, 1977) can perhaps be attributed to its early introduction there from South America.

Variation in alien species in Australia
In a country where many alien species occur in profusion over wide areas it is to be expected that considerable variation in individual seed-producing species would be evident. Few species, however, have been studied in detail. Perhaps the most well-known work is that concerned with *Trifolium subterraneum* (Aitken & Drake, 1941; Morley, 1961; Morley & Katznelson, 1965), where a great variety of forms differing in vegetative and floral characters and flowering times have been observed. Similar studies have been made on *T. glomeratum* (Woodward & Morley, 1974) where numerous variations have been noted. Much variation in growth habit, flowering times and spikelet characters have been observed in *Avena fatua*, *A. ludoviciana* and *A. barbata* (Whalley & Burfitt, 1972; Paterson, 1976). There is little evidence, however, that distinct Australian forms have developed. Rather, it is supposed that the

variation in Australian material is the direct result of a wide range of introductions.

Three forms of *Chondrilla juncea*, differing in the shape of rosette leaves, inflorescence morphology and fruit characters, were described by Hull & Groves (1973). One of the forms closely resembles plants collected from southern Italy and it is likely that the other two forms of this apomictic species could be matched with forms occurring in Europe.

Little variation occurs in the major occurrences of the tristylic species *Oxalis pes-caprae* throughout southeastern Australia. These occurrences are of a short-styled pentaploid clone, originally introduced as a garden plant which reproduces only by vegetative means. There are, however, other forms, all tetraploid and with all three style lengths, naturalised especially in Western Australia (Michael, 1964). Where tetraploid forms of different style lengths occur together viable seed may be produced, encouraging further variation in the species. The freely cross-breeding species, *Lolium perenne*, *L. multiflorum* and *L. rigidum* are very difficult to distinguish from each other in areas in Australia where any two or all three species have been introduced.

Proper attention to studies of variation in certain broad species groups has led, in recent years, to the definition of a number of distinct species within these groups, for example, in *Solanum nigrum sens. lat.* (Henderson, 1974), *Rubus fruticosus* sp. agg. (Amor & Miles, 1974) and *Amaranthus hybridus sens. lat.* (Michael, 1977).

Conclusions

Many of the studies on Australian alien species have been prompted by their economic significance, both favourable and unfavourable. Successful and desirable species, such as *Trifolium subterraneum*, *Stylosanthes humilis*, *Lolium rigidum* and *Chloris gayana* appear to share certain biological attributes with undesirable weeds, such as *Cirsium vulgare*, *Hyptis suaveolens*, *Hordeum* spp. and *Xanthium occidentale*. Both groups are able to establish themselves, often forming extensive populations, without man's deliberate action, and tend to be aggressive, competitive and adaptable. Efficient reproduction and the ability to survive under temporarily unfavourable conditions are held in common. We return, thus, to Thellung's definition of naturalised species given earlier.

There are big gaps in our knowledge of the alien flora as a whole. Problems of identity are foremost. It is to be hoped that recent taxonomic treatments like those of *Solanum nigrum* and related species

(Henderson, 1974), *Rubus fruticosus* sp. agg. (Amor & Miles, 1974) *Adonis* (Kloot, 1976), *Hordeum* (Cocks *et al.*, 1976), *Datura* (Haegi, 1976*a*), *Lycium* (Haegi, 1976*b*), *Paronychia* (Aston, 1977), *Amaranthus* and *Conyza* (Michael, 1977) and *Echium* (Piggin, 1977) will be extended to cover a much wider range of genera.

Work on alien genera is fraught with many difficulties, notably the general paucity of specimens collected in Australia and the difficulties in acquiring collections and literature from other countries. Co-operation with taxonomists in other parts of the world is indispensable.

There are few precise data on distribution of alien species in Australia. Campbell (1977), in contributing to our knowledge of *Nassella trichotoma* and *Hypericum perforatum* var. *angustifolium*, pointed to the difficulties in assessing area and distribution of weeds and suggested that new grid survey techniques as outlined by Brook (1976) could be used. In studies on Australian aliens artificial state boundaries should be ignored.

In view of the recent interest in methods of predicting the potential spread of species already established in Australia, as exemplified by Medd & Smith (1978) for *Carduus nutans* and Doley (1977) and Williams & Groves (1980) for *Parthenium hysterophorus*, familiarity with alien and weedy native species recorded for other parts of the world, but not yet recorded in Australia, is desirable. *Nassella trichotoma* (under the name *Stipa trichotoma*) was recorded by Coste (1906) as being completely naturalised on the banks of the River Orba in the Piedmont, Italy and *Parthenium hysterophorus* was recorded by Hauman (1928) as a garden weed in Tucumán, Argentina, at about 27°S. It is especially important that close attention be paid to the alien and weedy floras of South America and sub-tropical South Africa.

References

Aitken, Y. & Drake, F.R. (1941). Studies of the varieties of subterranean clover. *Proceedings of the Royal Society of Victoria, 53*, 342–93.

Ali, S.I. (1969). *Senecio lautus* complex in Australia. V. Taxonomic interpretations. *Australian Journal of Botany, 17*, 161–76.

Allan, H.H. (1940). *A Handbook of the Naturalized Flora of New Zealand.* Department of Scientific and Industrial Research Bulletin, Number 83.

Amor, R.L. & Miles, B.A. (1974). Taxonomy and distribution of *Rubus fruticosus* L. agg. (Rosaceae) naturalized in Victoria. *Muelleria, 3*, 37–62.

Anderson, R.H. (1939). The naturalised flora of New South Wales (excluding Gramineae). *Contributions of the New South Wales National Herbarium, 1*, 16–33.

Armstrong, T.R. (1978). Herbicidal control of *Parthenium hysterophorus*. *Proceedings of the First Conference of the Council of Australian Weed Science Societies*, pp. 157–64. Melbourne: Council of Australian Weed Science Societies.

Aston, H.I. (1977). The species of *Paronychia* (Caryophyllaceae) in Victoria. *Muelleria*, *3*, 209–14.

Auld, B.A. (1977). The introduction of *Eupatorium* species to Australia. *Journal of the Australian Institute of Agricultural Science*, *43*, 146–7.

Bailey, F.M. (1883). *A Synopsis of the Queensland Flora*. Brisbane: Government Printer.

Bailey, F.M. (1890). *A Synopsis of the Queensland Flora*, 3rd suppl. Brisbane: Government Printer.

Bailey, F.M. (1899–1902). *The Queensland Flora*. Brisbane: H.J. Diddams.

Bailey, F.M. (1913). *Comprehensive Catalogue of Queensland Plants*. Brisbane: Government Printer.

Bentham, G. (1863–78). *Flora Australiensis*. London: Lovell Reeve & Co.

Blakely, W.F. (1923). A census of the weeds and naturalised plants of New South Wales (Abstract). *Proceedings of the Pan-Pacific Science Congress*, *Australia*, Vol. 1, p. 330. (The list of plants was never published but what appears to be the final draft is held in the Library of the National Herbarium of New South Wales.)

Britten, J. (ed). (1906). Introduced plants at Sydney, 1802–4. *Journal of Botany (British and Foreign)*, *44*, 234–5.

Brook, A.J. (1976). A biogeographic grid system for Australia. *Search*, *7*, 191–5.

Burbidge, N.T. (1963). *Dictionary of Australian Plant Genera*. Sydney: Angus & Robertson.

Campbell, M.H. (1977). *Assessing the area and distribution of Serrated Tussock* (Nassella trichotoma), *St. John's Wort* (Hypericum perforatum *var.* angusti-folium) *and Sifton Bush* (Cassinia arcuata) *in New South Wales*. New South Wales Department of Agriculture Technical Bulletin, Number 18.

Candolle, A.L.P.P. de (1855). *Géographie Botanique Raisonnée*, Vol. 2. Paris

Chippendale, G.M. (1971). Check list of Northern Territory plants. *Proceedings of the Linnean Society of New South Wales*, *96*, 207–67.

Clapham, A.R., Tutin, T.G. & Warburg, E.F. (1952). *Flora of the British Isles*. London: Cambridge University Press.

Clifford, H.T. & Ludlow, G. (1972). *Keys to the Families and Genera of Queensland Flowering Plants*. Brisbane: University of Queensland Press.

Cocks, P.S., Boyce, K.G. & Kloot, P.M. (1976). The *Hordeum murinum* complex in Australia. *Australian Journal of Botany*, *24*, 651–62.

Coste, H.J. (1906) *Flore descriptive et illustrée de la France*, Vol. 3. Paris: Librairie des Sciences et des Arts.

Cronquist, A. (1968). *The Evolution and Classification of Flowering Plants*. Boston: Houghton Mifflin Co.

Cross, D.O. & Vickery, J.W. (1950). List of the naturalised grasses in New South Wales. *Contributions from the New South Wales National Herbarium*, *1*, 275–80.

Darlington, C.D. (1963). *Chromosome Botany and the Origins of Cultivated Plants*, 2nd edn. London: George Allen & Unwin.

Doing, H., Biddiscombe, E.F. & Knedlhans, S. (1969). Ecology and distribution of the *Carduus nutans* group (nodding thistles) in Australia. *Vegetatio*, *17*, 313–51.

Doley, D. (1977). Parthenium weed (*Parthenium hysterophorus* L.): Gas exchange characteristics as a basis for prediction of its geographical distribution. *Australian Journal of Agricultural Research*, *28*, 449–60.

Everist, S.L. (1960). Strangers within the gates. *Queensland Naturalist*, *16*, 49–60.

Ewart, A.J. & Tovey, J.A. (1909). *The Weeds, Poison Plants and Naturalized Aliens of Victoria. Pt. II. Census of the Naturalized Aliens and Introduced Exotics of Victoria.* Melbourne: Government Printer.

Good, R. (1974). *The Geography of the Flowering Plants*, 4th edn. London: Longman.

Gray, M. (1976). Miscellaneous notes on Australian plants. 2. *Chrysanthemoides* (Compositae). *Contributions from Herbarium Australiense, 16*, 1–5.

Haegi, L. (1976*a*). Taxonomic account of *Datura* L. (Solanaceae) in Australia with a note on *Brugmansia* Pers. *Australian Journal of Botany, 24*, 415–35.

Haegi, L. (1976*b*). Taxonomic account of *Lycium* (Solanaceae) in Australia. *Australian Journal of Botany, 24*, 669–79.

Haseler, W.H. (1976). *Parthenium hysterophorus* in Australia. *PANS, 22*, 515–17.

Hauman, L. (1928). Les modifications de la flore Argentine sous l'action de la civilisation. *Mémoires de l'Academie Royale de Belgique Classe des Sciences 2s, 9*, 3–95.

Healy, A.J. (1944). Some additions to the naturalised flora of New Zealand. *Transactions of the Royal Society of New Zealand, 74*, 221–31.

Henderson, R.J.F. (1974). *Solanum nigrum* L. (Solanaceae) and related species in Australia. *Contributions from the Queensland Herbarium*, Number 16.

Hilliard, O.M. (1977). *Compositae in Natal.* Pietermaritzburg: University of Natal Press.

Hisauchi, K. (1940). *Naturalised Plants* (in Japanese). Tokyo: Kagakutosho-shuppansha.

Hooker, J.D. (1860). On some of the naturalized plants of Australia. In *The Botany (of) the Antarctic Voyage*, part III. *Flora Tasmaniae*, vol. 1. London: Lovell Reeve.

Hull, V.J. & Groves, R.H. (1973). Variation in *Chondrilla juncea* L. in southeastern Australia. *Australian Journal of Botany, 21*, 113–35.

Jones, D.L. & Clemesha, S.C. (1976). *Australian Ferns and Fern Allies.* Sydney: A.H. & A.W. Reed.

Kloot, P.M. (1976). The species of *Adonis* L. naturalized in Australia. *Muelleria, 3*, 200–7.

Kuptsov, A.I. (1955). Geographical distribution of cultivated flora and its historical development (Russian). *Geograficheskoe obshchestvo, Izvestiya, 87*, 220–31.

McBarron, E.J. (1955). An enumeration of plants in the Albury, Holbrook and Tumbarumba districts of New South Wales. *Contributions from the New South Wales National Herbarium, 2*, 89–247.

Macknight, C.C. (1976). *The Voyage to Marege.* Melbourne: Melbourne University Press.

McMillan, C. (1975). The *Xanthium strumarium* complexes in Australia. *Australian Journal of Botany, 23*, 173–92.

McVean, D.N. (1966). Ecology of *Chondrilla juncea* L. in southeastern Australia. *Journal of Ecology, 54*, 345–65.

Maiden, J.H. (1916). Weeds at Sydney in 1802–4. *Agricultural Gazette, New South Wales, 27*, 40.

Maiden, J.H. (1920). *The Weeds of New South Wales*, Part 1. Sydney: Government Printer.

Mann, J. (1970). *Cacti Naturalised in Australia and their Control.* Brisbane: Department of Lands.

Martin, A.R.H. & Noel, A.R.A. (1960). *The Flora of Albany and Bathurst.* Grahamstown: Department of Botany, Rhodes University.

Medd, R.W. (1977). Some aspects of the ecology and control of *Carduus nutans* L. (Nodding Thistle) on the Northern Tablelands, New South Wales, Ph.D. thesis, University of New England, N.S.W.

Medd, R.W. & Smith, R.C.G. (1978). Prediction of the potential distribution of *Carduus nutans* (nodding thistle) in Australia. *Journal of Applied Ecology, 15,* 603–12.

Michael, P.W. (1964). The identity and origin of varieties of *Oxalis pes-caprae* L. naturalized in South Australia. *Transactions of the Royal Society of South Australia, 88,* 167–74.

Michael, P.W. (1970). Weeds of grasslands. In *Australian Grasslands,* ed. R.M. Moore, pp. 349–60. Canberra: Australian National University Press.

Michael, P.W. (1972). The weeds themselves – early history and identification. In *Symposium: The History of Weed Research in Australia. Proceedings of the Weed Society of New South Wales, 5,* 3–18.

Michael, P.W. (1977). Some weedy species of *Amaranthus* (amaranths) and *Conyza/Erigeron* (fleabanes) naturalized in the Asian-Pacific region. *Proceedings of the Sixth Asian-Pacific Weed Science Society Conference,* Vol. *1,* pp. 87–95.

Mitchell, A.S. (1978). An historical overview of exotic and weedy plants in the Northern Territory. *Proceedings of the First Conference of the Council of Australian Weed Science Societies,* pp. 145–53.

Moore, R.M. (1959). Ecological observations on plant communities grazed by sheep in Australia. In *Biogeography and Ecology in Australia,* (Monographiae Biologicae, VIII), ed. A. Keast, R.L. Crocker & C.S. Christian, pp. 500–13. The Hague: Junk.

Moore, R.M. (1967). The naturalisation of alien plants in Australia. *International Union for Conservation of Nature and Natural Resources Publications,* New Series, Number 9, 82–97.

Morley, F.H.W. (1961). Subterranean Clover. *Advances in Agronomy, 13,* 57–123.

Morley, F.H.W. & Katznelson, J. (1965). Colonization in Australia by *Trifolium subterraneum* L. In *The Genetics of Colonizing Species,* ed. H.G. Baker & G.L. Stebbins, pp. 269–82. London: Academic Press.

Paterson, J.G. (1976). Vernalisation and photoperiod requirement of naturalized *Avena fatua* L. and *Avena barbata* Pott. ex Link in Western Australia. *Journal of Applied Ecology, 13,* 265–72.

Piggin, C.M. (1977). The herbaceous species of *Echium* (Boraginaceae) naturalized in Australia. *Muelleria, 3,* 215–44.

Ponert, J. (1977). Ergasiophygophytes and xenophytes of East Asiatic origin in Adjaria. A stimulus to new terminology, especially for ergasiophygophytes. *Folia Geobotanica and Phytotaxonomia, 12,* 9–22.

Robbins, W.W. (1940). *Alien plants growing without cultivation in California.* California Agricultural Experiment Station Bulletin, Number 637.

Rousseau, C. (1968). Histoire, habitat et distribution de 220 plantes introduites au Quebec. *Naturaliste canadienne, 95,* 49–169.

Ryves, T.B. (1977). Notes on wool-alien species of *Lepidium* in the British Isles. *Watsonia, 11,* 367–72.

Salisbury, F.S. (1919) Naturalised plants of Albany and Bathurst. *Records of the Albany Museum, 3,* 161–77.

Sauer, J.D. (1967). The grain amaranths and their relatives: a revised taxonomic and geographic survey. *Annals of the Missouri Botanical Garden, 54,* 103–37.

Select Committee on the Scotch Thistle and Bathurst Burr Report (1852). Sydney: Government Printer.

Texas Plants (1962). *Texas Plants – a Checklist and Ecological Summary*. Texas Agricultural Experiment Station, MP–585.

Thellung, A. (1912). La flore adventice de Montpellier. *Mémoires de la Société Nationale des Sciences Naturelles et Mathematiques de Cherbourg, 38*, 57–728.

Vavilov, N.I. (1926). Studies on the Origin of Cultivated Plants. *Trudy po prikladnoî botanike, genetike i selektsii, 16* No. 2 (Russian with English translation pp. 139–245).

Vavilov, N.I. (1935). Phytogeographic basis of plant breeding. (Tr. by K.S. Chester from original Russian in Vavilov, N.I. (ed.) *Theoretical Bases of Plant Breeding*, Vol. 1. Moscow & Leningrad: State Agricultural Publishing House. (*Chronica Botanica, 13*, 14–54, 1949/50).

Watson, H.C. (1847). *Cybele Britannica*, Vol. 1. London: Longman & Co.

Westman, W.E., Panetta, F.D. & Stanley, T.D. (1975). Ecological studies on reproduction and establishment of the woody weed groundsel bush (*Baccharis halimifolia* L.: Asteraceae). *Australian Journal of Agricultural Research, 26*, 855–70.

Whalley, R.D.B. & Burfitt, J.M. (1972). Ecotypic variation in *Avena fatua* L., *A. sterilis* L. (*A. ludoviciana*) and *A. barbata* Pott. in New South Wales and Queensland. *Australian Journal of Agricultural Research, 23*, 799–810.

Widder, F.J. (1923). Die Arten der Gattung *Xanthium*. *Feddes Repertorium* Beihefte Bd. 20.

Williams, J.D. & Groves, R.H. (1980). The influence of temperature and photoperiod on growth and development of *Parthenium hysterophorus* L. *Weed Research, 20*, 47–52.

Willis, J.C. (1973). *A Dictionary of Flowering Plants and Ferns*, revised by H.K Airy Shaw, 1897, 8th edn. London: Cambridge University Press.

Woodward, R.G. & Morley, F.H.W. (1974). Variation in Australian and European collections of *Trifolium glomeratum* L. and the provisional distribution of the species in southern Australia. *Australian Journal of Agricultural Research, 25*, 73–88.

MAJOR VEGETATION TYPES

Preamble

The concept that vegetation types change along a continuum or gradient runs counter to the classification of vegetation types into neat categories. This conflict became apparent in planning this book and in writing individual chapters, especially on forests and woodlands and those distinctively Australian scrublands called 'mallees'.

The allocation of chapters in this section has been based on the most recent classification of Australia's major vegetation types by Specht (see Table 18.1, p. 397). But no effort has been made to constrain individual authors to the terminology of that classification. Whether a chapter refers to 'tall open-forest' or to the 'wet sclerophyll forest' of earlier writers matters little, provided the individual author conveys to the reader something of the intrinsic structure, floristic composition and ecological relationships of his area. The authors and the editor have tried to describe the vegetation types in their own ways without adhering to rigid definitions.

The chapters in the next section (Vegetation of Extreme Habitats), on the other hand, are based on readily identifiable regions of Australia, such as the alpine region, and together describe vegetation of a number of types. There is therefore some inevitable overlap between the sections.

I hope that there are sufficient cross-references to help the reader through the fundamental difficulty of viewing vegetation as a continuum when reading about it in separate chapters written by highly individualistic Australian botanists.

R.H. Groves

4

The rainforests of northern Australia

L.J. WEBB & J.G. TRACEY

Despite incursions by industrial man over the last century, the remaining Australian rainforest vegetation has a unique ecological status, taxonomic interest, and aesthetic appeal that continue to attract increasing attention from scientific specialists and the community at large. The original area of rainforest when European man arrived was relatively small (less than one per cent of the surface of the continent) and its distribution pattern resembled an 'archipelago of habitats' (Herbert, 1967).

The rainforest habitats preserve a remarkable wealth of endemic and, in some areas, primitive biota, as well as exhibiting strong affinities at the generic level with surrounding countries that were continuous with the Australian land mass in Gondwanic time. Although the processes of evolution and community development responsible for the patterns of Australian rainforests are being unravelled only now, evidence already forthcoming indicates a need for revision of traditional concepts in Australian phytogeography that previously regarded the floristic elements of the northern rainforests as alien and invasive (Webb & Tracey, 1980).

The earliest descriptions of Australian rainforests were taxonomic, and the northern types were soon recognised as 'of high interest for the genesis of the Australian flora' (translation of Diels, 1906). Ecological studies did not begin until about fifty years ago, at first near Sydney and Melbourne, and expanded northwards and then to the northwest (for references, see Webb & Tracey, 1980).

Except for the scattered pockets of monsoon forest in northern Australia, many of the remaining rainforests along the eastern coast are relatively accessible. In this chapter we shall describe briefly the widely differing types, their distribution and dynamics, and offer some interpretations of them.

Ecological characteristics
Australian rainforests have several striking characteristics that readily set them apart from the rest of Australian vegetation.

Definition
The canopy is closed and the trees densely spaced, in contrast to the open and generally scattered sclerophyllous vegetation that covers most of the forested area of the continent in the moister coastal and sub-coastal zones, but mainly in the east.

In tropical and sub-tropical rainforests there are three or more tree layers, with or without emergents. The tree layers become reduced to two and eventually to one distinct layer at higher latitudes and altitudes, i.e. temperate and montane forest types that are floristically impoverished. Rainforests are also distinguished from other forests by combinations of characteristic life forms, e.g. epiphytes, lianes, certain root and stem structures, certain tree ferns and palms, absence of annual herbs on the forest floor. Species composition is most complex and the interdependence of the different niches of the forest is most complete under optimal environmental conditions and on relatively large ground surfaces of long stability. Lowering of moisture and temperature, decrease in soil nutrient availability and aeration, and reduction in habitat size and stability are accompanied by lower species diversity. All rainforests, especially the most complex tropical ones, are composed of a mixture of species representing different stages of succession following different kinds of disturbance. These successions ensure the maximum saturation of an area in time as well as in space. Disturbances should be understood as an integral factor of the rainforest environment, and they strongly influence species diversity.

Rainforests, including monsoonal types of drier northern areas, vary in the proportion of evergreen and deciduous species in the canopy, and belong to the category of closed-forests and closed-scrubs, as opposed to open-forests, woodlands and scrubs (Specht, 1970). The closed-forests correspond to the '*forêts denses*', referred to mainly in the African and French literature (e.g. Aubréville, 1965*a*), and to the '*ombrophilous*' forests of the UNESCO world classification (UNESCO, 1969).

The term rainforest is well entrenched in the literature of Australia (e.g. Diels, 1906; Francis, 1929; Fraser & Vickery, 1938; Baur, 1957;

Webb, 1959) and elsewhere (e.g. Schimper, 1903; Richards, 1952; Odum & Pigeon, 1970; Walter, 1971; Whitmore, 1975) so that there are advantages in retaining it. One of the most cogent botanical reasons is that tropical, sub-tropical, monsoonal and temperate types of rainforest in Australia resemble formation-types in other countries, with which they have many biological affinities and can be readily compared. This is in spite of the lack of uniformity in terminology for rainforests in Africa, Asia and America noted by Letouzey (1978). Within Australia, seemingly disparate types of rainforest occur throughout a wide geographical range (see Figure. 4.1) stretching in an arc from the northwest to the southeast through some 6000 km, but they all form part of a floristic continuum, especially when other plants besides trees are included.

The spelling 'rainforest' as a single word adopted here follows Baur (1968), and denotes a series of formations that are generally independent

Figure 4.1. Distribution of Australian rainforests. From Webb & Tracey, 1980. The arrows indicate the approximate centre of distribution ('core area') of each climatic type but the boundaries of each type overlap because of past climatic changes.

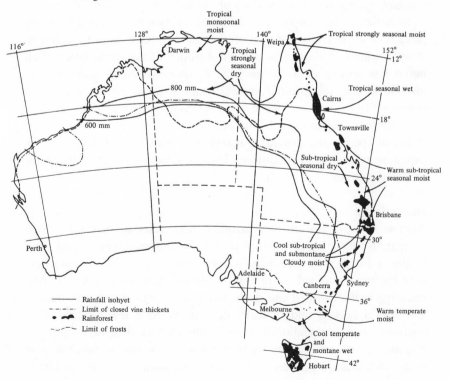

from other forest types. The spelling also avoids undue emphasis on rain as the sole determining environmental factor (Baur, 1968). It is, however, considered that the term rainforest should be restricted to the general formation-type, and that further subdivisions, equivalent to formations and subformations which can be broadly correlated with climate and soils, are best described as structural types following the nomenclature of Webb (1959, 1968, 1978). Floristic provinces or elements (see below), whose distribution among structural types and their habitats is the result of climatic sifting and historical factors, are most conveniently designated as climatic types in relation to their 'core' areas at the present time. These are assumed, perhaps arguably, to indicate the particular climates under which the floristic elements originally evolved.

To avoid circular reasoning, rainforest is strictly defined to exclude species that do not regenerate within a well-developed or slightly disturbed canopy of closed-forest. A corollary is that such excluded species are usually integral members of other formation-types such as eucalypt forests. This definition excludes, for example, all species of *Eucalyptus*, *Casuarina* and *Melaleuca*; all but one species of *Callitris* (*C. macleayana*); all but a few species of *Acacia* (e.g. *A. bakeri*, *A. fasciculifera* in the dry sub-tropics; and debatably *A. melanoxylon* in attenuated patches of warm temperate rainforest in the southeast, and *A. mangium*, *A. aulacocarpa*, *A. cincinnata* in the wet tropics of the northeast); and all but a few species of *Tristania* (e.g. *T. pachysperma* in semi-swampy lowland tropical rainforest; *T. exiliflora* and *T. laurina* mostly in riparian situations in tropical and temperate regions respectively; and arguably, *T. conferta* in the moist sub-tropics). There is most difficulty with the status of species and genera in marginal communities under generally extreme environmental conditions. These are discussed further in a later section (Interspersions, p. 92).

Discontinuity

Another striking feature of Australian rainforests is their small area compared with that of sclerophyll forests and woodlands, and their scattered distribution, especially in drier inland areas of the north where they become restricted to fire-proof niches. Rainforest 'pockets' representing well-integrated and distinctive biotic communities may occupy less than a hectare and typically do not exceed a few hectares. Discontinuity and fragmentation are especially characteristic of a rainforest floristic element towards its limits, for instance, monsoon rainforests in permanent soakage pockets in the northwest; warm temperate forests in fire-protected moist gullies in the southeast; warm sub-tropical forests on cool cloudy summits with increasing latitudes. More continuous distributions occur in more favourable and extensive habitats, e.g. the Cooktown–Ingham massif in northern Queensland (approx. 360 km by 80 km), and highland segments with orographic rainfall and frequent clouds along the eastern coast, e.g. the Border Ranges (Macpherson Range) of Queensland and New South Wales, and the scarps and foothills of basaltic plateaux farther southwards in New South Wales. However, the small-scale maps available tend to disguise the ecological fact that the scattered 'islands' of rainforests have tenuous connections by frequent outliers in gullies, along riparian alluvia, on cloudy mountain peaks and residual basaltic caps, and bouldery outcrops and rocky screes that function as 'fire shadows'. This virtual continuity of rainforest habitats is of great biological significance, as for bird-migration routes (Kikkawa, 1968) and preservation of related taxa of insects (Darlington, 1961; Parsons & Bock, 1977) and other organisms discussed in Keast (1980).

Airphoto patterns strikingly confirm impressions gathered from ground surveys that the distribution and composition of rainforest pockets reflect historical processes of climatic–edaphic–topographic sifting which predated the use of fire by Aboriginal man, and that many of the rainforest fragments were more continuous in earlier and more favourable climatic periods. Geomorphological and palynological evidence for these inferences, and for the profound influence of past climatic changes on the stability and trend of contemporary rainforest patterns, is discussed later in this chapter.

Segregation
Finally, there are few affinities between the floras of the rainforests and adjacent sclerophyll vegetation. This extraordinary ecological situation, so different for example from the rainforests and *campos cerrados* of Brazil or the evergreen, moist-dry deciduous forest series of India, prompted Aubréville (1965b) on a visit to northern Queensland to exclaim: 'les flores paraissent issues de deux mondes différents' (the floras seem to be derived from two different worlds). The segregation of the two floras is nevertheless incomplete, varying in extent between north and south and east and west. Interspersion of the floras may occur in ecotonal communities, among sclerophyll woodlands on suitable soils in the tropics and sub-tropics, and in temperate and montane situations. Additionally, a few taxa are common to both well-developed rainforests and genuinely sclerophyll vegetation types such as the scrubs and heaths on nutrient-poor sands, as well as strand and mangrove communities.

The significance of these shared taxa in the phylogenetic development of Australian vegetation communities is not properly understood (Webb & Tracey, 1980), but a list of examples at the level of families includes:

> Capparidaceae, Celastraceae, Combretaceae, Dilleniaceae, Ebenaceae, Euphorbiaceae, Fabaceae, Flindersiaceae, Meliaceae, Mimosaceae, Myrtaceae, Pittosporaceae, Proteaceae, Rhamnaceae, Rubiaceae, Rutaceae, Solanaceae, Sterculiaceae, Verbenaceae.

Each of these families has one or more genera shared between closed- and open-forest communities, and their distributions are presumably the result of sifting and co-adaptation in evolutionary time.

Distribution and area

The main areas of distribution are shown in Figure 4.1, but as already noted, allowance should be made for many small interconnecting 'islands' (see Figure 4.2) between the larger massifs. Xeric replace mesic types in 'dry corridors' along the northeastern coast, e.g. Coen–Cooktown, Ingham–Bowen, and Sarina–Gladstone areas in Queensland, as well as in rain-shadows in the lee of coastal ranges. The amount of rainfall is broadly correlated with the presence of coastal mountains which receive orographic rain from the southeastern trade winds and where precipitation is augmented by cloud-drip. Discontinuous patches

Figure 4.2. Typical 'islands' of rainforest in sclerophyll vegetation. Land unit 48 is microphyll vine thicket on stony basaltic soils, surrounded by 47 (*Eucalyptus orgadophila* grassy woodland on basaltic cracking clays) and 16 (*E. maculata* open-forest on lateritized sediments). From Gunn & Nix, 1977, p. 85. Photo: CSIRO.

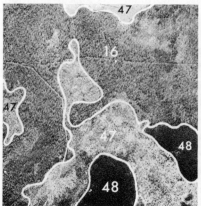

of rainforest extend away from the coast in northeastern Australia (north of severe frosts at latitude approx. 30°S) to about the 600 mm isohyet, and in northern Australia to about the 800 mm isohyet.

The geographic tropic has little significance in the differentiation of the climatic types. Structural and floristic classifications (see next sections) indicate that tropical and sub-tropical types extend southwards to the Illawarra district south of Sydney, e.g. to Minna Murra Falls. Frosts extend north of the tropic in Queensland on the lowlands to about Ingham (latitude 19°S). The Cardwell Range near Ingham, the adjacent mangroves of Hinchinbrook Channel, and the palm swamps just to the north near Tully all contribute to a distinct tropical ecological boundary for many plant and animal taxa.

There are also outliers northwards within the strictly tropical zone of northern Queensland of temperate and submontane floristic elements, and outliers southwards along the Queensland coast of northern monsoonal types.

Correlations of climatic factors and particular rainforest types are therefore not clear-cut and are valid only for core areas, outside which there are many extraneous floristic elements as the result of past climatic changes.

It is estimated that the total area of rainforests remaining in coastal eastern Australia is approximately two million hectares, representing between 1/4 and 1/3 of the original area. In addition there were several hundred thousand hectares of rainforests (the so-called 'Bottletree Scrubs') interspersed among the country dominated by brigalow (*Acacia harpophylla*) in sub-coastal Queensland and northern New South Wales. This totalled approximately five million hectares before clearing, which is now practically complete. The scattered patches of monsoon forests in northern and northwestern Australia average from two to twenty ha for each stand, but there are many biologically significant fragments less than one hectare. Amalgamation of the larger stands may be imagined to total a few thousand hectares.

Structural classification

An intuitive classification for Australian rainforests was developed by Webb (1959, 1968, 1978) following the use of physiognomic–structural features by earlier workers, notably Beard (1944, 1955). This classification was later elaborated into a methodology by the numerical analysis of data collected in an 'open-ended' *pro forma* (Webb, Tracey, Williams & Lance, 1970; Webb, Tracey & Williams, 1976). The method is suitable for systematic and large-scale surveys in tropical and sub-tropical rain-forest areas to establish vegetation types and associated environmental types, and does not require taxonomic knowledge.

The intuitive classification provides a useful perspective, from the synecological point of view, of the main kinds of rainforest at local, regional, and continental levels. A field key based on this classification is given in Webb (1978). The major structural types are *vine forests* (tropical and sub-tropical), *fern forests* and vine–fern forests (sub-montane and warm temperate), and *mossy forests* and fern–moss forests (montane and temperate). These are further subdivided by periodicity of leaf fall, leaf size, and structural complexity to provide an empirical–inductive classification of types. Specific environmental relationships may then be deduced. The striking contrast in physiognomy and struc-ture between these forests and adjacent sclerophyll vegetation is unique to Australia, so that the use of special life forms to differentiate them is singularly apt. In other countries, however, where rainforest vegetation is a continuum, life forms such as vine, fern, or moss are universal within a given climatic zone and such nomenclature is not applicable.

As with any 'pigeon-hole' classification, there are many misfits and overlaps, but the classification is flexible enough to allow prefixes to be omitted and different nodal types to be linked to allow for sites that fall 'somewhere in between' the key units. The environmental relationships of the structural types are summarised in Figure 4.3, and the role of edaphic factors is discussed subsequently.

Floristic classification

Community rank
Unlike structure and physiognomy, which reflect the integrated impact of the physical environment on vegetation, floristic patterns are also the result of phytosociological processes and climatic–edaphic–topographic sifting on a variety of time scales. Definition of structural attributes, although lacking precise cut-off points, does not have the formidable taxonomic difficulties of species identification in complex forests, where even in Australia many taxa remain undescribed. For example, out of specimens representing 765 species collected from 140 rainforest sites in northern Queensland, 117 cannot be satisfactorily placed, and many of

Figure 4.3. Environmental relationships of the structural types of Australian rainforest vegetation. After Webb, 1968.

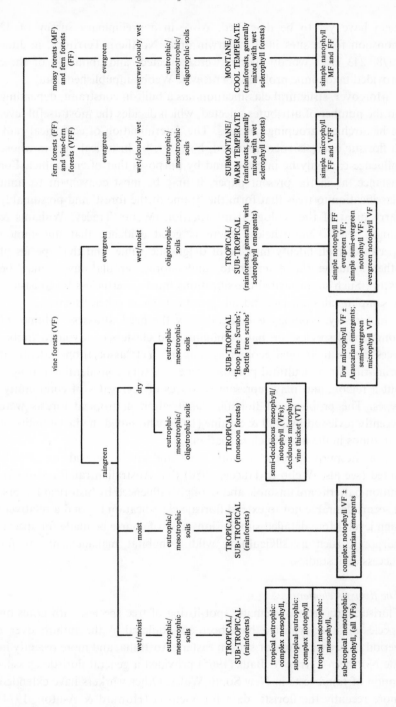

them have yet to be described. Also, in a preliminary survey of 42 monsoon forest sites in the Darwin area, Northern Territory, in June 1978, 243 erect woody species were collected, and about 40 of these provided taxonomic problems (Tracey & Webb, unpublished data).

Moreover, structural classification has a built-in constraint, depending on the number of attributes selected, which decides the most useful level in hierarchical groupings of sites. The determination of ecological rank in floristic analysis is complicated, however, by historical factors whose influence can only be inferred, and by purposes that often conflict. For instance, as in the present paper, it may be most convenient to limit classification to trees that form the 'frame of the forest' and presumably carry most of the ecological information (Webb, Tracey, Williams & Lance, 1967). Nevertheless, there is good evidence that uncommon species of high fidelity may be of diagnostic value, and that species of other synusiae (liane, epiphyte, understorey, ground layer) may be useful. Such sophisticated classifications must nevertheless be consistent in sampling all synusiae, and this is not feasible in general surveys.

In forestry, typology is controlled by the need to assess volumes of commercial species, so that dominants and clusters of favoured timber trees create their own ecological ranking. Yet 'associations' identified primarily for silvicultural purposes generally vary considerably throughout a region, and may represent mixtures of segregates of community-types. The problems of floristic classification in tropical forests were recently reviewed by Hall & Swaine (1976) who noted the inevitability of variations in the units of classification.

For the purposes of a synoptic view, and given the evidence already noted (see also Webb & Tracey, 1980) that Australian rainforest distribution patterns are unstable and strongly influenced by historical factors, it seems desirable not to extend floristic classification beyond a relatively high level. More detailed subdivisions may of course be made for special purposes, such as silviculture, wildlife habitat management, or for successional studies.

The floristic elements
Floristic data from systematic spot-listing of tree species, with notes on species of other synusiae, have been accumulated by the authors over a period of twenty years for sites in eastern Australia, and more recently in the Northern Territory. Baur (1965) provided a general floristic classification of rainforests in New South Wales. Other workers have extended more recently the floristic data for Victoria (Howard & Ashton, 1974;

Ashton & Frankenberg, 1976; Busby & Bridgewater, 1977; Parsons, Kirkpatrick & Carr 1977; D.M. Cameron, personal communication); Tasmania (Jackson, 1972; Busby & Bridgewater, 1977); New South Wales (Bowden & Turner, 1976; A.G. Floyd, personal communication); Queensland (Lavarack & Stanton, 1977; J.P. Stanton, personal communication; G.C. Stocker & B.P.M. Hyland, personal communication; W.J.F. McDonald, personal communication); and northwestern Australia (George & Kenneally, 1975; J.S. Beard, 1976, personal communication; Kabay, George & Kenneally, 1977; Smith, 1977).

Although the data are still patchy and incomplete, especially for rainforest pockets in remote areas, it is now possible to attempt a general floristic classification. Numerical analysis of 561 sites representing 1316 tree species was undertaken by W.T. Williams and produced the groupings of sites (8-group level) shown in the dendrogram of Figure 4.4 (L.J. Webb, J.G. Tracey & W.T. Williams, unpublished). The distribution of

Figure 4.4. Dendrogram of numerical analysis showing grouping of sites into floristic regions and provinces (elements).

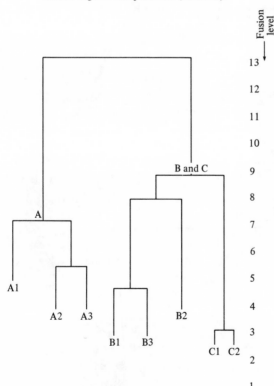

the sites is given in the map of Figure 4.5. Each site-grouping has a broad geographical reality, although there are many areas of overlap, both latitudinally and altitudinally, and one site-grouping is differentiated within the tropical, strongly seasonal zone by its occupation of nutrient-poor sands. The densest and most characteristic aggregation of sites is interpreted as a core area for a particular floristic element. It is convenient to regard the distribution of the elements at the 3-group level as floristic regions, and at the 8-group level as floristic provinces. It should be noted, however, that these are established from floristic data from specific habitats and community-types, and not as in classical floristic geography from the boundaries of combinations of species with approximately similar ranges.

Figure 4.5. Distribution of sites in floristic regions and provinces of different types of Australian rainforests.

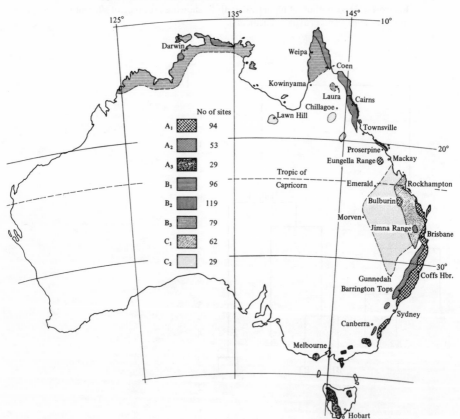

Environmental relationships

Each floristic element or province is defined primarily as a climatic type roughly correlated with its core area. The distribution of the elements, the structural types within them, and their broad environmental relationships are shown in Table 4.1. The meteorological data available from recording stations remote from rainforest patches, (which are, in addition, widely scattered and 'out of phase') do not include important factors such as mountain fogs and cloud-drip, so that the climatic parameters inferred from one or two stations for each province are very approximate.

There is broad agreement between the climatic zones correlated with major structural types (see Figure 4.3), and the climatic types correlated with the core areas of the different floristic provinces. Within each climatic zone there may be considerable variation in structural type depending on altitude, rainfall, soil fertility, and soil drainage. Habitat diversity is highest for the tropical, seasonally-wet province, and least for

Table 4.1. *Environmental relationships of the floristic provinces (climatic types) and the structural types, using selected meteorological data*

Floristic element or province (climate type)	Structural type (on optimal site)	'Average' climate station	Rainfall			Temperature	Soil fertility
			mean annual (mm)	mean annual raindays	mean for driest six consecutive months (mm)	mean minimum, coldest month (°C)	
Tropical seasonal wet	complex mesophyll vine forest	Innisfail, Qld	3644	155	760	15.1	High
Tropical strongly seasonal moist	semi-deciduous mesophyll vine forest	Iron Range, Qld	2049	202	215	18.4	High/medium often riverine alluvium
Tropical strongly seasonal dry	deciduous microphyll vine forest	Kowanyama, Qld	1222	71	36	14.8	High/medium/ low
Tropical extreme seasonal (monsoonal) moist	semi-deciduous notophyll vine forest	Darwin, NT	1534	109	110	18.9	Medium generally enriched coast sand
Warm sub-tropical seasonal moist	complex notophyll vine forest ± araucarians	Condong, NSW	1722	142	550	5.8	High
Sub-tropical seasonal dry	semi-evergreen microphyll vine thicket	Biloela, Qld	699	75	187	5.1	High/medium
Cool sub-tropical cloudy moist	notophyll vine forest	Cloud's Creek, Qld	1397	138	413	−0.3	High/medium
Submontane cloudy moist	microphyll fern forest	Wentworth Falls, NSW	1374	149	405	1.5	High/medium
Warm temperate moist	microphyll vine-fern forest	Cann River, Victoria	1004	146	448	2.0	High/medium
Cool temperate wet	nanophyll fern or moss forest	Waratah, Tasmania	2201	252	820	1.5	High/medium/ low

Figure 4.6. Profile diagrams illustrating selected structural types of rainforest. *a.* complex notophyll vine forests (Cannabullen, Queensland, map reference Tully 540430). 1, *Myristica muelleri;* 2, *Rockinghamia angustifolia;* 3, *Austromyrtus dallachyana;* 4, *Endiandra sankeyana;* 5. *Acronychia haplophylla;* 6, *Xanthophyllum octandrum;* 7, *Argyrodendron peralatum;* 8, *Synima cordieri;* 9, *Doryphora aromatica;* 10, *Sloanea australis;* 11, *Tetrasynandra pubescens;* 12, *Endiandra muelleri;* 13, *Dysoxylum klanderi;* 14, *Beilschmiedia bancroftii;* 15, *Ochrosia poweri;* 16, *Endiandra tooram;* 17, *Hylandia dockrillii;* 18, *Siphonodon membranaceum;* 19, *Dysoxylum oppositifolium;* 20, *Apodytes brachystylis;* 21, *Flindersia acuminata;* 22, *Daphnandra repandula;* 23, *Acronychia vestita;* 24, *Toechima erythrocarpum;* 25, *Alangium villosum* ssp. *tomentosum;* 26, *Alphitonia whitei;* 27, *Elaeocarpus michaelii;* 28, *Cardwellia sublimis;* 29, *Gmelina fasciculiflora.*

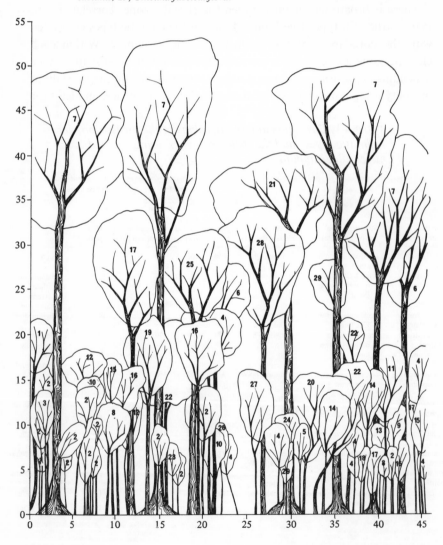

Fig. 4.6 *b*. Semi-deciduous mesophyll vine forest (Dowlings hill, near Heienvale, Queensland, map reference Helenvale 180225). 1, *Bombax ceiba;* 2, *Ficus albipila;* 3, *Albizia toona;* 4, *Vitex acuminata;* 5, *Garuga floribunda;* 6, *Terminalia sericocarpa;* 7, *Canarium australianum;* 8, *Argyrodendron polyandrum;* 9, *Acacia polystachya;* 10, *Cupaniopsis anacardioides;* 11, *Linociera ramiflora;* 12, *Harpullia pendula;* 13, *Polyalthia nitidissima;* 14, *Randia cochinchinensis;* 15, *Mimusops elengi;* 16, *Antidesma ghaesembilla;* 17, *Wrightia pubescens;* 18, *Aglaia elaeagnoidea;* 19, *Glycosmis pentaphylla;* 20, *Ixora klanderana.*

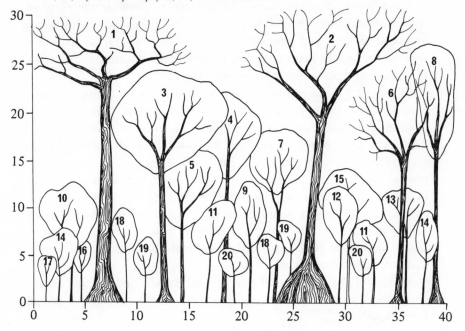

Fig. 4.6 *c*. Simple microphyll vine-fern thicket (Bellenden Ker, Queensland, map reference Bartle Frere ca 786912). 1, *Cinnamomum propinquum;* 2, *Dracophyllum sayeri;* 3, *Leptospermum wooroonooran;* 4, *Trochocarpa laurina;* 5, *Drimys membranea;* 6, *Rapanea achradifolia;* 7, *Rhodomyrtus sericea;* 8, *Alyxia orophila;* 9, *Eugenia apodophylla;* 10, *Planchonella singuliflora;* 11, *Balanops australiana;* 12, *Wilkiea macooraia;* 13, *Austromyrtus metrosideros;* 14, *Polyscias bellendenkerensis;* 15, *Flindersia unifoliolata;* 16, *Orites fragrans;* 17, *Laccospadix australasicus;* 18, *Linospadix* sp.

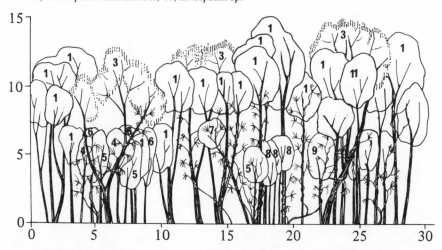

the tropical, strongly seasonally-dry areas. To show the structural types possible within one province would make Table 4.1 too complicated, so that only the optimal type is shown, i.e. for conditions of highest temperature, rainfall, and soil fertility.

A few profile diagrams have been selected to illustrate the range of stratification, canopy height, tree density, and occurrence of life forms such as tree ferns and palms (Figure 4.6). The topographic sequences selected for two transects in northern Queensland illustrate the diversity of vegetation types within this province (Figure 4.7).

Soils exert a strict control on structural type within a given climatic zone (Tracey, 1969; Webb, 1969a), although at the extremes of the tropical monsoonal dry type and the cool temperate wet type, soil differences tend to be overwhelmed by climate. Soil drainage, whether excessively free to produce droughty sites or seasonally impeded to produce swampy sites, is generally obvious enough. Soil nutrient status is, however, of great importance in differentiating simple evergreen and complex raingreen structural types in the tropical and sub-tropical moist zones. Nutrient availability may be inferred from the mineral composition of the parent rocks, provided that climatic and topographic conditions are not extreme (Webb, 1968).

An idea of the complicated interplay of climatic and edaphic factors correlated with the structural types listed in Table 4.1 may be gained from Figure 4.3.

Fig. 4.6 *d*. Simple notophyll vine-fern forest (Mt Spurgeon, Queensland, map reference Mossman 120720). 1, *Elaeocarpus largiflorens;* 2, *Tieghemopanax murrayi;* 3, *Pullea stutzeri;* 4, *Rhodomyrtus trineura;* 5, *Apodytes brachystylis;* 6, *Opisthiolepis heterophylla;* 7, *Cryptocarya* sp. (cf. *C. cinnamomifolia* BRI. 160375); 8, *Ceratopetalum virchowii;* 9, *Sarcopteryx martyana;* 10, *Cryptocarya corrugata;* 11, *Orania appendiculata;* 12, *Cinnamomum laubatii;* 13, *Podocarpus ladei;* 14, *Rapanea achradifolia;* 15, *Carnarvonia* sp. BRI, 160356; 16, *Antirhea tenuiflora* (Hairy Form) BRI. 10528; 17, *Eugenia* sp. BRI, 104967; 18, *Eugenia hemilampra;* 19, *Sphalmium racemosum;* 20, *Endiandra montana;* 21, *Symplocos cyanocarpa;* 22, *Flindersia pimenteliana;* 23, *Laccospadix australasicus;* 24, *Rhodamnia blairiana;* 25, *Garcinia brassii;* 26, *Eugenia cormiflora;* 27, *Planchonella* sp. BRI. 126143; 28, *Cryptocarya angulata;* 29, *Diospyros* sp. BRI. 153554; 30, *Xylopia* sp. BRI. 126112; 31, *Beilschmiedia* sp. (cf. *B. obtusifolia* BRI. 160361); 32, *Sphenostemon lobosporus;* 33, *Wilkiea macooraia;* 34, *Aceratium ferrugineum;* 35, *Alphitonia petriei;* 36, *Planchonella macrocarpa;* 37, *Elaeocarpus* sp. (cf. *E. stellularis* BRI. 174055).

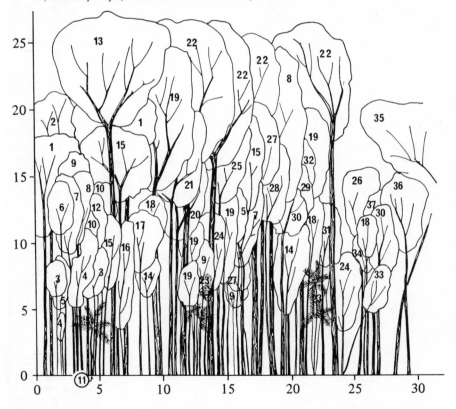

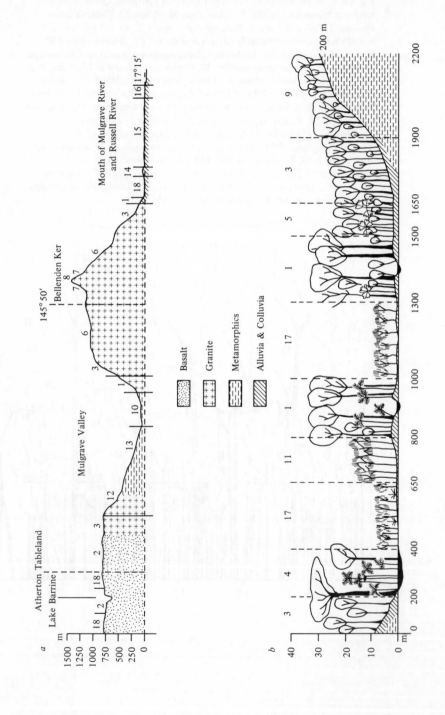

Figure 4.7. Topographic sequences at different scales in northern Queensland (Sugarcane Creek, Tully – Mission beach road, map reference approx. 17°55′S, 146°00′E.) Vertical scale is exaggerated as shown. *a*, transect of 40 km; *b*, transect of 2200 m. Key to vegetation types.
Rainforest: 1, complex mesophyll vineforest on wet lowlands, alluvial and colluvial soils; 2, complex mesophyll vineforest on wet cloudy uplands, basalt; 3, mesophyll vineforest on lowlands and foothills, granite and schist; 4, mesophyll vineforest with dominant feather palms (*Archontophoenix alexandrae*) lowland swamps; 5, mesophyll vineforest with dominant fan palms (*Licuala ramsayi*) lowland swamps; 6, simple notophyll vineforest on cloudy wet uplands, granite; 7, simple microphyll vinefern forest on cloudy wet highlands, granite; 8, simple microphyll vine-fern thicket on exposed cloudy wet highlands, granite. *Rainforest with sclerophyll emergents and co-dominants:* 9, mesophyll vineforest with *Acacia mangium, A. aulacocarpa* emergents on lowlands, schist;

10, mesophyll vineforest with *Eucalyptus grandis, Acacia melanoxylon, A. aulacocarpa* emergents on foothills, uplands, and highlands, granite. *Open sclerophyll forests and woodlands:* 11, medium open-forest *Melaleuca quinquenervia*, lowland swamps; 12 medium layered woodland with *Syncarpia glomulifera* and *Eucalyptus intermedia*, upland granite; 13, medium layered woodland with *Eucalyptus tereticornis, E. tessellaris* and *E. intermedia*, lowland schists. *Vegetation complexes and mosaics:* 14, swampy coastal plains complex with *Melaleuca quinquenervia* as main component; 15, saline littoral complex with Mangrove forest and scrub as main component; 16, coastal beach ridge and swale complex with medium layered woodland *Eucalyptus tereticornis, E. tessellaris, E. pellita* as main component; 17, coastal plains (impeded drainage) complex with stunted paperbark forest *Melaleuca viridiflora* as main component. *Cleared land:* 18, mainly sugarcane on lowlands and mixed farming and grazing on uplands.

Biogeographic significance

Structural patterns

The distribution of the main structural types provides a general climatic typology for Australian rainforests, based primarily on latitude and altitude (temperature), and secondarily on rainfall and topography (moisture) as in Table 4.1, and as discussed by Webb (1959, 1968). At the tertiary level, and sometimes even higher, edaphic factors including climatic–edaphic compensation become important.

It is clear, however, that the elucidation of phylogenetic processes that are central to biogeography requires the use of taxa.

Floristic patterns

At the 3-group level (Figure 4.4) the floristic regions denote hot wet, hot dry and cool wet forests, as in the broad structural classification (Figure 4.3) of tall mostly evergreen vine forests, low mostly deciduous vine thickets, and tall fern or mossy forests respectively.

The primary separation of the hot forests from the cool forests indicates that they are relatively remote phylogenetically, suggesting relatively long periods of independent evolution. It does not support the traditional view (e.g. Herbert, 1960) that the tropical lowland rainforest of northern Queensland simply becomes attenuated southwards in response to a gradient of decreasing temperature.

The wet and the dry forests of the tropics are also relatively remote phylogenetically, at least at the species level. There is no continuous gradient of impoverishment westwards from the northeastern coast along a classical formation-series of decreasing moisture as idealised, for example, by Beard (1944) and Webb (1968). The humid eastern rainforest massif of northern Queensland ends abruptly as a wall, and only at distances generally beginning hundreds of kilometres to the north and west do small pockets of dry deciduous forests appear. On the contrary, in the sub-tropics of southern Queensland, there is a more or less continuous gradation westwards from the wet coastal types to the increasingly fragmented dry semi-evergreen types. This suggests a less disruptive climatic history in recent times than in northern Queensland. This interpretation may be relevant to late Quaternary changes described by Kershaw (see Chapter 2, Walker & Singh, 1981, for references). Except for *Agathis* and *Podocarpus*, these climatic changes practically obliterated the conifer forests that flourished before about 30 000 B.P. on the Atherton Tableland, northern Queensland.

The sub-tropical dry floristic region tapers southwards into sub-coastal New South Wales, suggesting much earlier extensions southwards of the dry monsoonal climates of the north and northwest of the continent. Ancient and widespread floristic elements in the dry tropics of the Old and New Worlds are recognised as paleotropical or archeotropical (e.g. Schnell, 1970).

Examples of common tree species exclusive to each of the three regions are:

Tropical wet

> *Acmena graveolens, Agathis microstachya, Aglaia argentea, Backhousia bancroftii, Cardwellia sublimis, Caryota rumphiana, Cryptocarya oblata, Darlingia ferruginea, Elaeocarpus bancroftii, E. ferruginiflorus, Endiandra palmerstonii, Flindersia ifflaiana, Hypsophila halleyana, Idiospermum australiense, Maniltoa lenticellata, Neorites kevediana, Normanbya normanbyi, Placospermum coriaceum, Toechima erythrocarpum, Tristania pachysperma.*

Tropical dry

> *Alstonia constricta, Backhousia angustifolia, Bauhinia cunning-hamii, Brachychiton australe, B. rupestris, Cochlospermum gregorii, Croton arnhemicus, Dolichandrone heterophylla, Ehretia membranifolia, Erythroxylon australe, Flindersia collina, F. maculosa, Gyrocarpus americanus, Notelaea microcarpa, Planchonella cotinifolia, Strychnos lucida, Terminalia aridicola, T. oblongata, Ventilago viminalis, Wrightia saligna.*

Sub-tropical and temperate wet

> *Anopterus macleayanus, Archontophoenix cunninghamiana, Argyrodendron actinophyllum, Atherosperma moschatum, Ceratopetalum apetalum, Cryptocarya meisnerana, Doryphora sassafras, Elaeocarpus holopetalus, Eugenia brachyandra, E. moorei, Flindersia bennettiana, Geissois benthamii, Nothofagus cunninghamii, N. moorei, Orites excelsa, Polyosma cunninghamii, Sarcopteryx stipitata, Schizomeria ovata, Sloanea woollsii, Synoum glandulosum.*

It is significant that the fairly clear-cut geographical separation of the three regions occurs only at the species level. At the generic level, the boundaries of the regions (and more so those of the provinces) are blurred (Webb & Tracey, 1980). The distribution among the regions of endemic and non-endemic species of the total of species sampled in the present analysis is given in Figure 4.8.

At the 8-group level (Figure 4.4) the floristic provinces can be correlated with climatic types as in Table 4.1. Despite the desirability of following a standard international classification (Letouzey, 1978) it is not possible to match many of the Australian rainforest types with those in the UNESCO world classification (UNESCO, 1969).

The floristic elements are no longer neatly segregated geographically, and extend in various directions outside the core areas of the provinces. The extent of overlap by these outliers is shown in Figure 4.5.

It would have been preferable to identify the different floristic elements taxonomically as well as climatically, but this becomes too complicated for other than the simple types, e.g. *Nothofagus–Atherosperma* (cool temperate), *Nothofagus–Ceratopetalum* (warm temperate and submontane moist), *Bombax–Gyrocarpus* (tropical monsoonal dry).

Species vary in their fidelity for a particular floristic element and province. Besides exclusive species which are completely or almost completely confined to one element, there are species with decreasing degrees of fidelity: selective, preferential, indifferent, and accidental (Braun-Blanquet, 1932; Mueller-Dombois & Ellenberg, 1974). Certain of

Figure 4.8. Distribution of *a*, 103 endemic rainforest genera; *b*, 442 non-endemic rainforest genera, and *c*, 145 rainforest families, among the Australian rainforest floristic regions A, B, and C (from Webb & Tracey, 1980).

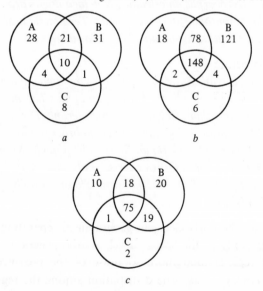

these less exclusive species are highly vagile in relation to seed dispersal mechanisms.

The community-types

The term 'community-type' is adopted as a neutral term and denotes, in this chapter, a characteristic and widespread species group recognisable at a high level in the hierarchy of phytosociological units provided by numerical analysis of site floristic data. One or more community-types characterise a particular core area of a floristic element or province. The number is greatest in the large rainforest massifs which have the greatest habitat diversity, such as the tropical seasonal wet province of northern Queensland. However, even in this complex area, the number of community-types may be limited to those situations where there is obvious floristic discontinuity, i.e. to distinctly different habitats, such as mountain summits, swamps, and parent rocks within a wide altitudinal zone in the high rainfall belt. It may be argued that the community-type is definable at the point where emphasis changes from discontinuities between stand groups to continuities within them (Dale & Webb, 1975). Unless it is necessary to subdivide them for a particular purpose, differences in frequency of tree species within one extensive and continuously varying phytosociological unit are treated as 'segregates' in the sense of Braun (1950) and Schulz (1960).

The segregates at different places are characterised by variations in frequency of leading species which may become rare, absent, or abundant. Variations that are obviously related to local site differences, generally soils, are regarded as 'lociations'. Where the variations in dominants over a geographical or altitudinal range are not correlated with obvious site differences, they are termed 'faciations' and interpreted as the result of historical changes, such as climatic sifting. This interpretation of a widespread community-type segregating in deterministic or probabilistic ways (cf. Webb, Tracey & Williams, 1972; Dale & Webb, 1975; Diamond, 1975) emphasises the floristic continuity of rainforest patches that are widely scattered, and recognises the dynamic nature of communities. It implies that the boundaries between different floristic regions and provinces are not definitive or fixed, although it may be convenient to regard a concentration of sites of a particular community-type as a core area. It is to be expected that many examples of extraneous floristic elements occur outside core areas. These elements may be interdigitated in different niches in other core areas, or interspersed as community-fragments with other community-types. Both occurrences reflect the historical changes discussed below.

A detailed catalogue of community-types and their segregates is outside the scope of this chapter, and the data are still being assembled for northern Australia (e.g. Tracey & Webb, unpublished data).

Dynamics of the rainforests

The role of refugia

There are many examples of disjunct, narrowly endemic, relict, and vicarious species within one or more provinces, some of which are listed by Webb & Tracey (1980). There are also examples, less readily identified in the absence of a comprehensive framework of community-types for all the provinces, of segregates and fragmentary communities which occur outside the province typified by a particular community-type. As already noted, these outliers, whether individual species or recognisable community fragments, are interpreted most satisfactorily as the result of historical changes.

There are many examples of disjunctions between north and south in eastern Australia, but a much greater number of southern species occur as relicts in the uplands of northern Queensland than do northern species in the south. This implies that southern floristic elements previously extended during cooler periods to the north. There are also outliers in northern New South Wales of temperate flora now common in Tasmania (see Chapter 5, Howard, 1981).

Relict and fragmentary tropical strongly seasonal communities also occur scattered farther south along the coast of central Queensland, suggesting earlier extension of dry seasonal climates southwards. The scattered occurrence of grasslands characterised by relict montane and sub-alpine species in northern New South Wales (e.g. in the Deervale area, east of the New England scarp) and in southern Queensland (e.g. Emuvale, Bunya Mts) may be interpreted as indicating a previous cool and dry steppe climate (Webb, 1964).

There is also considerable evidence for the relict distribution of organisms other than plants along the discontinuous tract of eastern Australian rainforest which is summarised by various authors in Keast (1980).

The evolution of contemporary rainforest patterns, and the degree of instability implied by them, are conveniently considered on three time scales: geological, prehistorical, and historical (Webb & Tracey, 1980). Historical changes in the strict sense of written history since the arrival of industrial man occupy only two centuries. Unlike other countries, where natural and human disturbances of vegetation occupying many thousands of years are confused, man's recent impacts on Australian rainforests are readily identified, but for reasons of space will not be pursued here (see e.g. Seddon & Davis, 1976, for references).

On a geological scale, evidence from modern patterns of plant distribution, fossils, and geomagnetic studies indicates that Australia was once part of the Gondwanaland continent in the Cretaceous (Keast, 1980). Of 436 non-endemic genera of woody rainforest plants listed by Webb & Tracey (1980), substantial percentages are found today in what were parts of Gondwanaland, e.g. Australasia (New Zealand, New Hebrides, New Caledonia, Fiji, Lord Howe Is., Norfolk Is.) and Indo-Malayan (syn. Malesian) (28%); Africa (26%); world tropics (23%); New Caledonia (21%); and India (21%). It now seems reasonable to suppose that what were traditionally regarded as Indo-Malayan immigrant elements by Hooker (1860) and his followers originated mainly *in situ*, and differentiated since Gondwanic times. At least 103 genera of woody rainforest plants are endemic. Of 98 rainforest genera of primitive angiosperms and gymnosperms listed by Specht, Roe & Boughton (1974), more than 73 (allowing for recently described Proteaceae not included) are found in the tropical seasonal wet province, and 42 in the sub-tropical moist province. Of 14 primitive angiosperm families listed by Takhtajan (1969), eight occur in Australasia. This suggests that these refugia are of great antiquity, dating back to the Cretaceous and possibly earlier for taxa, such as certain conifers and cycads, that now have very narrow and presumably relict distributions in northeastern Australia. Of course the possibility of seed dispersal over moderate, and more rarely, long distances cannot be excluded, as we shall discuss subsequently.

On a prehistorical time scale, there is now direct evidence from palynological studies on the Atherton Tableland, northern Queensland (for details, see Chapter 2, Walker & Singh, 1981). On a world scale, additional evidence suggests that the distribution patterns of tropical rainforests at the present day are less than 4000 to 12 000 years old (Synnott, 1977).

Viewed in the perspective of these now well-documented climatic changes, and allowing for the influences of topography, soils, wildfires, and in historical times European man, the archipelago of rainforest habitats that characterises the eastern and northern Australian landscape becomes an *archipelago of refugia*.

Australian rainforest refugia were classified into four main types by Webb & Tracey (1980): (1) large, relatively wet areas such as lowland gorges and cloudy wet summits and gullies on slopes of coastal mountains, (2) small, relatively dry and fireproof topographic isolates such as bouldery outcrops, (3) small, often narrow edaphic isolates such as riverine alluvia and perennial spring-fed soaks, and calcareous coastal dune systems, and (4) topographic–edaphic–climatic isolates in lower rainfall areas such as residual basaltic caps of coastal areas and mountains.

Given the extent of past climatic changes and their interactions on different time scales, and recognising the role of refugia of different types and ages in the ebb and flow of rainforest vegetations, the overlap between the floristic elements, the faciations, and the consistent patchiness of rainforest patterns away from previous refugia become explicable.

Interspersions

The use of structure (i.e. closed canopy) to separate rainforests from the rest of Australian vegetation loses much of the biogeographical significance of the floristic elements. As already noted (p. 87) there is a limited but significant floristic continuity between closed- and open-forests under extreme environmental conditions, notably lowered temperature and rainfall, and low soil fertility.

Three kinds of interspersions may be recognised:

(1) Boundaries and ecotones. These are narrow in the tropics and subtropics of northern Australia, and wide (forming tall open-forests or 'wet sclerophyll forests', see Chapter 6, Ashton, 1981) in warm and cool temperate regions of the southeast. Differences in ecotones are related to different fire behaviour under particular seasonal climatic, topographic, and edaphic conditions. Surface fires in late winter–spring, and running crown fires in summer–autumn, characterise the tropical and temperate regions respectively (Webb, 1968).

Where the rainforest communities become fragmented at the limits of their range, the boundaries become involuted, forming a mosaic of ecotones in what was originally a rainforest matrix (see Figure 4.2).

It should also be noted that certain species belonging to sclerophyll and rainforest genera are exclusive to ecotones, which accordingly deserve consideration as ecological entities in their own right.

(2) Scattered understorey in grassy eucalypt woodland and in acacia open-forest. In the former case the floristic elements are often deciduous and characterise tropical mixed deciduous woodland on well-drained soils of higher fertility in northern Australia. The species, unlike those of the closed rainforests, are fire-tolerant.

There is a remarkable convergence ('epharmosis') of leaf size and shape in certain taxa to narrow linear and often falcate to resemble sclerophylls such as eucalypts, e.g. *Eugenia eucalyptoides* (Webb & Tracey, 1980).

(3) Mixed communities in at least two different ecological situations. The intermingling of floristic elements from closed rainforests and open sclerophyll forests is so intimate that sample plots of 20 m^2 or less cannot avoid both floristic types. The mixing is so complete that the elements do not appear to represent fragmentary associations, and the species are widespread and characteristic of similar habitats, implying long periods of co-adaptation. They form an exception to the strict definition of rainforest given earlier.

(a) Nutrient-poor sands in coastal areas of northern Australia support vegetation gradients from sclerophyll heath and scrub through layered open-forest to vine forest in which sclerophylls are never absent.

(b) Submontane and montane exposed situations on eastern Australian highlands and in Tasmania.

Vagility
The distribution of rainforest patches and the configuration of floristic elements at the present day depend partly on the suitability of habitat factors and past climatic–edaphic sifting, and partly on the rate and distance of dispersion of diaspores, i.e. on the vagility of the plants.

Dispersal of seeds is least effective by gravity, and increasingly effective by wind, mammals, water, birds and bats. The subject is, however, controversial and the data are meagre, for example, length of time that viable seeds are retained in the digestive tract of birds in relation to distances traversed, migration routes, and habitats visited (see, e.g. Proctor, 1968; van Steenis, 1972; McKey, 1975; Whitmore, 1975; Regal, 1977; Webb & Tracey, 1980). In the absence of firm and comprehensive evidence, it is assumed that the efficiency of long-distance dispersal (to be reckoned in hundreds and thousands of kilometres) remains questionable and its operation rare, especially when the sizes of many rainforest seeds (10–20 mm diameter or more) are considered.

It has also been argued that rainforests composed of many biotically interdependent synusiae and involving entire phylogenetic series must migrate communally (Takhtajan, 1969), which further decreases the possibility of transfer of adequate breeding populations of plants over large distances. Perhaps there are also ecological barriers to ecesis in new habitats even when seeds arrive (Regal, 1977). Given the restricted vagility of many rainforest species, and accepting the role of refugia, the interpretation of scattered occurrences of populations as relict seems more reasonable.

Rainforest as a habitat for birds

Northern Australian rainforests form a closed community in which, as in other humid tropical regions of the world, plants and animals are closely associated, and animals often act specifically as pollinators and seed dispersers. The 'island effect' of rainforest distribution is also paralleled by high fidelity of many faunal elements that have been ecologically trapped as the result of aridity (Kikkawa 1974; Kikkawa, Monteith & Ingram, 1980). The following is a summary of circumstantial evidence for these trends as seen in the birds of Australian rainforests.

With increasing latitudes and along mesic–xeric gradients the species richness of land birds generally decreases over the Australian continent (Kikkawa 1968; Kikkawa & Pearse, 1969; Recher, 1971). In the sub-tropical region of eastern Australia the rainforest birds, compared with birds of semi-arid habitats, are characterised by stronger association of species, relative abundance of tree-nesting frugivores and small clutch size (Kikkawa, 1974). The same tendencies are more pronounced in the tropical rainforests compared with sub-tropical rainforests within Australia (Kikkawa & Webb, 1967; Kikkawa, 1968) and in the lowland rainforests compared with the montane rainforests in New Guinea (Kikkawa & Williams, 1971a, b). In the sub-tropical region of Australia the species diversity of birds in the rainforest is not as high as in the surrounding open-forests (wet sclerophyll forests) but in northeastern Australia this relation is reversed. The guilds of birds that contain additional elements in the tropical rainforests are (1) tree-feeding her-bivores and omnivores, (2) small predators and (3) ground-feeding in-sectivores, indicating greater niche volumes of these guilds. For example, nectar feeders are almost absent from sub-tropical rainforests whilst they are abundant in tropical rainforests. In the sub-tropical region most species of honeyeaters (Meliphagidae) occur in sclerophyll vegetation; only one species occurs regularly in rainforests. On the other hand, most tropical species occur inside the rainforest rather than outside (functional response). In the case of lorikeets (Loriidae) the species abundant in tropical rainforest vegetation (numerical response) are found in sclero-phyll vegetation in the sub-tropical region. Since these birds are resi-dents, both functional and numerical responses to increased resources reflect a sustained supply of nectar and other food sources in tropical rainforests, hence interdependence of nectar-feeders and bird-pollinated plants (e.g. Proteaceae). In the case of rainforest meliphagids most nectar-feeders are facultative in their diet and consume a large amount of insects as well. However, their foraging activity is restricted to the rainforest habitat. Fruit-eating birds (mostly pigeons) also show both numerical and functional responses suggesting their increased role as seed dispersers in tropical rainforests. Two species of lowland rainforest fruit-eaters perform regular long-distance migration to New Guinea and adjacent islands reflecting the seasonality of fruiting. Their role in long-distance seed dispersal is therefore potentially greater than that of resident species. Two ground species of lowland rainforests are also migrants from New Guinea and arrive only after the wet season begins locally (Kikkawa, 1976). The guild of foliage-feeding insectivores on the other hand is increased in tropical rainforests when fruit-eaters and ground-feeders meet adverse conditions in the lowlands. The increase is due to an influx of southern migrants escaping the winter conditions of sub-tropical rainforests. There are at least three species of rainforest flycatchers (Muscicapidae) in this category. Since these birds are associated with rainforest vegetation wherever they occur the patches of rainforests on the route of migration play an important role for their survival. Finally, association of birds with vegetation structure and floristics within tropical rainforests has been demonstrated, showing parallel trends of ecology and biogeography between the northern rainforests and the birds that inhabit them (Webb, Tracey, Kikkawa & Williams, 1973).

Social values and conservation status

For those who have seen something of the vast, life-teeming evergreen and deciduous forests – where they survive – in other tropical countries, a first impression of northern Australia must be surprise and even disbelief that the areas of rainforests are so meagre. Yet, as we have suggested, it was ever so for man. The tropical rainforest flora, whose evolution is rooted in Gondwanaland, has not been extensive since Tertiary times when the sclerophylls became rampant. For the Aboriginal immigrants some forty thousand years ago who had resource utilisation patterns adapted to a tropical rainforest environment, botany began anew in sclerophyll Australia (Webb, 1973).

The fragmented and restricted distribution of the remaining rainforest communities highlights the need to conserve them. As a rare resource they are precious for a variety of reasons. The patterns of distribution and organisation of separate communities are themselves of great bio-geographical and ecological interest, and provide keys to evolutionary processes and niche specialisation. An understanding of these processes is of more than academic interest, and is essential for successful agricul-tural and forestry practice. For example, the pockets of rainforest serve as 'markers' for particular kinds of biotic environments or habitat types that can be matched elsewhere in the world tropics which lack structural and floristic counterparts of the all-pervasive Australian sclerophyll vegetation. Habitat-matching based on similar structural types would provide a more discriminate method for reciprocal introduction of tree species, as well as of potential pasture species that inhabit forest fringes (cf. Webb, 1966).

The Australian rainforests are of international significance as ancient and isolated reservoirs of plant and animal taxa, many of them primitive or endemic, that have so far received little scientific attention (Hope, 1974). Even at the descriptive level of plant taxonomy, about two hundred tree species in northern Australia that have been collected cannot be placed satisfactorily, and many of them are probably new to botany.

Although the Australian rainforests were exploited for a variety of cabinet woods from the earliest times (the first log of Red Cedar (*Toona australis*) was exported seven years after the settlement at Sydney) they yielded very few foodstuffs, spices, medicines, etc., compared with the forests of Malesia. *Macadamia* (Proteaceae) is the only edible nut from a native plant to be developed commercially. Although the Aborigines, as hunter–gatherers, used a wide range of leaves, fruits, seeds and tubers from the tropical rainforest and monsoon forest flora in northern Australia, these plants did not enjoy millennia of selection and culti-vation as in, say, India and Thailand. Some of the species would undoubtedly qualify after selection as edible fruits, e.g. *Antidesma, Elaeagnus, Elaeocarpus, Parinari, Securinega, Syzygium*. Some of the Aboriginal bush medicines and poisons have also been shown by phyto-chemical analysis to contain physiologically active compounds, e.g. *Alstonia, Barringtonia, Duboisia, Isotoma, Nicotiana, Stephania, Strychnos, Tylophora* (Webb, 1969*b*, *c*). Australian rainforests have also contributed plants to world horticulture, such as flowering species of Proteaceae (silky oak, scrub waratah, wheel of fire), lillypillies (*Syzygium* spp.), native frangipani (*Hymenosporum flavum*), and many ornamental forms such as palms, ferns, vines, and epiphytes.

Patches of rainforest in gullies and gorges often with spectacular waterfalls, on isolated mountain summits, and fringing rivers and beaches represent exceptional environments among the dominant sclerophyll vegetation. They have become increasingly popular as 'destinations' for people seeking recreation and variety. Most of the rainforests that survive in eastern Australia, but to a less extent in northern Australia, are now located in National Parks and State Forests. Nearly all the tall rainforests that have not been cleared south of Cape Tribulation (north of Cairns) in eastern Australia have been logged. Unlogged areas are generally on steep upper slopes and skeletal soils of mountains that support trees of poor form for sawmilling. These areas, if they do not contain minerals, tend by default to National Park status, or remain under the administration of State forestry authorities as 'protection' (i.e. catchment protection) forests.

The increased mobility of urban populations along the eastern seaboard has meant increasing pressures for recreation and amenity in the accessible rainforest areas, and increasing scrutiny of intensified forestry operations. This has generated considerable public debate and an unfortunate polarisation between the foresters and wood-using industries on the one hand, and the conservationists on the other, who reflect growing international concern to manage in the most conservative manner the remaining rainforests (Hope, 1974). This concern is the result of the virtual disappearance of the lowland rainforests in many tropical regions, and the extinction of many of their species. It is especially relevant to the Australian rainforest communities that are vestigial in evolutionary time, a non-renewable resource when subjected to extensive clearing, yet one of the most fascinating elements in our moist tropical and sub-tropical landscapes.

The section on rainforest as a habitat for birds was kindly written by Dr. J. Kikkawa of the Zoology Department, University of Queensland, St. Lucia.

References

Ashton, D.H. (1981). Tall open-forests. In *Australian Vegetation*, ed. R.H. Groves, pp. 121–151. Cambridge University Press.

Ashton, D.H. & Frankenberg, J. (1976). Ecological studies of *Acmena smithii* (Poir.) Merrill & Perry with special reference to Wilson's Promontory. *Australian Journal of Botany*, 24, 453–87.

Aubréville, A. (1965a). Principes d'une systématique des formations végétales tropicales. *Adansonia*, 5, 153–96.

Aubréville, A. (1965b). Instabilité de l'équilibre biologique des forêts de l'Australie tropicals orientale et de la Nouvelle-Calédonie. *Comptes Rendu Academie de Sciences Paris*, 261, 3463–6.

Baur, G.N. (1957). Nature and distribution of rainforests in New South Wales. *Australian Journal of Botany*, *5*, 190–233.

Baur, G.N. (1965). *Forest types in New South Wales*. Forestry Commission of New South Wales Research Note No. 17.

Baur, G.N. (1968). *The Ecological Basis of Rainforest Management*. Sydney: Forestry Commission of New South Wales.

Beard, J.S. (1944). Climax vegetation in tropical America. *Ecology*, *25*, 127–58.

Beard, J.S. (1955). The classification of tropical American vegetation-types. *Ecology*, *36*, 89–100.

Beard, J.S. (1976). The monsoon forests of the Admiralty Gulf, Western Australia. *Vegetatio*, *31*, 177–92.

Bowden, D.C. & Turner, J.C. (1976). *A preliminary survey of stands of temperate rainforest on Gloucester Tops*. University of Newcastle, N.S.W., Research Papers in Geography Number 10.

Braun, E.L. (1950). *Deciduous Forests of Eastern North America*. New York: Hafner Publ. Coy.

Braun-Blanquet, J. (1932). *Plant Sociology*. New York & London: McGraw-Hill.

Busby, J.R. & Bridgewater, P.B. (1977). Studies in Victorian vegetation. II. A floristic survey of the vegetation associated with *Nothofagus cunninghamii* (Hook.) Oerst. in Victoria and Tasmania. *Proceedings of the Royal Society of Victoria*, *89*, 173–82.

Dale, M.B. & Webb, L.J. (1975). Numerical methods for the establishment of associations. *Vegetatio*, *30*, 77–87.

Darlington, P.J. (1961). Australian carabid beetles. V. Transition of wet forest faunas from New Guinea to Tasmania. *Psyche, Journal of Entomology*, *68*, 1–24.

Diamond, J.M. (1975). Assembly of species communities. In *Ecology and Evolution of Communities*, ed. M.L. Cody & J.M. Diamond, pp. 342–444. Cambridge, Massachusetts & London: Belknap Press.

Diels, L. (1906). *Die Pflanzenwelt von West-Australian südlich des Wendekreises*. Leipzig: Wilhelm Engelmann.

Francis, W.D. (1929). *Australian Rain Forest Trees*. Brisbane: Government Printer.

Fraser, L. & Vickery, J.W. (1938). The ecology of the Upper Williams River and Barrington Tops districts. II. The rainforest formations. *Proceedings of the Linnean Society of New South Wales*, *63*, 139–84.

George, A.S. & Kenneally, K.F. (1975). Flora. In *A Biological Survey of the Prince Regent River Reserve, North-west Kimberley, Western Australia*, ed. J.M. Miles & A.A. Burbidge, pp. 31–68. Wildlife Research Bulletin of Western Australia No. 3.

Gunn, R.H. & Nix, H.A. (1977). Land units of the Fitzroy region, Queensland. *CSIRO, Australia, Land Research Series No. 39*.

Hall, J.B. & Swaine, M.D. (1976). Classification and ecology of closed-canopy forest in Ghana. *Journal of Ecology*, *64*, 913–51.

Herbert, D.A. (1960). Tropical and sub-tropical rainforest in Australia. *Australian Journal of Science*, *22*, 283–90.

Herbert, D.A. (1967). Ecological segregation and Australian phytogeographic elements. *Proceedings of the Royal Society of Queensland*, *78*, 110–11.

Hooker, J.D. (1860). Introductory essay. In *The Botany (of) the Antarctic Voyage*, part III. *Flora Tasmaniae*, vol. 1. London: Lovell Reeve.

Hope, R.M. (1974). *Report of the National Estate*. Canberra: Australian Government Publishing Service.

Howard, T.M. (1981). Southern closed-forests. In *Australian Vegetation*, ed. R.H. Groves, pp. 102–120. Cambridge University Press.

Howard, T.M. & Ashton, D.H. (1974). The distribution of *Nothofagus cunninghamii* rainforest. *Proceedings of the Royal Society of Victoria, 86*, 47–75.

Jackson, W.D. (1972). Vegetation of the Central Plateau. In *The Lake Country of Tasmania*, ed. M.R. Banks, pp. 61–85. Hobart: Royal Society of Tasmania.

Kabay, E.D., George, A.S. & Kenneally, K.F. (1977). The Drysdale River National Park Environment. In *A Biological Survey of the Drysdale River National Park, North Kimberley, Western Australia*, ed. E.D. Kabay & A.A. Burbidge, pp. 13–30. Wildlife Research Bulletin of Western Australia No. 6.

Keast, A. (ed.) (1980). *Ecological Biogeography in Australia*. The Hague: Junk.

Kikkawa, J. (1968). Ecological association of bird species and habitats in eastern Australia: similarity analysis. *Journal of Animal Ecology, 37*, 143–65.

Kikkawa, J. (1974). Comparison of avian communities between wet and semi-arid habitats of eastern Australia. *Australian Wildlife Research, 1*, 107–16.

Kikkawa, J. (1976). The birds of Cape York Peninsula. *Sunbird, 7*, 25–41, 81–106.

Kikkawa, J., Monteith, G.B. & Ingram, G. (1980). Cape York Peninsula: the major region of faunal interchange. In Keast (1980), pp. 1695–1742.

Kikkawa, J. & Pearse, K. (1969). Geographical distribution of land birds in Australia – a numerical analysis. *Australian Journal of Zoology, 17*, 821–40.

Kikkawa, J. & Webb, L.J. (1967). Niche occupation by birds and the structural classification of forest habitats in the wet tropics, north Queensland. *Proceedings of the XIV Congress of the International Union of Forest Research Organizations* (26), pp. 467–82.

Kikkawa, J. & Williams, W.T. (1971a). Altitudinal distribution of land birds in New Guinea. *Search, 2*, 64–5.

Kikkawa, J. & Williams, W.T. (1971b). Ecological grouping of species for conservation of land birds in New Guinea. *Search, 2*, 66–9.

Lavarack, P.S. & Stanton, J.P. (1977). Vegetation of the Jardine River catchment and adjacent coastal areas. *Proceedings of the Royal Society of Queensland, 88*, 39–48.

Letouzey, R. (1978). Floristic composition and typology. In *Tropical Forest Ecosystems, A State-of-knowledge Report Prepared by UNESCO/UNEP/FAO*, pp. 91–111. Paris: UNESCO.

McKey, D. (1975). The ecology of co-evolved seed dispersal systems. In *Co-evolution of Animals and Plants*, ed. L.E. Gilbert & P.H. Raven, pp. 159–91. Austin: University of Texas Press.

Mueller-Dombois, D. & Ellenberg, H. (1974). *Aims and Methods of Vegetation Ecology*. New York: John Wiley & Sons.

Odum, H.T. & Pigeon, R.F. (1970). *A Tropical Rain Forest*. Springfield, Virginia, USA: Atomic Energy Commission.

Parsons, P.A. & Bock, I.R. (1977). Australian endemic *Drosophila*. I. Tasmania and Victoria, including descriptions of two new species. *Australian Journal of Zoology, 25*, 249–68.

Parsons, R.F., Kirkpatrick, J.B. & Carr, G.W. (1977). Native vegetation of the Otway region, Victoria. *Proceedings of the Royal Society of Victoria, 89*, 77–88.

Proctor, V.W. (1968). Long-distance dispersal of seeds by retention in digestive tract of birds. *Science, 160*, 321–2.

Recher, H.F. (1971). Bird species diversity: a review of the relation between species number and environment. *Proceedings of the Ecological Society of Australia, 6*, 135–52.

Regal, P.J. (1977). Ecology and evolution of flowering plant dominance. *Science, 196*, 622–9.

Richards, P.W. (1952). *The Tropical Rain Forest*. Cambridge University Press.

Schimper, A.F.W. (1903). *Plant Geography upon a Physiological Basis*. Oxford University Press.

Schnell, R. (1970). *Introduction à la Phytogéographie des Pays Tropicaux: Les Problèmes Généraux*. I. Paris: Gauthier-Villars Editeur.

Schulz, J.P. (1960). Ecological studies on rainforest in northern Suriname. Verhandelingen der Koninkluke Nederlandse Akademie van Wetenschappen, AFD. Natuurkunde, 2 sect., *53,* 1–367.

Seddon, G. & Davis, M. (ed.) (1976). *Man and Landscape in Australia*. Australian UNESCO Committee for Man and the Biosphere, Publication No. 2, pp. 1–373. Canberra: Australian Government Publishing Service.

Smith, F.G. (1977). Vegetation map of the Drysdale River National Park. In *A Biological Survey of the Drysdale River National Park, North Kimberley, Western Australia*, ed. E.D. Kabay & A.A. Burbidge, pp. 31–78. Wildlife Research Bulletin of Western Australia No. 6.

Specht, R.L. (1970). Vegetation. In *The Australian Environment*, 4th edn. (rev.), ed. G.W. Leeper, pp. 44–67. Melbourne: CSIRO, Australia & Melbourne University Press.

Specht, R.L., Roe, E.M. & Boughton, V.H. (1974). Conservation of major plant communities in Australia and Papua New Guinea. *Australian Journal of Botany, Supplementary Series* No. 7.

Steenis, C.G.G. J. van (1972). *The Mountain Flora of Java*. Leiden: E.J. Brill.

Synnott, T.J. (1977). Monitoring tropical forests: a review with special references to Africa. *MARC Report, Monitoring and Assessment Research Centre, University of London* (1977), *No. 5*, 1–45. Oxford: University Department of Forestry.

Takhtajan, A. (1969). *Flowering Plants: Origin and Dispersal*. Edinburgh: Oliver & Boyd.

Tracey, J.G. (1969). Edaphic differentiation of some forest types in eastern Australia. I. Soil physical factors. *Journal of Ecology, 57*, 805–16.

UNESCO (1969). *A framework for a classification of world vegetation*. Paris: UNESCO.

Walker, D. & Singh, G. (1981). Vegetation history. In *Australian Vegetation*, ed. R.H. Groves, pp. 26–43. Cambridge University Press.

Walter, H. (1971). *Ecology of Tropical and Sub-tropical Vegetation*. Edinburgh: Oliver & Boyd.

Webb, L.J. (1959). A physiognomic classification of Australian rainforests. *Journal of Ecology, 40*, 551–70.

Webb, L.J. (1964). An historical interpretation of the grass balds of the Bunya Mountains, South Queensland. *Ecology, 45*, 159–62.

Webb, L.J. (1966). An ecological comparison of forest-fringe grassland habitats in eastern Australia and eastern Brazil. *Proceedings IX International Grassland Congress, Brazil*, pp. 321–30.

Webb, L.J. (1968). Environmental relationships of the structural types of Australian rainforest vegetation. *Ecology, 49*, 296–311.

Webb, L.J. (1969*a*). Edaphic differentiation of some forest types in eastern Australia. II. Soil chemical factors. *Journal of Ecology*, *57*, 817–30.

Webb, L.J. (1969*b*). The use of plant medicines and poisons by Australian Aborigines. *Mankind*, *7*, 137–46.

Webb, L.J. (1969*c*). Australian plants and chemical research. In *The Last of Lands*, ed. L.J. Webb, D. Whitelock & J. Le Gay Brereton, pp. 82–90. Brisbane: The Jacaranda Press.

Webb, L.J. (1973). 'Eat, die, and learn' – the botany of the Australian Aborigines. *Australian Natural History*, *17*, 290–5.

Webb, L.J. (1978). A general classification of Australian rainforests. *Australian Plants*, *9*, 349–63.

Webb, L.J. & Tracey, J.G. (1980). Australian rainforests: patterns and change. In *Ecological Biogeography in Australia*, ed. A. Keast, pp. 605–94. The Hague: Junk.

Webb, L.J., Tracey, J.G., Kikkawa, J. & Williams, W.T. (1973). Techniques for selecting and allocating land for nature conservation in Australia. In *Nature Conservation in the Pacific*, ed. A.B. Costin & R.H. Groves, pp. 39–52. Canberra: Australian National University Press.

Webb, L.J., Tracey, J.G. & Williams, W.T. (1972). Regeneration and pattern in the sub-tropical rainforest. *Journal of Ecology*, *60*, 675–95.

Webb, L.J., Tracey, J.G. & Williams, W.T. (1976). The value of structural features in tropical forest typology. *Australian Journal of Ecology*, *1*, 3–28.

Webb, L.J., Tracey, J.G., Williams, W.T. & Lance, G.N. (1967). Studies in the numerical analysis of complex rainforest communities. II. The problem of species sampling. *Journal of Ecology*, *55*, 525–38.

Webb, L.J., Tracey, J.G., Williams, W.T. & Lance, G.N. (1970). Studies in the numerical analysis of complex rainforest communities. V. A comparison of the properties of floristic and physiognomic-structural data. *Journal of Ecology*, *58*, 203–32.

Whitmore, T.C. (1975). *Tropical Rain Forests of the Far East*. Oxford: Clarendon Press.

5

Southern closed-forests

T. M. HOWARD

Definition of community

Southern closed-forests can be segregated from the northern closed-forests (rainforests) of the previous chapter (Webb & Tracey, 1981) by selecting communities dominated primarily by species of *Nothofagus*, the Antarctic beech. Closed-forests were defined by Specht (1970) as communities consisting of single-stemmed trees ranging in average height from 5 to 30 m and having a projective foliage cover of between 70 and 100%. These communities can be further subdivided on the basis of average tree height (Specht, 1970), with those over 30 m high being designated 'tall' and those between 5 and 10 m as 'low'. In this chapter I shall consider all those communities dominated by *Nothofagus* species and not just the closed-forests dominated by *Nothofagus* in the sense of Specht (1970). Those communities dominated by *Nothofagus gunnii* in highland parts of Tasmania are related to closed-scrub communities and will not be considered further.

Closed-forests dominated by *Nothofagus* have been described previously as 'cool temperate rain forests' (Webb, 1959; Wood & Williams, 1960) or as 'nanophyll moss forests' by Webb (1959, 1968) in his physiognomic classification of eastern Australian rainforests.

Nothofagus species and their distribution

The southern closed-forests of Australia are dominated by two species of *Nothofagus*, namely *N. cunninghamii* and *N. moorei*. From the general distribution of these two species (Figure 5.1) it can be seen that the species do not overlap geographically. Both species form self-regenerating forests of trees up to 30 m or more high and belong to the '*menziesii*' group of species, based on their pollen morphologies (Cranwell, 1939; Cookson & Pike, 1955).

Other communities containing Nothofagus

Nothofagus occurs not only in mature self-regenerating forests, but also as young plants in the understorey of tall open-forest (see Chapter 6, Ashton, 1981). Indeed, the latter may be interpreted as part of a succession to closed-forest with increasing time from the last fire. Mature to overmature trees may persist as individuals or groups within northern closed-forests, and may be interpreted as relics of a previous climatic regime, especially one with slightly lower temperatures.

Distribution of Nothofagus forests

Nothofagus, a genus in the Fagaceae, is restricted in its distribution to the Southern Hemisphere. Living species occur in New Caledonia, New Guinea, Australia (including Tasmania), New Zealand and South America; fossil remains have been found in Antarctica. The present Australian distribution (Figure 5.1) has contracted very markedly since the Tertiary, when *Nothofagus* species were widespread in southern Australia, and included not only members of the *menziesii* group, but also of the *'fusca'* and *'brassii'* groups (Cookson, 1958). *Nothofagus gunnii* is the only living member of the *fusca* group in Australia and there are no Australian representatives of the *brassii* group extant.

Constraints on living Australian species of Nothofagus

The present constraint on the distribution of *Nothofagus* in Australia is overwhelmingly climatic, with fire, competition from plants of other genera, and soil fertility interacting to form local patterns. Each of these constraints will now be considered briefly.

In common with many facies of northern closed-forests (see Chapter 4, Webb & Tracey, 1981) the southern closed-forest is able to develop only in climates with a high annual rainfall, which is evenly distributed and highly reliable throughout the year. In Victoria and Tasmania *N. cunninghamii* appears to be able to form self-regenerating forests only in areas receiving more than 1500 mm rain per year (Howard, 1970, and unpublished), although riparian stands may form forests in areas with a rainfall of 1375 mm per year if supplemented by ground-water. The limited information available for forest sites of *N. moorei* suggests that the distribution of this species is constrained similarly, although minimum effective annual rainfalls may be higher for the more northerly occurrences of this species.

Nothofagus is able to grow over a very wide temperature range. In Victoria, *N. cunninghamii* grows over an altitudinal range from sea-level, with a mean annual temperature of 14°C, to 1370 m, where the mean annual temperature is 8.5°C, and in Tasmania it grows from sea-level to the tree-line. Seeding plants of *N. cunninghamii* failed to grow in Sydney where the mean annual temperature is 17.3°C (Howard, unpublished). *Nothofagus moorei* probably has a maximum mean temperature limit of less than 17.3°C for growth, based on my observations on seedlings planted outside its natural range. This boundary is, however, obscured by competitive interactions with species from northern

Figure 5.1. Distribution of southern closed-forest types dominated either by *Nothofagus moorei* (m) or by *N. cunninghamii* (hatched areas), although the latter species may not always be present in this forest type.

closed-forests. As for *N. cunninghamii* and the New Zealand species, *N. solandri*, *N. moorei* may be capable of extending its distribution to the tree line, although this is conjecture.

Failure of *N. cunninghamii* to establish and grow at high altitudes appears to be related to an insufficiently long frost-free period, in which flowering and seed development can take place (at least three months is required), so that natural regeneration becomes impossible (Howard, 1970). Failure to establish in warm climates appears to be related to problems of early seedling growth rather than to reproduction.

Nothofagus forests are not present in many areas which appear to be climatically suitable. This could be due in part to difficulties of seed redistribution to suitable sites from existing enclaves, as the seed is poorly designed for wind distribution (Howard, 1970). Although it floats well and germinates after 15 days' immersion in fresh water, this enables only downstream distribution within an individual catchment, and not lateral spread between catchments. Whether the initial retreat of *Nothofagus* into limited enclaves is due to past climatic change, and/or to widespread forest fires, present expansion is very much restricted by regular stand 'trimming' by forest fires.

A mature closed *Nothofagus* forest is remarkably fire-resistant, in that it creates an internal microclimate which is cool, moist and almost free of understorey, because of its high level of foliage cover. Surrounding tall open-forests, with their more open canopies and denser understoreys dry out more rapidly and carry fire much more efficiently. Lateral spread of fire into *Nothofagus* forest through wind eddies is also restricted by the dense canopy which extends to ground level and impedes draughts where the two forest types meet. Individual seedlings and saplings of *Nothofagus* established within the tall open-forest lack the organised resistance to fire of the mature *Nothofagus* stand, so that they are destroyed in a fire, and the potential lateral extensions of closed-forest are cut back to the boundaries of the original stand.

Both *N. moorei* and *N. cunninghamii* come into direct contact with the northern closed-forests or rainforests described in the previous chapter. Many species of the northern closed-forest appear to have higher minimum temperature requirements for growth, so that at temperatures below these, *Nothofagus* does not face direct competition. On the other hand, *Nothofagus* is able to grow successfully in a range of temperatures which overlaps those of species of the northern closed-forest, but it is rarely found in such situations.

Unfortunately, very little is known about the light compensation points for species of northern closed-forests, but based on field observations and one such measurement for *Acmena smithii* (Ashton & Frankenberg, 1976), it is apparent that seedlings of the northern closed-forest species are able to grow successfully at extremely low light intensities, even less than for *Nothofagus*. Seedlings of *N. cunninghamii* had a light compensation point of 1.15% of full sunlight in summer (Howard, 1973a) compared with a level of 0.1% for *A. smithii* (Ashton & Frankenberg, 1976).

Relict stands of *N. moorei* on the Coast Range in northern New South Wales consist entirely of mature to overmature trees, often as a second or third generation of trunks on the same rootstock, and are completely surrounded by species typical of the northern closed-forest. These *N. moorei* trees still set large quantities of viable seed each year, but seedlings were observed to establish only along forestry tracks and around log-loading platforms where the overstorey of the northern closed-forest had been removed, and mineral soil exposed. Seedlings and saplings of northern closed-forest species, however, exist throughout the mature forest, and amongst the relict *N. moorei*.

Where conditions suitable for establishment of both groups of species are available within the one catchment area, as in the Bellangry area, near Wauchope, in New South Wales for *N. moorei* and in eastern Victoria for species associated with *N. cunninghamii*, the northern closed-forest interdigitates smoothly with the southern closed-forest with increase in elevation. Where the northern closed-forest element is absent, as in southern Victoria and Tasmania, forests of *N. cunninghamii* occur at sea-level.

Soils are, in general, not an important constraint on *Nothofagus* forest development. However, as outlined by Jackson (1968), there is a strong interaction between fire frequency and forest regeneration on areas of inherently nutrient-poor bedrock, e.g. quartzite. On such rock types, the proportion of nutrients held in the standing biomass is very high compared with that in the soil, or with that available for replacement from the bedrock. In the self-perpetuating *Nothofagus* forest, these nutrients are rapidly recycled by litter decomposition or by re-absorption before leaves fall (Ashton, 1976), but if the area is burnt, particularly at short intervals, much of the nutrient capital will be leached or washed out of the root zone; thus prolonging *Nothofagus* re-establishment to the same time-scale as that necessary for rebuilding the soil nutrient pool. As the replacing vegetation is more fire-prone than the *Nothofagus* forest, this may mean that the nutrient capital is never restored, as the soil fertility continues to decline with further fires. In mainland Australia, where suitable climatic conditions are generally much less widespread than in Tasmania, *Nothofagus* forests are effectively restricted to high-nutrient bedrocks, such as granite, basalt, dacite, selected sedimentaries and alluvium. The soils developed on these bedrocks are generally deep (2 m or more), friable kraznozems.

Structure and floristics
Southern closed-forests can be subdivided most readily into two distinct structural types on the basis of average tree height; tall closed-forests and closed-forests which range from 10 to just over 30 m in height, and low closed-forests from 5 to 10 m high. Only *N. cunninghamii* at its highest altitudes (over 1300 m in Victoria) forms low closed-forest; mountains of sufficient elevation are lacking within the geographical range of *N. moorei* for the development of this type. Within the tall closed-forest and closed-forest type, many local variants have been distinguished on the basis of computer analyses of floristics, some being delineated by co-dominant or subdominant tree species (Busby & Bridgewater, 1977), and others by variations in the suite of understorey species (Howard & Ashton, 1973).

Low closed-forest
This forest type is restricted to the highlands of Victoria and Tasmania and the following description is based mainly on a limited number of stands described by Howard & Ashton (1973) from Victoria.

Low closed-forest contains a large number (2113 stems ha^{-1}) of small (36.7 m^2 ha^{-1} basal area), often multi-stemmed trees of *N. cunninghamii* and a similar number of *Leptospermum grandifolium* trees (1840 stems ha^{-1}, 2411 m^2 ha^{-1} basal area). The average height of both species is 9 m, although towards the upper altitudinal limit of this community *L. grandifolium* becomes dominant in height and density, with *N. cunninghamii* averaging only 3.3 m in height. The total crown cover of the stand exceeds 75% foliage projective cover, and the trees are smooth-trunked and light-limbed, with the crowns occupying a third to half of the total tree height.

Low closed-forest is characterised further by a large number of species of mosses and lichens which cover trunks, dead wood and soil, often forming a soft spongy carpet over the entire forest floor. Sparse shrubs of species such as *Drimys lanceolata*, a few tree-ferns (*Dicksonia antarctica*) and occasional tall shrubs of *Atherosperma moschatum* and

Podocarpus alpina also contribute to the intermittent shrub stratum. Ground ferns are few, *Polystichum proliferum* and *Blechnum penna-marina* being the most common, and the only filmy fern present is *Hymenophyllum peltatum*. Small herbs are very common in the moss mat, particularly around streams. A species list (excluding mosses, lichens and liverworts) is given in Table 5.1 for one typical stand of low closed-forest from Victoria. In low closed-forest in Tasmania additional endemic shrubs make up a significant part of the understorey.

Tall closed-forest and closed-forest

These forest types form the major stronghold of *Nothofagus* today in Australia, and include not only forests dominated by the two different species, *N. moorei* and *N. cunninghamii*, but also a wide range of different associated species reflecting variations in local environment and the wide latitudinal range of this forest type. Forest structure, however, remains relatively constant, local variations generally reflecting the age of the stand rather than environmental variations.

Structure. The mature self-regenerating closed-forest of *Nothofagus* consists typically of trees from 10 to more than 30 m in height, generally single-stemmed, with canopies occupying half to two-thirds of total tree height. *Nothofagus* trees of all size classes are present, but large mature to overmature trees predominate. Seedlings, saplings and poles are not randomly scattered throughout the stand, but are aggregated in clusters under gaps in the canopy, generally caused by the death or destruction of mature trees, remnants of which can usually be distinguished. Associated species may be co-dominant (e.g. *Athrotaxis selaginoides*, *Doryphora sassafras*, *Ceratopetalum apetalum*, *Atherosperma moschatum*) or sub-dominant (e.g. *Phyllocladus rhomboidalis*, *Atherosperma moschatum*) depending on the history of and conditions at the site. The presence of long-lived seral species, such as *Eucalyptus regnans*, *E. delegatensis*, *E. saligna* and *E. fastigata*, as an emergent stratum, or of *Acacia melano-xylon* as a co-dominant, reflect the comparative youth of such a stand (less than 400 years old), as such species are unable to regenerate further within the mature closed-forest. *Nothofagus* basal area is generally high and density comparatively low, and may at times be exceeded by that of the co-dominants (Table 5.2).

Nothofagus trees themselves vary considerably in their form. Whilst most trees have single stems, multi-stemmed trees are common in stands which result from coppicing after fire (*N. cunninghamii* in Victoria) or in

Table 5.1. *Species composition of a low closed-forest at Lake Mountain, Victoria (altitude 1470 m). CD, co-dominant; +, casual species, intruding from surrounding vegetation; EF, epiphytic fern; FF, filmy fern; GF, ground fern; H, herb; S, shrub; T, tree*

Angiosperms

Caryophyllaceae
Nertera depressa H

Compositae
Astelia alpina H
Cotula filicula H
Gnaphalium sp. H
?*Lagenifera stipitata* H

Cyperaceae
Uncinia tenella H
Carex appressa H

Epacridaceae
Wittsteinia vacciniacea S

Ericaceae
Gaultheria appressa S

Fagaceae
Nothofagus cunninghamii T, CD

Geraniaceae
Geranium pilosum H

Gramineae
Hierochloe redolens H

Iridaceae
Libertia pulchella H

Labiatae
Prostanthera cuneata S, +

Myrtaceae
Baeckea utilis var. *latifolia* S, +
Leptospermum grandifolium T, CD

Orchidaceae
Chiloglottis sp. H

Ranunculaceae
Ranunculus plebeius H

Rosaceae
Acaena anserinifolia H

Rubiaceae
Coprosma nitida S

Stylidiaceae
Stylidium graminifolium H

Umbelliferae
Hydrocotyle javanica H

Violaceae
Viola hederacea H

Winteraceae
Drimys lanceolata S

Ferns and fern allies

Aspidiaceae
Polystichum proliferum GF

Blechnaceae
Blechnum fluviatile GF
B. penna-marina GF
B. wattsii GF

Grammitidiaceae
Grammitis billardieri EF

Hymenophyllaceae
Hymenophyllum peltatum FF

Lycopodiaceae
Lycopodium scariosum
L. selago

Table 5.2. *Basal area and density values at five selected sites of tall closed-forest and closed-forest, showing density and basal area values for dominant (D) and subdominant (SD) species*

Site and Species	Density (no. stems ha^{-1})	Basal area (m^2 ha^{-1})
1. Gloucester Tops, NSW[a]		
Nothofagus moorei (D)	309	60.8
Doryphora sassafras (SD)	107	1.6
Elaeocarpus holopetalus (SD)	202	6.3
2. Maits Rest, southwestern Victoria		
Nothofagus cunninghamii (D)	247	379.9
3. Mt Donna Buang, Victoria		
Nothofagus cunninghamii (D)	865	73.5
Atherosperma moschatum (SD)	284	20.7
4. Surrey Hills, northwestern Tasmania		
Nothofagus cunninghamii (D)	543	72.8
Atherosperma moschatum (SD)	123	15.8
5. Errinundra Plateau, eastern Victoria		
Atherosperma moschatum (D)	679	27.5

[a]from Bowden & Turner (1976), based on 50 plots, 0.0201 ha in area, excluding emergent *Eucalyptus fastigata* trees. These values are for all stems $\geqslant$ 10 cm girth; values at all other sites are for all stems.

relict stands of *N. moorei* which can only reproduce from coppice shoots because of seedling competition from species of northern closed-forests. Individual trunks are rarely smooth, particularly in *N. cunninghamii* in lowland northwestern Tasmania and Victoria, and in *N. moorei* at its junction with areas of northern closed-forests. In these situations, the epicormic burl which is formed in all situations is very large, developing as a corrugated gnarled bulge up to twice the basal stem diameter from below the soil level up to 2 to 3 m up the trunk. From this woody mass of dormant buds, roots and coppice shoots grow at random even when the crown is intact. Beneath the major forest canopy there may be an ill-defined layer of subdominant tree species, and a few straggling tall shrubs, but the primary understorey is non-angiospermous and made up of tree ferns, ground ferns, mosses, lichens and liverworts. Epiphytic ferns (including filmy ferns), orchids and one dicotyledon, *Fieldia australis*, are common in *N. moorei* forests, and in lowland *N. cunninghamii* forests. Mosses, lichens and liverworts occur mainly on trunks, branches, fallen wood and stones, the general forest floor being covered with a mosaic of patches of ground ferns, deep leaf litter and bare mineral soil. In Tasmania, where lyrebirds (*Menura novaehollandiae*) are not present to cultivate the topsoil in their search for food, a litter layer of 10 cm is common between trees, and around the bases of trunks may build up to 60 cm deep. Such litter accumulation is common also around the bases of trees of *N. moorei*.

Floristics. Some life-form and floristic data for angiosperm, gymnosperm and fern species occurring at a selection of sites of tall closed-forests and closed-forests are listed in Table 5.3. Although the data are compiled from species lists which represent only a small part of the information available from selected sites in eastern Australia (Fraser & Vickery, 1938; Gilbert, 1959; Howard, 1970; Howard & Hope, 1970; Howard & Ashton, 1973; Bowden & Turner, 1976; Busby & Bridgewater, 1977; Kirkpatrick, 1977; Howard, unpublished) they do show some of the general characteristics of all such forests so far described.

The number of spermatophyte families is very low, especially of tree species, and this distinguishes this forest type from the northern closed-forests described in the previous chapter. The diversity of spermatophyte families is highest at those sites where the forest overlaps northern closed-forest, and less where the stands occur on bedrock of inherently low nutrient status or in areas of high fire frequency. At those sites where stands are limited in area, and especially if the site has a history of infrequent fires, the stand may retain a permanently ecotonal aspect, with scattered emergent eucalypts, acacias and other species more typical of surrounding tall open-forest stands being present. Shrubs and herbs generally are very poorly represented throughout southern closed-forest; both groups are more diverse and individually more abundant in low closed-forests.

Lianes are poorly represented generally, except in the northernmost stands which overlap with northern closed-forest. They tend to be comparatively slender and inconspicuous, compared with the abundant, ropy lianes of adjacent stands of northern closed-forest.

Ferns are well represented in lowland *N. cunninghamii* and all *N. moorei* stands, imparting much of the character to southern closed-forests when viewed from the ground (Figure 5.2). Tree-ferns are particularly abundant in breaks in the permanent canopy, such as along streams, and ground ferns may form large patches on the forest floor. Epiphytic ferns follow the same pattern of distribution as lianes and are most abundant in lowland *N. cunninghamii* and all *N. moorei* stands.

Table 5.3. *Some life form and floristic data for seven selected sites of southern closed-forest (excluding mosses, lichens and liverworts)*

Site	Nichols Creek, Tasmania (150 m)	Basils Road, Tasmania (502 m)	Cradle Mountain, Tasmania (822 m)	Maits Rest, Victoria (196 m)	Mt Donna Buang, Victoria (1005 m)	Mt Boss, NSW (1066 m)	Bar Mt NSW (1080 m)
Number of species: trees	5	3	3	2	4	7	6
shrubs	1	1	5	2	2	6	2
herbs	1	0	2	3	9	3	0
lianes	0	0	0	1	0	6	8
ferns	3	10	5	18	8	9	8
TOTAL	10	14	15	26	23	31	24
Number of epiphytic species	1	6	3	10	3	6	6
Total number of families	10	12	11	19	19	27	22
Number of non-fern families	7	4	6	8	12	18	15

Vascular epiphytes are few in number, are mainly orchids, and are more prominent in stands of *N. moorei*.

Tall closed-forest and closed-forest not dominated by Nothofagus *species*
There are at least three types of tall closed-forest and closed-forest which are closely related floristically or structurally to *Nothofagus*-dominated stands. These are forests dominated by species either usually

Figure. 5.2. Characteristic appearance of tall closed-forest and closed-forest at Mt Donna Buang, Victoria (1005 m altitude), viewed from the roadside. Note the character imparted by the tree-fern *Dicksonia antarctica* in the foreground, the varied size of the trunks of *Nothofagus cunninghamii* trees and the lack of a shrub stratum. A clump of subdominant *Atherosperma moschatum* is seen on the left.

subdominant or co-dominant, or essentially seral (e.g. *Acacia*), or floristically similar to the understorey of tall closed- and closed-forest. Stands dominated solely by *Atherosperma moschatum* occur in gullies of the highlands of eastern Victoria and in the Dandenong Ranges, near Melbourne. They are otherwise indistinguishable from the forests of *N. cunninghamii* and *A. moschatum* of the central highlands of Victoria, and of Tasmania. The absence of *N. cunninghamii* from these areas may be because of past climatic change and/or the creation of new environments by geological changes and subsequent failure of seed redistribution of *N. cunninghamii*. (*Atherosperma moschatum* seed, on the other hand, is ideally designed for wind distribution (Howard, 1970).) Pure stands of *Ceratopetalum apetalum* occur within stands of *N. moorei* in the Bellangry area of northern New South Wales; these most probably arose from random interactions of fire, seed distribution and the availability of seed for initial establishment.

Seral species, such as *Acacia elata*, *A. melanoxylon* and *A. dealbata*, form localised stands having a typical closed-forest structure and flora, and appear to be one end of a spectrum of post-fire communities of closed-forest. They are unique in that the dominant species have been entirely eliminated.

Suites of understorey species of closed-forest type occur in localised habitats, such as deep gullies, regardless of the type of surrounding vegetation. Although these are not strictly southern closed-forest they tend to intergrade with fire-induced stands of *Acacia* species. *Acacia melanoxylon* trees may form the overstorey to such species and similar assemblages of species often occur in the understorey to tall open-forest.

The understorey species of southern closed-forest are not tied necessarily to the *Nothofagus*/*Atherosperma*/*Ceratopetalum*/*Acacia* type of overstorey to closed-forest, but they are able to grow wherever local conditions provide adequate shade and a continuously moist environment. Individual fern species are prominent in the tall closed-forest and complete assemblages of understorey species may occur within any rainfall zone. In Stradbroke Chasm in eastern Victoria, for example, such an assemblage occurs in a rain-shadow area (450 mm annual rainfall) in a deep narrow defile where the surrounding vegetation is low woodland dominated by *Callitris rhomboidea*.

Regeneration, succession and the present status of southern closed-forest
Some attempts have been made in the past to describe the interaction of environmental variables with the biology of *N. cunninghamii* forests in

Tasmania (Jackson, 1968) and Victoria (Howard & Ashton, 1973). Practising foresters in Tasmania, where the eucalypt forests are economically important and yet occupy potential *N. cunninghamii* sites, have paid considerable attention to the effect of *N. cunninghamii* invasion of such stands, with a view to excluding the undesirable *N. cunninghamii* because it reduces the productivity of or actually kills the eucalypt overstorey (Ellis, 1964, 1971). No such attempt at integration of forest management has yet been attempted for *N. moorei* forests.

Nothofagus cunninghamii *in Victoria*
In areas such as Victoria, where forests of *N. cunninghamii* are limited in area and surrounded by tall open-forest dominated by eucalypts, fire is the single most important threat to their continued existence. *Nothofagus cunninghamii* is now largely restricted to soils of high fertility on bedrocks of high nutrient status and to the most sheltered, moist positions, such as deep gullies or southerly aspects. Lateral expansion of such stands is still occurring, aided by the ability of *N. cunninghamii* to regrow from the basal burl as well as by the spread of seed from undamaged stands.

The widespread fires of 1939 in Victoria reduced the distribution of mature closed-forest to the most sheltered enclaves. Subsequent regeneration by coppicing, particularly when regeneration of *Atherosperma moschatum* from seed also occurred, has led after 40 years to local canopy closure of the *N. cunninghamii* understorey within the tall open-forest. These trees of *N. cunninghamii* are now flowering and filling in of the stratum by seedling regeneration is also taking place. At least another 30 years may be necessary, however, for the establishment of a continuous canopy of closed-forest species beneath the eucalypts, during which time the re-establishing closed-forest is still highly vulnerable to destruction by fire. At least two generations of fire-produced coppice shoots may be produced from an epicormic burl, which suggests that regeneration may be able to recur several times after fire. A fire within this next 30-year period may not necessarily eliminate *N. cunninghamii* from these sites.

Nothofagus cunninghamii *in Tasmania*
In Tasmania, where *N. cunninghamii* forms very extensive forests, fire is still important, especially as it interacts with tree death induced by the *Platypus* beetle on sites of low nutrient status. Tasmania has extensive areas climatically suited to growth of *N. cunninghamii* and forests have

been able to develop not only on high-nutrient bedrocks but also on those lower in nutrients. In the latter, repeated fires can eliminate *N. cunninghamii* and subsequent regeneration may be prevented by nutrient depletion and increased waterlogging of the site. Carried to its extreme, other species may invade a site with each alteration in the environment through a series of reverse successional changes to produce finally a closed-sedgeland, composed mainly of tussocks of *Gymnoschoenus sphaerocephalus* (Jackson, 1968). Such a 'disclimax' is not only extremely flammable, but it is able to perpetuate itself with or without fire. It is debatable whether such a site is ever able again to pass through the successional stages back to closed-forest, even if complete protection from fire were possible, as soils are shallower, lower in nutrients and waterlogged seasonally.

Extensive areas of *N. cunninghamii* in Tasmania also occur on soils and bedrocks of high nutrient status. Within these areas the closed-forest is under pressure from two main forces: firstly, a desire to harvest commercially eucalypts typical of tall open-forest; and secondly, an interaction between a species of *Platypus* beetle, which may cause tree death naturally, and forestry activity in opening up the vegetation by means of roads and logging practices.

To harvest eucalypts of tall open-forest succesfully, areas which would be capable of supporting *N. cunninghamii* must be kept free of this species, both because of its direct competitive effect with eucalypts for soil nutrients and also because the development of a closed-forest understorey may lead to death of the eucalypt component long before the end of the natural life span of the eucalypts. This effect may arise because of a reduction in soil temperatures produced by the dense crowns of *N. cunninghamii* and associated trees (Ellis, 1971). Forestry practice is to replace logged-over virgin closed-forest with eucalypts and, if necessary, to burn off or slash the developing understorey of *N. cunninghamii*.

The role of the Platypus *beetle in Tasmania*
Several species of *Platypus* beetle occur in the *Nothofagus* forests of New Zealand (Milligan, 1972), where they play much the same role as the one species found in Tasmania (Howard, 1973*b*). It belongs to the family Platypodinae and is an ambrosia beetle, that is, it carries spores of a fungus into its nest sites in the wood of living trees, which then grow in the tunnel systems of the nest and feed the developing larvae. The fungus is often lethal to the host trees; for example, as in the case of *Scolytus* ambrosia beetles in *Ulmus* in the Northern Hemisphere.

Under natural conditions in undisturbed mature forests of *N. cunninghamii* a very high proportion of the trees are overmature and provide suitable sites for the beetle to inhabit. Young trees appear to withstand the attack by beetles, perhaps because of a lower ratio of sapwood to heartwood or perhaps because they provide insufficient volume of wood for the species-specific nest pattern to be bored. Gaps in the forest canopy may be found locally throughout the forest and these are caused by the death of concentric rings of trees, with the most decayed at the centre and newly dead trees occupied by *Platypus* at the periphery. Seedling, sapling and pole trees of *N. cunninghamii* are present in the gap, thereby suggesting that the *Platypus* beetle in the natural state fits neatly into the regeneration pattern of *N. cunninghamii*.

Platypus spread is greatly accelerated by forestry practices such as logging or by the development of access roads, and extensive tree deaths may occur, particularly in dry summers. Coppicing stumps are also killed, so that forest re-establishment depends solely on seedling regeneration.

Regeneration after fire may be similarly affected. Coppicing burls may be completely destroyed by *Platypus* beetles and, as seed is often not available or is very limited in amount (Howard, 1974), re-establishment of *N. cunninghamii* becomes dependent on lateral spread of seed from unburnt areas of adjoining forest.

The absence of the *Platypus* bettle from some areas of marginal rainfall in Tasmania and from mainland Australia and the consequent importance of coppice regeneration may best be explained by limitations on the supply of host material, especially if at some time in the past extensive forest fires reduced all the closed-forest in an area to young plants and destroyed all the large trees suitable as hosts for the *Platypus* beetle.

Nothofagus moorei *in New South Wales and Queensland*

Forests of *N. moorei* are restricted generally to soils of high nutrient status and are constrained in much the same way as forests of *N. cunninghamii* in Victoria when surrounded by tall open-forests of eucalypts. Only one really extensive area of *N. moorei* forest still exists, in the Bellangry area of New South Wales, and whilst some eucalypts are removed from it for timber, it is generally recognised that it is a unique forest type and considerable areas of it are preserved. It is, however, still liable to damage from fire and from the incursion of logging roads.

Whereas forests of *N. cunninghamii* are able to develop over a wide altitudinal range, *N. moorei* is successfully outcompeted by elements of the northern closed-forest at the lower end of its potential altitudinal range. Only Mt Nothofagus (28°17′S; 152°37′E) on the border of New South Wales and Queensland is of sufficient elevation for *N. moorei* to persist as a self-regenerating forest free of competition from elements of the northern closed-forest. Enclaves of *N. moorei* forest in Wiangarie State Forest, New South Wales, and Lamington National Park, Queens-

Figure 5.3. A comparatively small individual of *Nothofagus moorei* from Wiangarie State Forest, northern New South Wales (1080 m altitude), where northern closed-forest species now preclude seedling regeneration of *Nothofagus*. Note the enormously enlarged basal burl and epicormic shoots. Evidence of a previous generation trunk was visible between the two main trunks; the trunk of sapling size could represent a replacement for the larger trunk, the crown of which was mostly dead.

land, are no longer able to regenerate from seed, as discussed previously, but the species persists as individual trees by coppice replacement from the epicormic burl (Figure 5.3). In Wiangarie State Forest (28°23'S; 153°06'E) recent development of roads for logging has led to some windthrow of *N. moorei* as well as limited seedling regeneration on areas free from competition, such as disused snig-tracks and log-loading areas.

The conservation status of southern closed-forest
Southern closed-forest dominated by *N. moorei* appears to be in the greatest danger of extinction because of its limited extent and the scarcity of sites suitable for its development. The existence of some reservations indicates that its unique status is recognised but these reserves may need to be more actively protected from fire.

Forests of *N. cunninghamii* in Victoria are limited in distribution but, because of recognition of this, together with the attractive character of such forests, a large number of sites have been preserved. Protection from both fire and over-use by visitors will be essential to their survival.

In Tasmania, where *N. cunninghamii* is very widespread, comparatively small areas are protected and a detailed inventory of forest sites is required as a basis for choosing areas to be preserved. Species composition of Tasmanian closed-forest is very varied, presumably because of variations in soil type, altitude and rainfall. Further studies such as those of Kirkpatrick (1977) and Busby & Bridgewater (1977) may help to provide a sounder foundation on which such a re-appraisal could be made.

References

Ashton, D.H. (1976). Phosphorus in forest ecosystems at Beenak, Victoria. *Journal of Ecology*, *64*, 171–86.

Ashton, D.H. (1981). Tall open-forests. In *Australian Vegetation*, ed. R.H. Groves, pp. 121–151. Cambridge University Press.

Ashton, D.H. & Frankenberg, J. (1976). Ecological studies of *Acmena smithii* (Poir.) Merrill & Perry with special reference to Wilson's Promontory. *Australian Journal of Botany*, *24*, 453–87.

Bowden, D.C. & Turner, J.C. (1976). *A preliminary survey of stands of temperate rainforest on Gloucester Tops*. University of Newcastle, NSW, Research Papers in Geography No. 10.

Busby, J.R. & Bridgewater, P.B. (1977). Studies in Victorian vegetation. II. A floristic survey of the vegetation associated with *Nothofagus cunninghamii* (Hook.) Oerst. in Victoria and Tasmania. *Proceedings of the Royal Society of Victoria*, *89*, 173–82.

Cookson, I.C. (1958). Fossil pollen grains of *Nothofagus* from Australia. *Proceedings of the Royal Society of Victoria*, *71*, 25–30.

Cookson, I.C. & Pike, K.M. (1955). The pollen morphology of *Nothofagus* Bl. subsection *Bipartitae* Steen. *Australian Journal of Botany, 3*, 197–206.

Cranwell, L.M. (1939). Southern beech pollens. *Records of the Auckland Institute & Museum, 2*, 176–96.

Ellis, R.C. (1964). Dieback of alpine ash in north-eastern Tasmania. *Australian Forestry, 28*, 75–90.

Ellis, R.C. (1971). Dieback of alpine ash as related to changes in soil temperature. *Australian Forestry, 35*, 152–63.

Fraser, L. & Vickery, J.W. (1938). The ecology of the Upper Williams River and Barrington Tops district. II. The rainforest formations. *Proceedings of the Linnean Society of New South Wales, 63*, 139–84.

Gilbert, J.M. (1959). Forest succession in the Florentine Valley, Tasmania. *Papers and Proceedings of the Royal Society of Tasmania, 93*, 129–51

Howard, T.M. (1970). The ecology of *Nothofagus cunninghamii* Oerst. Ph.D. thesis, University of Melbourne.

Howard, T.M. (1973a). Studies in the ecology of *Nothofagus cunninghamii* Oerst. I. Natural regeneration on the Mt Donna Buang massif, Victoria. *Australian Journal of Botany, 21*, 67–78.

Howard, T.M. (1973b). Accelerated tree death in mature *Nothofagus cunninghamii* Oerst. forests in Tasmania. *Victorian Naturalist, 90*, 343–5.

Howard, T.M. (1974). *Nothofagus cunninghamii* ecotonal stages. Buried viable seed in North West Tasmania. *Proceedings of the Royal Society of Victoria, 86*, 137–42.

Howard, T.M. & Ashton, D.H. (1973). The distribution of *Nothofagus cunninghamii* rainforest. *Proceedings of the Royal Society of Victoria, 86*, 47–76.

Howard, T.M. & Hope, G.S. (1970). The present and past occurrence of beech (*Nothofagus cunninghamii* Oerst.) at Wilson's Promontory, Victoria, Australia. *Proceedings of the Royal Society of Victoria, 83*, 199–209.

Jackson, W.D. (1968). Fire, air, water and earth – an elemental ecology of Tasmania. *Proceedings of the Ecological Society of Australia, 3*, 9–16.

Kirkpatrick, J.B. (1977). Biological environment. Native vegetation of the west coast region of Tasmania. In *Landscape and Man*, ed. M.R. Banks & J.B. Kirkpatrick, pp. 55–80. Launceston: Royal Society of Tasmania.

Milligan, R.H. (1972). A review of beech forest pathology. *New Zealand Journal of Forestry, 17*, 201–11.

Specht, R.L. (1970). Vegetation. In *The Australian Environment*, 4th edn (rev.), ed. G.W. Leeper, pp. 44–67. Melbourne: CSIRO Australia & Melbourne University Press.

Webb, L.J. (1959). A physiognomic classification of Australian rainforests. *Journal of Ecology, 47*, 551–70.

Webb, L.J. (1968). Environmental relationships of the structural types of Australian rainforest vegetation. *Journal of Ecology, 49*, 296–311.

Webb, L.J. & Tracey, J.G. (1981). The rainforests of northern Australia. In *Australian Vegetation*, ed. R.H. Groves, pp. 67–101. Cambridge University Press.

Wood, J.G. & Williams, R.J. (1960). Vegetation. In *The Australian Environment*, 3rd edn (rev.), pp. 67–84. Melbourne: CSIRO Australia & Melbourne University Press.

6

Tall open-forests

D.H. ASHTON

The tall open-forests in Australia are unique. At maturity, the eucalypts which dominate them may rise to heights of over 100 m, their shaft-like trunks supporting light, open crowns of pendant leaves. They are the supreme expression of the genus *Eucalyptus sensu lat.* The tall open-forest form includes all stands which, at maturity, are greater than 30 m tall and whose foliage covers 30–70% of the area. Where the canopy exceeds 40 m, the understorey is frequently luxuriant and consists of mesomorphic broad-leaf trees, shrubs and ferns. These forests, which occur in the most mesic sites, have been called 'wet sclerophyll forest' (Beadle & Costin, 1952) or 'sclerophyll fern forest' (Webb, 1959).

Tall open-forests as they occur in Australia are variable. In colder or less mesic sites tree height is reduced to between 30 and 60 m and the understorey is simplified. On more fertile soils, tussock grasses predominate with or without a shrub stratum. On poorer soils in south-eastern Australia, however, highly sclerophyllous shrubs, sedges and climbing ferns may be found. In Arnhem Land, some specially favoured sites may produce tall open-forests with a layered understorey of rank, tall grasses and shrubs. Along the flood plains of the Murray River, first-quality forests may be found where flooding is regular, but not severe. Elite trees of *E. camaldulensis* grow up to 42 m high in what must be classed as an exceptional form of tall open-forest. Here, the understorey consists of hydrophytic forbs and grasses. In this chapter most discussion will centre on the commercially valuable wet sclerophyll type of tall open-forest, rather than on the variants described above.

In many places the mature wet sclerophyll eucalypts emerge above a well-developed rainforest (or closed-forest), see the previous two chapters (Webb & Tracey, 1981; Howard, 1981). Since the closed-forests are self-perpetuating and the eucalypts light-demanding, this forest-in-forest structure is clearly successional. Such stands have been called 'mixed' forests (Gilbert, 1959). The canopies of some of the eucalypts on the extreme southwest coast of Western Australia bear unusually heavy crowns and this may have led Gardner (1942) to regard these wet forests as a form of temperate rainforest.

In Australia, the tall open-forests of the wet sclerophyll type are distributed in a discontinuous arc of high rainfall country from southern Queensland (latitude 25°S) to southern Tasmania (latitude 42°S) then across a low rainfall gap of 2100 km from western Victoria to the southwest of Western Australia (latitude 35°S). Over this enormous range, a wide floristic variation occurs, not only in the eucalypt component but also in the understorey species (Tables 6.1, 6.2). A marked floristic discontinuity is evident between southwestern Australia and the eastern coast. In the latter region two trends are evident: firstly, a gradual replacement of species dominance with decreasing latitude; and secondly, the extension of 'southern' species into northern latitudes where compensating altitudinal increases exist.

Three broad groups of wet sclerophyll eucalypt are discernable: those of northern New South Wales and southern Queensland (Figure 6.1), *Eucalyptus cloeziana*, *E. microcorys*, *E. pilularis*, *E. saligna* and *E. grandis*; those of Victoria (Figure 6.2) and Tasmania (and some in the highlands of northern NSW) *Eucalyptus regnans*, *E. viminalis*, *E. obliqua*, *E. globulus*, *E. fastigata*, *E. delegatensis*, *E. cypellocarpa*, *E. dalrympleana* and *E. nitens*; and those in southwest Western Australia (Figure 6.3), *Eucalyptus diversicolor*, *E. calophylla*, *E. guilfoylei*, and *E. jacksonii*.

About 11 species of *Eucalyptus* are more or less confined to the wet sclerophyll forests, and another 27 to 30 species occur in them at the wetter ends of the range. It is likely that these variable species, e.g. *E. pilularis*, *E. obliqua*, are made up of numerous ecotypes which enable them to occur also in dry sclerophyll and grassy forests, woodland or as scattered low trees in heathland.

Of the eucalypts exclusive to wet sclerophyll forests, six are smooth-barked, three are completely fibrous and two are 'half-barks'. These eucalypts are amongst the tallest trees in the world, although today only vestiges of the once magnificent forests remain. The tallest *E. regnans* surveyed in Victoria last century was 110 m (Hardy, 1935); today the tallest tree stands at 98.75 m in the Styx Valley, southwestern Tasmania (Australian Newsprint Mills, personal communication). The tallest *E. diversicolor* in Western Australia is 83.3 m. A recently discovered, tall *E.*

Table 6.1. *Characteristic understorey species of certain wet sclerophyll forests in five regions of Australia*

	Sub-tropics (southern Queensland and northern NSW)	Warm Temperate (central and southern NSW)	Cool Temperate (central and southern Victoria)	Cool Temperate (southern Tasmania)	Warm Temperate (southwest Western Australia)
Tall trees	E. cloeziana	E. saligna	E. regnans	E. regnans	E. diversicolor
	E. grandis	E. paniculata	E. viminalis	E. obliqua	E. calophylla
	E. pilularis	E. pilularis	E. obliqua	E. ovata	E. jacksonii
	E. microcorys	E. smithii	E. cypellocarpa	E. delegatensis	E. guilfoylei
Small trees and shrubs	Syncarpia glomulifera	Syncarpia glomulifera	Acacia melanoxylon	Acacia melanoxylon	Casuarina decussata
	Tristania conferta	Acacia binervata	Acacia dealbata	Acacia dealbata	Agonis flexuosa
	Casuarina torulosa	Casuarina torulosa	Pomaderris aspera	Pomaderris apetala	Trymalium spathulatum
	Callicoma serratifolia	Callicoma serratifolia	Olearia argophylla	Acacia pentadenia	Acacia urophylla
	Rapanea howittiana	Rapanea howittiana	Bedfordia arborescens	Bedfordia salicina	Lasiopetalum floribundum
	Dodonaea triquetra	Dodonaea triquetra	Hedycarya angustifolia	Hedycarya angustifolia	Bossiaea laidlawiana
	Persoonia attenuata	Persoonia linearis	Zieria arborescens	Zieria arborescens	Oxylobium lanceolatum
	Elaeocarpus reticulatus	Elaeocarpus reticulatus	Prostanthera lasianthos	Phebalium squameum	Logania vaginalis
	Brachychiton acerifolium	Zieria arborescens	Correa lawrenciana	Drimys lanceolata	Hibbertia amplexicaulis

Table 6.1 (*continued*)

	Sub-tropics (southern Queensland and northern NSW)	Warm Temperate (central and southern NSW)	Cool Temperate (central and southern Victoria)	Cool Temperate (southern Tasmania)	Warm Temperate (southwest Western Australia)
Small trees and shrubs (*continued*)	*Backhousia myrtifolia*	*Tieghemopanax sambucifolius*	*Tieghemopanax sambucifolius*	*Cenarrhenes nitida*	
Pachycauls	*Archontophoenix cunninghamiana* *Livistona australis* *Cyathea australis* *Cyathea leichardtiana* *Lepidozamia peroffskyana*	*Livistona australis* *Cyathea australis* *Macrozamia communis*	*Cyathea australis* *Dicksonia antarctica*	*Dicksonia antarctica*	*Macrozamia riedlei*
Lianes	*Marsdenia rostrata* *Smilax australis* *Flagellaria indica* *Cissus hypoglauca*	*Marsdenia rostrata* *Smilax australis* *Clematis aristata* *Cissus hypoglauca*	*Clematis aristata* *Parsonsia brownii* *Pandorea pandorana* *Billardiera longiflora*	*Clematis aristata*	*Clematis aristata* *Billardiera floribunda* *Hardenbergia comptoniana*

	Parsonsia brownii *Pandorea pandorana* *Morinda jasminoides* *Celastrus australis* *Ripogonum discolor*	*Parsonsia brownii* *Pandorea pandorana* *Morinda jasminoides* *Geitonoplesium cymosum*			
Ground ferns	*Culcita dubia* *Blechnum cartilagineum* *Hypolepis rugosula* *Asplenium australasicum*	*Culcita dubia* *Blechnum cartilagineum* *Adiantum hispidulum* *Pteridium esculentum*	*Polystichum proliferum* *Blechnum nudum* *Blechnum wattsii* *Histiopteris incisa* *Pteridium esculentum* *Culcita dubia*	*Polystichum proliferum* *Blechnum nudum* *Blechnum wattsii* *Histiopteris incisa* *Pteridium esculentum*	*Adiantum aethiopicum* *Pteridium esculentum*
Herbs	*Oplismenus aemulus* *Imperata cylindrica* *Viola betonicifolia*	*Oplismenus aemulus* *Imperata cylindrica* *Hydrocotyle hirta*	*Australina muelleri* *Hydrocotyle hirta* *Viola hederacea*	*Australina pusilla* *Hydrocotyle hirta* *Viola hederacea*	*Hydrocotyle hirta* *Veronica calycina*

Table 6.2. *Forest-in-forest communities in eastern Australian rainforest with eucalypt emergents, a late stage in secondary succession*

	Southern Queensland and Northern NSW	Southern NSW and Eastern Victoria	Central Victoria	Southern Tasmania
Very tall trees (emergents)	*E. microcorys* *E. cloeziana* *E. grandis*	*E. botryoides* *E. maidenii* *E. viminalis* *E. elata*	*E. regnans* *E. viminalis*	*E. regnans* *E. obliqua*
Tall trees Rainforest	*Casuarina torulosa* *Syncarpia glomulifera* *Tristania conferta*	*Acacia melanoxylon*	*Acacia melanoxylon* *Nothofagus cunninghamii*	*Nothofagus cunninghamii*
Trees	*Argyrodendron actinophyllum* *Cryptocarya rigida* *Litsea reticulata* *Ackama paniculata* *Callicoma serratifolia* *Cinnamomum oliveri* *Rhodamnia trinervia*	*Acmena smithii* *Eucryphea moorei* *Rapanea howittiana* *Acronychia oblongifolia*	*Atherosperma moschatum* *Pittosporum bicolor* *Hedycarya angustifolia*	*Atherosperma moschatum* *Phyllocladus aspleniifolius* *Pittosporum bicolor* *Anodopetalum biglandulosum*

Shrubs	*Citriobatus multiflorus* *Wilkiea huegeliana* *Eupomatia laurina*	*Elaeocarpus reticulatus* *Eupomatia laurina*	*Coprosma quadrifida*	*Drimys lanceolata* *Coprosma nitida* *Cenarrhenes nitida*
Pachycauls	*Archontophoenix cunninghamiana* *Livistona australis* *Cyathea leichardtiana*	*Livistona australis* *Cyathea australis* *Cyathea leichardtiana*	*Cyathea australis* *Cyathea cunninghamii* *Dicksonia antarctica*	*Dicksonia antarctica*
Lianes	*Ripogonum discolor* *Cissus hypoglauca* *Morinda jasminoides* *Piper novaehollandiae* *Celastrus australis* *Smilax australis*	*Marsdenia rostrata* *Celastrus australis* *Parsonsia brownii* *Smilax australis*	*Clematis aristata* *Parsonsia brownii*	

viminalis in northwestern Tasmania has been authentically measured to 89.9 m (Institute of Foresters of Australia, 1976).

Similar groupings of the understorey components could also be made. Tree-ferns are a conspicuous feature of the east coast forests, especially in the south, but are totally absent in Western Australia. Palms, however, are characteristic of the warmer coastal forests in New South Wales and Queensland. In the northern forests, the understorey of tall shrubs

Figure 6.1. Tall open-forest of *Eucalyptus pilularis* near Taree, New South Wales (reproduced by kind permission of the Forestry Commission of New South Wales).

Figure 6.2. Mature tall open-forest of *Eucalyptus regnans* at Wallaby Creek, Victoria, with a *Pomaderris aspera* understorey.

Figure 6.3. Mature tall open-forest of *Eucalyptus diversicolor*, near Pemberton, Western Australia, with *Trymalium floribundum* in the understorey.

and vines includes notophyllous and microphyllous forms such as *Callicoma serratifolia, Dodonaea triquetra, Elaeocarpus reticulatus* and *Brachychiton acerifolium.* 'Small trees' (10 to 40 m) include the myrtaceous *Syncarpia glomulifera* and *Tristania conferta,* as well as *Casuarina torulosa.* Palms (*Livistona australis*) and vines are common. In the southeast the understorey has similar leaf-size classes and includes *Pomaderris aspera* and *P. apetala,* the composites, *Olearia argophylla* and *Bedfordia* spp. and the tree-ferns *Cyathea australis* and *Dicksonia antarctica.* Small trees (20 to 30 m) consist of acacias such as *A. dealbata* and *A. melanoxylon.* In southwest Western Australia the appearance of the undergrowth is astonishingly similar to some of the southeastern forests, except for the absence of tree-ferns and the low diversity of ground ferns. The understorey includes such tall shrubs as *Trymalium spathulatum, Lasiopetalum floribundum* and *Bossiaea laidlawiana.* Small trees include *Agonis flexuosa* and *Casuarina decussata.* Perhaps the latter species may prove to form a climax stand in the prolonged absence of fire in over-mature *E. jacksonii* forest. No rainforest trees of Malesian or Antarctic affinities now occur in this wet southwest corner of Australia.

The environment

Climate
The outstanding features of the environment are high, reliable rainfall and shelter from the worst desiccation of the fire-promoting winds. Since wet sclerophyll forest is potentially capable of developing in most rainforest sites under appropriate fire regimes, annual rainfalls may exceed 1500 to 2000 mm and that of the driest month may exceed 50 mm. However, wet sclerophyll forest will extend over large areas of mountainous terrain receiving between 1000 and 1500 mm per annum. In the drier sites, the annual total may be compensated for by protection in deep gullies or by extremely equable maritime conditions. In areas receiving only 25 to 30 mm in the driest month, rainforest elements are lacking or rare; such areas, which are found in western Victoria and southwestern Australia, are likely to be more fire-prone. Along all the coastal mountains benefits are likely to accrue from the incidence of low cloud and fog-drip. This may add significantly (15–20%) to the total precipitation in some mature *E. regnans* forests at altitudes of 600 to 700 m (Brookes, 1949). In addition, the great height of the mature forest may increase orographic rainfall, since a rise in altitude of 100 m may increase annual

precipitation by about 150 mm on the Great Dividing Range in central Victoria.

The seasonal distribution of rainfall follows the general latitudinal trend in which winter-dominant rains occur in the southwest and south, uniform distributions occur in the east, and summer-dominant rains occur in the north. Droughts are a recurrent feature of the Australian continent and rainfalls less than half the average total may occur at intervals of 20 to 50 years. When this situation developed in Victoria in 1967, wet sclerophyll forests were killed or severely damaged only in dense stands on rocky ridge-tops and northern slopes, or where streams dried out as water tables fell. In some cases tree and shrub death could have been due to a synergistic combination of droughts and fungal or insect parasites. The most erratic climate is that encountered in the northern range of the wet sclerophyll forests in southern Queensland where the coefficient of variability of annual rainfall is of the order of 40%. The most reliable rainfalls, on the other hand, are in the western half of Tasmania and southwest Western Australia where the coefficient of variability is only 10% (Leeper, 1970).

The temperature regimes of wet sclerophyll forest areas are cool to mild and warm since they encompass all climates from sub-tropical in southern Queensland to sub-montane in Victoria, Tasmania and New England in New South Wales. In coastal southern Queensland and in southwest Western Australia, frosts and snowfall are uncommon, whereas in the cooler and higher areas in southeastern Australia these phenomena may be occasional to frequent. Relative humidities of wet sclerophyll forest areas are controlled by the temperature and the source of the winds. Because their habitats are situated along mountain ranges and plateaux which fringe the oceans, the inflow of maritime air contributes greatly to the amelioration of the climate. The generally brief periods dominated by desiccating winds radiating from the Australian interior constitute a climatic hazard which encourages the spread of fires. Above all else, this factor is of the greatest importance to the survival and perpetuation of these forests.

Fires

There can be no doubt that fire is an integral part of the environment of these wet eucalypt forests. Fire requires dry fuel, oxygen, low humidities and high temperatures for its unimpeded progress (Luke & McArthur, 1978). The high forests provide an enormous potential fuel supply, hence fires are dependent on the combinations of meteorological conditions for which the whole of the so-called wet sclerophyll belt is notorious. The onset of prolonged, erratic, dry summer conditions following dry winters can provide explosive conditions in these areas. The most hazardous conditions arise when slow-moving high-pressure systems direct hot, dry, desert winds outwards to the coast. Temperatures reach and exceed 40°C, relative humidities fall to between 10 and 20% and winds may gust up to 70 to 80 km hr^{-1}. In general terms the lower, drier forests ignite most readily and therefore are burnt more frequently than the high forests. Forests in the wettest climates are least likely to dry out and thus least likely to be disturbed by fire; ridges and spurs exposed to dry winds will tend to carry fire more easily and constitute 'fire paths'. In heavily dissected terrain, fires will tend to burn more severely on the windward slopes than on the lee slopes and gullies may be 'jumped' and constitute 'fire shadows' (Webb, 1970). Fires are notoriously capricious in these forests and are under the control of wind, slope and above all the flammability, quantity and compaction of fuel.

In the summer of 1938–9 enormous fires razed 1 380 000 ha of mountain forests in Victoria (in total, 13.5% of the state). Similar holocausts occurred at the turn of the century and before that, in the early gold mining times of 1851. A reasoned review by King (1963) suggested that small fires were more frequent in the times of Aboriginal man, and large damaging fires more characteristic of the period of European occupation.

The outstanding characteristic of the high eucalypt forests is their even-agedness. This, together with the presence of persistent soil charcoal is suggestive of catastrophic fires in the past centuries. The commonest fires in these forests on mainland Australia are the surface fires which consume the ground-stratum of ferns and grasses and kill or ignite the shrub and sensitive small tree strata. Under severe conditions of higher wind speeds, fires sweep up into the tree canopy. The crown fires which result may be free-running under explosive conditions, or dependent on the fuel of the undergrowth (Luke, 1961). As pointed out by Mount (1969), crown fires often totally destroy the forests, but permit dense regeneration to follow. In the wet country, such conditions originally did not recur more than once every one or two hundred years. Since the arrival of European man these types of fires have been of a frequency such that even eucalypts have been eliminated from large areas. In cooler climates of Tasmania and certain limited sites in Victoria humus accumulation in the top soil may allow slow smouldering humus fires to persist. These fires girdle giant eucalypt trees as well as rainforest trees

(Cremer, 1962) and set in train a whole sequence of recolonisation that may take centuries to complete.

In eucalypt forests the presence of deciduous bark streamers, rough fibrous bark of lichen epiphytes (e.g. *Usnea*), tend to carry fire up to the canopy under some conditions. These materials also serve to disseminate fire ahead of the main front, leading to a common Australian phenomenon of spot-firing (Luke, 1961). A chaotic melée of fires can develop anywhere up to 30 km ahead of the main fire front (Stretton, 1939).

The quality of the fuel is modified by nutrient content (Vines, 1975), so that poor soils supporting markedly sclerophyllous, flammable undergrowth are likely to be sites of severe fires. As Jackson (1968) and Ashton (1976a) have pointed out, the fertility of soils is likely to be depleted in areas of high rainfall subjected to repeated burning. In terms of ecosystem potential it is a 'downward spiral'.

Soils

The soils of the tall open-forest are variable and are derived from a wide spectrum of parent materials. Since they are relatively deep and their infiltration rates are good, their water relations are favourable for the sustenance of large and complex forests. In relatively 'dry' climates in Victoria (1000 mm per annum), fertile krasnozems from basalt support grassy tall open-forest of *Eucalyptus globulus*, *E. obliqua* and *E. viminalis*. On arkoses, podsolic soils support layered forest with an understorey of both shrubs and grasses. Tall open-forests of *E. delegatensis*, with grassy understoreys, also occur in sub-montane areas (1200–1500 m) where soils are humus-rich brown loams.

Where rainfalls in Victoria are higher (1500 mm or more per annum), krasnozems and related brown, acid, friable soils occur under wet sclerophyll forest on igneous (granites, granodiorites, dacites and basalts) and sedimentary rocks (mudstones and shales). In Tasmania, such forests occur on similar red-brown soils derived from dolerite. Yellow-grey, podzolic, silty clay soils support these forests on arkose in Victoria, and on limestone in Tasmania. A similar pattern of soils and parent materials may be found in wet sclerophyll forests in the sub-tropics. In southwestern Australia, forests of *E. diversicolor* occur on deep red-earth soils from gneiss. Although low in nutrients such soils are much more fertile than those of the *E. marginata* dry sclerophyll forests on nearby laterite.

In the wet, monsoonal climates of coastal Arnhem Land, layered tall open-forests occur on very poor soils and contain an understorey of tall, rank grasses and broad-leaf shrubs of Indo-Malayan affinities.

The influence of the litter cycle is indicated by the marked concentration of nutrients in the top soil. The litter fall of the wet sclerophyll forests is relatively high by world standards in relation to climate (Bray & Gorham, 1964). In *Eucalyptus regnans* forests in central Victoria it is 8.1 t ha^{-1} in mature forest and 6.8 t ha^{-1} in pole-stage forests. The *E. regnans* leaf component, which makes up about half of the total, is greatest in half-grown spar-stage forests 40–50 years old. The amount of leaf material dropped annually varies with the climate: it is greatest in sub-tropical *E. pilularis* forests and least in the mediterranean-type climate of southwest Western Australia (Table 6.3).

The contribution of nutrients to the top soil by means of leaf fall is relatively high and similar in the *E. pilularis* and *E. regnans* forests. However the return of calcium by *E. diversicolor* forest is much greater than that of the eastern forests studied. In the southwest, the level of nitrogen is relatively high, and that of phosphorus is low. The contribution of *Pomaderris aspera* to the calcium return in *E. regnans* forest is out of all proportion to the weight of leaf matter dropped. Thus 28% of the leaf fall contributes 56% of the calcium (Ashton, 1975*b*). Potassium values are likely to be underestimated because of pre-fall leaching. In general, the whole of the nutrient return is somewhat underestimated due to the persistent and sometimes substantial attacks on the canopy foliage by insects. Much of this cycling would thus occur by means of their frass.

The nitrogen levels of these forests are most likely to have been

Table 6.3. Eucalytus *forest leaf litter (t ha^{-1}) and its nutrient contribution (kg ha^{-1} yr^{-1})*

Mature forest	Parent material	Total litter	Nitrogen	Phosphorus	Potassium	Calcium
E. pilularis[a]	rhyolite	6.0	40.0	1.3	8.1	24.2
E. regnans[b] (eucalypt component)	granodiorite	4.15	43.5	1.4	5.7	34.8
		2.84	23.9	0.9	3.4	14.9
E. diversicolor[c] (eucalypt component)	gneiss	3.54	31.5	0.7	10.9	53.6
		2.23	10.4	0.4	6.2	35.0

[a] Webb, Tracey, Williams & Lance (1969).
[b] Ashton (1975 *b*).
[c] A. O'Connell (personal communication).

enhanced by the nitrogen-fixing plants. Thus acacias, which are almost universally associated with eucalypts in Australia could be important factors, especially in the wake of fires. Thus in *Eucalyptus regnans* forests *Acacia dealbata, A. melanoxylon* and *A. obliquinervia* are common, in *E. pilularis* forest *A. binervata* and *A. silvestris*, and in *E. diversicolor* forest *A. pentadenia* and *A. urophylla*. In addition, *Casuarina* may be involved in this contribution, notably *C. torulosa* in the northern forests and *C. decussata* in those of southwestern Australia. The low availability of nitrogen is a frequent problem encountered in the plantation establishment of *E. regnans* (R.N. Oldham, personal communication).

The status of phosphorus is low, as is usual for most Australian soils (Chapter 7, Gill, 1981), yet rarely does it appear to be so low as to be a primary nutrient deficiency, although it has been shown to increase markedly growth in *E. diversicolor* plantations (Christensen, 1974).

The soils of the wet sclerophyll forests are acid, pH values range from 4 to 6. The general availability and potential of nutrient supply is greater than in the dry sclerophyll forest (Beadle, 1962; McColl, 1966; Ashton, 1976a). Undoubtedly, the rate of turnover of litter and dead roots is a major factor contributing to fertility. Ellis (1969, 1971) showed in Victoria that the respiratory activity in soils under wet sclerophyll forest was also greater than that under dry sclerophyll forest.

On mainland Australia, even in the montane zone, temperatures are warm enough to allow the active decomposition of litter and mull humus is the rule. In the east and southeast, the forest floor is vigorously cultivated into the top soil by foraging lyre-birds (*Menura novaehollandiae*). In the *Eucalyptus regnans* forests of central Victoria, the weight of the forest floor amounts to 22 t ha^{-1}, of which the leaf material component is very little more than the current year's fall. In Tasmania, however, cooler maximum temperatures result in the accumulation of fibrous and amorphous humus on the soil surface, which may amount to 160 t ha^{-1} (Frankcombe, 1966). Although the nutrient return of the higher-quality wet sclerophyll forests is about equal to the average conifer forest in North America (Lutz & Chandler, 1947), the rate of turnover of the relatively nutrient-rich leaf material in the former is very rapid, due to decomposition by fungi and bacteria and through disintegration by soil fauna. The abundant moisture, good aeration of soil and the rapid flux of nutrients via the litter cycle is probably responsible for a rapid utilisation of energy and very high growth rates.

Growth and development

The forest form

Some of the fastest growing eucalypts are found in the wet sclerophyll forests. For trees in their juvenile state, annual height growth increments of 2 m during the first 10 years are not uncommon. In *Eucalyptus regnans* forests, half of the final mature height is achieved in the first 25 to 35 years (Jackson, 1968; Ashton, 1976*b*). The greatest standing biomass may be achieved between 75 and 100 years, although the culmination of the annual increment may be achieved earlier in the spar-stage between 40 and 60 years. The wet sclerophyll forests characteristically commence as even-aged stands following intense fires. Mineral soil is exposed and the nutrient status and biological level stimulated by the heat treatment imposed on the organic horizon by the fires. Under these initially non-competitive conditions, light, water and nutrients are adequate for prodigious growth. Seeds of the understorey species are often stored in the soil, although some species may regenerate vegetatively from organs at or below ground level, or be dispersed into the site from outside sources. The acacias grow very fast and may, in some cases, temporarily outgrow the eucalypts. This may be an advantage in some mountain sites where radiation frosts in hollows can be lethal. At Noojee, Victoria 2.4×10^6 seedlings per hectare of both *E. regnans* and *A. dealbata* occurred immediately after the 1939 fires (Beetham, 1950). The thinning out of wet sclerophyll eucalypts with age is very rapid. In *E. regnans* forests, a marked change in physiognomy of the stand takes place after 25 to 40 years when most of the early-suppressed individuals finally die. During this period tree numbers may diminish from nearly 2000 ha^{-1} at 16 years to 380 ha^{-1} at 40 years. The rate of decrease is reduced thereafter and stabilises at about 40 to 80 ha^{-1} at maturity after 150 years. Similar kinds of changes are found in other wet sclerophyll forests, although the denser crowns of *E. microcorys* and *E. calophylla* usually have more or less horizontally disposed foliage and their capacity to survive low light under canopies is much greater (Jacobs, 1955). Suppression of weaker trees is likely to be considerably delayed if root fusion with more vigorous trees has taken place (Ashton, 1975*c*).

The generalisation that trees from wet sites tend to develop root systems at an initially slower rate than those from dry sites (Toumey & Korstian, 1947) seems to hold for those species of eucalypt studied

(Zimmer & Grose, 1958; Ashton, 1975c). Seedlings of *E. regnans* are relatively slow to develop both the juvenile shoot and root in the first few months, but thereafter growth is accelerated and continues unabated for several years. The vigorous growth in the dense pole-stages results in marked apical dominance and conical crowns confer on the stand a conifer-like appearance. Continued shedding of branches leads to the development of a clear bole. The mature crown becomes open and infilled of secondary branch systems (Jacobs, 1955). Not until late maturity does die-back of the apical portions persist and lead to stag-headedness. Epicormic growth on the trunks descends to lower heights in very old trees (300–400 years), especially if such trees have been damaged by fires (Jacobs, 1955).

Results of a study of the water relations of tall eucalypts by Connor, Legge & Turner (1977) have revealed that moisture tensions at the tops of trees are developed in accordance with a static hydraulic head of about -0.1 bar m^{-1}. Such water potential in mature *E. regnans* may reach -8 bars at 60 m height in winter. On hot afternoons the water potentials at the summits of mature and pole stage trees may be similar at about -17 to -20 bars. The gradient, however, is very much steeper in the young dense forests (-0.45 bars m^{-1}), compared with that of mature forests (-0.27 bars m^{-1}).

Not surprisingly, shoot growth at the tops of tall mature trees is very slow, and amounts to only about 10 cm per year. The leaves are consequently small and possess thick cuticles. In *E. regnans*, the uppermost leaves of mature trees 75 m high are about 8 to 9 cm^2 in area and 8 to 10 cm long. Spar-sized trees 50 m tall develop leaves 20 to 40 cm^2 in area, 20 cm long and with cuticles 12 to 15 μm thick. Saplings 5 m tall may have leaf areas of 50 to 60 cm^2 and cuticle thicknesses of only 7 to 9 μm. The average diameter of vessels in secondary xylem of roots is much larger than that of stems. For instance, twigs and roots 1.5 mm in diameter have vessels averaging 44 and 74 μm in diameter respectively, whereas those from similar organs 1 cm in diameter are 72 and 155 μm. N. Legge (personal communication) has recorded vessels nearly 1 mm in diameter in long ropy roots of *E. regnans*. Such features are likely to facilitate water movement. The ability of these species to utilise the water supply and sustain rapid growth in their early decades must in part be due to their capacity to develop sufficiently low tensions. The long growing season of 7 to 9 months (Ashton, 1975d; Cremer, 1975) of *E. regnans* ensures the maintenance of large leaf areas in the vigorous early stages of the development of the stand. Hopkins (1964) suggested from the results of seedling experiments that the greater growth of *E. regnans* in fertile topsoils is a self-imposed factor restricting this species to sites with a sustained water supply.

The catchments of wet sclerophyll forests near large cities are frequently utilised for their water yield. In central Victoria, *E. regnans* covers at least half of the area of the water supply catchments for Greater Melbourne.

Following the destruction of many of these forests by bushfires, especially those of 1939, young dense forests have grown up. In the succeeding decades the harvest of water has steadily declined, irrespective of the trends of annual precipitation. This is thought to be due to the increased water use by the dense young stands (Langford, 1974). The extent of the reductions of water yield in four catchments ranges from 31 to 13% of that yielded in the pre-fire period and is strongly correlated with the area occupied by young dense regeneration of *E. regnans* (62 to 20%). With the development of these stands beyond the phase of their maximum leaf area index, an improvement of water yield may be anticipated.

Regeneration
The failure of eucalypts to regenerate in the absence of fire is primarily related to low light levels. This increases the susceptibility of seedlings to fungal disease, insect attack and marsupial browsing. Under high light intensities, eucalypts recover from many such attacks with remarkable tenacity and vigour. The absence of the lignotuber in many of the wet sclerophyll eucalypts diminishes their chances of persistence. Some species, especially those with horizontally-oriented leaves (such as *Eucalyptus calophylla* and perhaps *E. microcorys*), are relatively persistent under moderate amounts of shade, and survive damage due to their effective lignotuber development. In forests of *E. regnans*, *E. obliqua*, *E. delegatensis* and many others, the harvesting of seed by insects is rife. Ants have been implicated (Jacobs, 1955; Cunningham, 1960*a*, *b*; Grose, 1963) as have also lygaeid bugs (Grose, 1960, 1963; Cremer, 1965*a*). In the *E. regnans* forests at Wallaby Creek (37°24′S; 145°15′E) more than 60% of seed in the annual seed fall of *E. regnans* may be removed, which in the warmer months involved nearly all the full, fertile seed (Ashton, 1979). In addition, growth of eucalypt seedlings in the mature forest soils is frequently relatively poor. In *E. pilularis* forest, Florence & Crocker (1962) ascribed this to a microbiological antagonistic effect. Evans, Cartwright & White (1967) reported a phyto-

toxin (nectroline) from some root surface isolates of the fungus *Cylindrocarpon radicicola*. Willis & Ashton (1978) suggest that part of the inhibitory property of mature forest soil *in situ* is due to an inhibition of the nitrifying bacteria by leachate concentration from the eucalypt leaves. Results of other work (Ashton, 1962) suggested that inhibitory substances may be involved with the close approximation of old, living root systems. The removal of all these problems by fire is dramatic. Inhibition is replaced by stimulation. This 'ash-bed' effect has been variously ascribed to the enhancement of phosphorus availability (Attiwill, 1962; Loneragan & Loneragan, 1964), to an increased surge of nitrogen availability (Hatch, 1960), and to the death of micro-organisms and the consequent release of their nutrients and, most importantly, to the replacement of the normal microbial flora by an entirely different one over the period of increased fertility (Renbuss, Chilvers & Pryor, 1972).

One of the significant features of the wet sclerophyll eucalypts is the virtual absence of hard, dormant seed. Usually they are virtually the only members of the forest association not represented as soil-stored seed (Gilbert, 1959; Cunningham, 1960*b*; Cremer, 1965*b*; Floyd, 1976; Ashton, 1979). The recent seed fall may be found occasionally in the uppermost 1 cm of soil but it is likely that such temporarily buried seed would be destroyed by the heat of the fire. The storage of eucalypt seed is in the capsules of the canopies. The persistence of capsules for three or more years usually ensures that at least the crop of one good flowering season is present. In *E. regnans* forest a twentyfold difference in flowering and seed-set may occur from a good to a bad season. After a peak flowering year many millions of seed are stored per hectare of canopy, and released over several years or after fires, apparently in one large fall (Cremer, 1965*a*; Christensen, 1971; Ashton, 1975*a*).

The ecology of the wet sclerophyll forests therefore hangs on an extraordinarily fine thread. It seems almost inconceivable that the small capsules of these species – often 1 cm or less long – can protect seed sufficiently long in the holocaust of raging crown fires. Mount (1969) suggested that, due to the peculiar phenology of many of the eucalypts, the moist capsules are situated below the current foliage. The heat therefore would tend to be swept upwards and away. Results of preliminary research by Webb (1966) indicated that seeds within large green *E. globulus* fruits were protected for about 9 min from a lethal rise in temperature when such capsules were subjected to intense red heat at 440°C in an electric furnace. Results of experiments by Gloury (1978) suggest that when green capsules of *E. regnans* were subjected to uniform heat in an oven the germination capacity of the enclosed seed was diminished below that of seed exposed nakedly under the same conditions. Thus, these attempts to gauge the 'protectiveness' of capsules of different ages of various species indicate the complexity of the problem.

Succession

Due to the vicissitudes of fire, a forest may be partially damaged or totally razed. Repetitive burning within the primary non-flowering period of an obligate seed-regenerating species can result in its complete elimination from a site. The variability of vegetation structure and floristics in burnt forests is therefore great, and the boundaries of the various communities are often dramatically sharp. Throughout the tall open-forest even-aged stands from pole to mature form occur, each being related to a specific fire history. However, partially-damaged forest may show two or even three distinct age-classes, the older ones being invariably butt-scarred. This characteristic damage is due to litter accumulation on uphill sides of trunks and to the chimney-flue effect of flame vortices in the lee of the trunk bases (Gill, 1975). Surface fires in the ground stratum and the shrub understorey may leave the eucalypts relatively undamaged yet initiate even-aged strata of undergrowth seedlings or coppice shoots. In southern Tasmania where the humus layers burn above the mineral soil, the entire forest community may be destroyed (Cremer, 1965a), the colonising species arriving by wind and bird dispersal. In such instances it has been estimated by Cremer (1965a) that eucalypt seed sources need to be within 200 m for replacement of this dominant life form to occur. In the half-barked *E. regnans* stands, the butt-bark appears to be sufficiently thick to protect the tree from heat, if surface fires of low intensity occur at intervals greater than about 25 years. Therefore, within this age limit, the whole forest is even-aged, but beyond this period the understorey may be either contemporaneous or younger than the overstorey.

If rainforest species are present in the general vicinity, they may be expected to invade the maturing understorey of the wet sclerophyll forest within 100 years, and completely replace it in the succeeding century. Within the following two centuries the eucalypts will disappear as their life span is reached. The higher the rainfall, the less likely is this sequence likely to be interrupted by fire. The Tasmanian maritime environment is less conducive to destructive fires than the more continentally influenced mainland areas in Victoria. Large areas of rainforest of *Nothofagus cunninghamii* and *Atherosperma moschatum* are therefore rare in

Victoria, although as shown by Howard (1973), the potential for such to occur is relatively great (see Chapter 5, Howard, 1981).

Where rainforest species are remote from the burnt stands, successions in the understorey types of the wet sclerophyll forests proceed along different lines. There is a tendency for shrubs and small trees to reach their life span after specific intervals. In *E. regnans* forests a light-crowned understorey of the composite, *Cassinia aculeata*, begins to senesce after 25 to 30 years. It is then replaced by *Pomaderris aspera*, *Olearia argophylla* and *Bedfordia arborescens*, a common trio of under-storey dominants in this forest type in central Victoria. After about 100 years, *P. aspera* becomes senescent and the ground-stratum of rosette ferns (*Polystichum proliferum*) or tussock grass (*Poa ensiformis*) increases in abundance. This may inhibit the regeneration of the understorey shrubs and trees unless fire occurs. Only those shrubs with vegetative reproduction (e.g. layering and root suckering) persist. Cunningham & Cremer (1965) suggested that certain seeds may be stored in the soil for one or two centuries or more. Fires at intervals longer than 100 years do not result in the renaissance of an understorey of *Pomaderris apetala*.

The big forests of both *E. diversicolor* and *E. regnans* possessed structurally and floristically variable understoreys at the time of dis-covery by Europeans. Persistent folk-lore relates both to the riding of a horse through the virgin forests and to the hacking of a path for progress (Select Committee, 1876; Galbraith, 1937; Underwood, 1973). Such variation can be interpreted in terms of persistent, patchy, fire regimes.

The first communities to arise after fires are derived from soil-borne seeds, from fire-resistant capsules and from rapidly dispersed dissemi-nules. Bryophytes may become particularly abundant in severely burnt soil in the first two years after fire but their lawn-like mats and cushions are eventually overcome by more vigorous and taller herbaceous and woody regrowth (Cremer & Mount, 1965). During the first few years the communities are often dominated by so-called fire weeds. These are usually composites, both native and introduced. In warmer climates, weeds from other families are also involved, such as the introduced *Phytolacca octandra* and *Lantana camara* (Floyd, 1966, 1976), and wiry lianes (Van Loon, 1969).

In southeastern Australia the invasion of rainforest species may be sufficiently delayed by fires or long dispersal distances so that large tracts of wet sclerophyll forest continue to persist. The climatic climax to much of the tall eucalypt forest along the east coast is derived from the so-called Indo-Malayan element (Hooker, 1860) which forms rainforests of the 'notophyll vine' and 'microphyll vine' forest (Webb, 1959) types. The climatic climax of the southern areas and the wet cool plateaux to the north are of the so-called Antarctic element which forms 'microphyll mossy' and 'microphyll fern' forests, epitomised by *Nothofagus*. In Victoria, variations of the temperate rainforest are climax to a whole range of wet sclerophyll forests from near sea-level to about 1450 m (Howard & Ashton, 1973). At the northernmost distribution of tall eucalypts (e.g. *E. torelliana*, *E. pellita* *E. cloeziana*) on the Atherton Plateau, the transition to rainforest occurs too quickly for the development of a recognisable wet sclerophyll stage (G.C. Stocker, personal communication). In southwestern Australia such potential climaxes appear to have been eliminated by climatic changes in the geological past (Crocker, 1959).

The limit of rainforest distribution in the complete cessation of fire is difficult to determine. Certainly, many of their members are more resistant to drought (Williams, 1978) than has hitherto been thought. Howard (1970), for example, was able to establish *N. cunninghamii* in dry sclerophyll forest in a ridge-top site, Mt Donna Buang, and in gullies in the Dandenong Ranges, central Victoria, where it is not native. The limits in drier country appear to be obscured by fire. What exactly is climax again becomes an academic question, the solution of which requires experimentation and long-term observations.

In *E. regnans* forests, two small trees of the family Compositae (Asteraceae) possess lignotubers. Frequently these species, *Olearia argophylla* and *Bedfordia arborescens*, coppice from this organ and rapidly dominate the ensuring understorey. In Victoria, two rainforest trees, *Nothofagus cunninghamii* and *Acmena smithii*, are also endowed with lignotuberous burls from which recovery is possible after moderately intense fires (Howard, 1973; Ashton & Frankenberg, 1976). Repeated surface or crown fires, however, are likely to eliminate these species if the heartwood of the original trunk is incompletely callused, since the stumps may be ignited and the living buds killed from within by smouldering fires.

Almost all of the understorey species are represented in the store of soil seed and are stimulated to germinate by the passage of the fire. In *E. regnans* forests, this particularly applies to *Pomaderris aspera*, *P. apetala*, *Acacia dealbata*, *A. melanoxylon*, *Prostanthera lasianthos* and *Cassinia aculeata*. Dense understoreys of these species may result, their composition depending on the longevity of seed and the original species composition. In the mature *E. regnans* forest at Wallaby Creek, Victoria,

Pomaderris aspera is ten to fifteen times more abundant than *Prostanthera lasianthos*; yet in regeneration on large cleared plots the proportion of these two species is about parity. That fires do not always result in the exact replication of the original understorey was noted by the exploiters of the great forest of South Gippsland at the turn of the century (Committee of the South Gippsland Pioneers' Association, 1920).

Since European man's activities, magnificent eucalypt forests have been reduced to scrub by top fires at intervals of less than 15 to 20 years. This has been discussed by Gilbert (1959) and Jackson (1965), who both point out that there is a selection for the more fire-tolerant species in the process. *Pomaderris aspera* will likewise be removed if surface fires recur at intervals of less than five to ten years. The ability of the older eucalypts to withstand the heat of an undergrowth fire is almost certainly related to the greater thickness of the bark in the butt region. Many wet sclerophyll eucalypts, particularly *E. regnans*, have very thin gum bark until the spar stage when a thicker persistent sub-fibrous bark develops and gradually extends up to a height of 10 m by maturity. In general, bark thickness increases with age and diameter of the trunk (Gill & Ashton, 1968; Vines, 1968) so that younger trees will be more likely to be damaged or killed by fire than older. Many of the 'true' wet sclerophyll eucalypts develop a 'half-bark' condition at maturity similar to that of *E. regnans*. This, and the isolation of the high canopy from the understorey, are probably factors permitting a wide expression of understorey types in mature stands.

Rainforest components in mixed forest may persist vegetatively, although fire-sensitive eucalypts are killed. Thus, in Victoria, coppice *Nothofagus cunninghamii* may be found under even-aged pole-stage *E. delegatensis*, *E. nitens* and *E. regnans* on Mt Donna Buang and Mt Baw Baw, and *Acmena smithii* under *E. obliqua* and *E. muellerana* at Wilson's Promontory (see Chapter 5, Howard, 1981).

In areas of moderate altitude (100–800 m) in central Victoria where mean annual rainfalls are 1100–1250 mm, a mixture of *Pomaderris aspera*, *Olearia argophylla* and *Bedfordia arborescens* tends to form a relatively stable understorey for 100 years or more, provided there is no easily available seed source from rainforest patches in gullies. Under this rainfall regime, fire frequencies are likely to maintain this condition.

The question of the effect of acacias on the whole nitrogen economy of the wet sclerophyll forests has often been verbally expressed by many colleagues. Detailed research on this problem is long overdue. Investigations by Shirrefs (1977) in stands of 1939 origin at Toolangi, Victoria, suggest that the nitrogen level of one-year-old foliage of *E. regnans* in the presence of *A. dealbata* is enhanced 50% (significant at $P = 7\%$), compared with those stands in which it is absent. If this work is substantiated with other sites and species combinations, an important effect of fire in Australian ecosystems may be established. One important effect of soil which cannot be overlooked in the ecology of these forests is the so-called 'ash-bed' effect. This has been shown by many workers (e.g. Pryor, 1963) to cause a two- to threefold increase in tree growth for 2–3 years following fire. It is undoubtedly important in allowing the eucalypt species to exploit their superior growth potential.

Evolution of the Tall Open-forest of Wet Sclerophyll type
It seems paradoxical that the forests which produce vast amounts of fuel and generate the most fires, consist of fire-sensitive species. That they need to be burnt down at some time in their seed-bearing life is axiomatic. The reason why these eucalypts frequently bear thin bark, lack lignotubers and coppice poorly may be related to the channelling of their energy resources towards the development of a superlative juvenile height growth rate.

At the moment it appears that the sheer flammability of the overstorey may be the secret of the success of the forest regeneration, a philosophy expressed by Mutch (1970). The thermal death point of tissues must be expressed in terms of both temperature and time of exposure. The protection offered by various tissues is chiefly concerned with delaying the penetration of heat to the vital areas. The flammability of the eucalypt crowns therefore may be expected to create conditions of explosive heat for a very short time. Under this regime one may expect capsules to protect seed for this critically short period. If this is the explanation, the achievement is almost a miracle of timing. In one sense, the tolerance of the high forests to severe fire once every one, two or three centuries is due to their low resistance to it.

It would indeed be intriguing to know the state of the wet sclerophyll component in Pleistocene times before the arrival of man. As pointed out by Jackson (1968), the incidence of fire must have been greater in the more arid post-Pleistocene times. Successional trends have been accelerated since the control of firing was imposed by Forestry Authorities 30–40 years ago. A similar successional picture is provided by the analogous forest fire regimes in the Pacific West Coast of North America. One would assume the fires in the drier times would have been initiated by lightning, aided in some areas by vulcanism or the exo-thermic reactions of cracked and desiccating peats. Possibly fires were a much rarer event and the wet sclerophyll forests relatively scarce eco-tones between rainforests and more drought-resistant eucalypt forests or woodlands. According to Webb (1970), the wet sclerophyll forests in the east, south and southwest of Australia are the special product of intense fires on a grand scale. In the northern areas, surface fires of lower intensity cause the gradual attrition of the rainforest margins. The boundaries in the former areas therefore are likely to be blurred; in the latter areas they are often unbelievably sharp.

The evidence of fire in Tertiary times in Australia has been deduced from fusain in lignites (Kemp, 1980). More recently, the identification of microscopic charcoal in Pleistocene deposits has revealed a long history of fire, albeit at relatively low levels of intensity. From about 40 000 years B.P. in the Lake George deposits near Canberra (Singh, Kershaw & Clark, 1980), there is a dramatic and sustained increase in charcoal frequency. This time coincides with the known earliest records of Aboriginal man in Australia. This is no mere coincidence, since these peoples were known to light fires, either accidentally in the course of their fire-stick culture (Mulvaney, 1969), or deliberately in their manage-ment and hunting of game. Some early writers at least were impressed with the skill with which they fired the country. Given the erratic unpredictable weather (even to modern meteorologists!) in southern regions of the continent, it must be assumed that fires were an integral part of much of the country from this time.

Since the arrival of European man and his drive for pastures and minerals, the severity and frequency of large fires has been a normal consequence of most long periods of dry weather. Huge areas of high forest lie in a ruin of secondary scrub and testify to this blatant mismanagement. Pollen grains of *Eucalyptus* and other genera typical of wet sclerophyll forests have been traced back well into the last glaciation period (130 000 year B.P.) in northwestern Tasmania (E.A. Colhoun, personal communication). The assemblages of the pollen record over this time span in both this area and Lake George, indicate oscillations compatible with wide climatic fluctuations involving both temperature and precipitation. Any decrease in precipitation is likely to result in a tendency for greater fire frequency and the shrinkage of rainforest areas to more sheltered niches. With the advent of man such shrinkages of rainforest need not necessarily be the result of climatic change, since fire frequency would be greater.

It seems likely that such eucalypts would be able to assert their growth supremacy only on good soils or on those whose fertility was sufficiently stimulated by fire. Only on the very poor quartzose soils in high rainfall areas of western Tasmania are both the complex rainforest and the sclerophyllous *E. nitida* forests of relatively poor stature.

Whatever the palaeo-fire regime in the wetter parts of Australia, it seems likely that it has been a major factor shaping the evolution of fast-growing, light-demanding eucalypts in close proximity to rainforest elements. Whether the arrival of man on the continent has accelerated the rate of evolution of the wet sclerophyll eucalypt species as a consequence of increased firing is a question to be pondered upon, but probably never answered adequately.

References

Ashton, D.H. (1962). Some aspects of root competition in *E. regnans*. In *Proceedings 3rd General Conference Institute of Foresters of Australia*. Melbourne: Institute of Foresters of Australia.

Ashton, D.H. (1975a). The flowering behaviour of *Eucalyptus regnans* F. Muell. *Australian Journal of Botany*, *23*, 399–411.

Ashton, D.H. (1975b). Studies on the litter of *Eucalyptus regnans* F. Muell. forests. *Australian Journal of Botany*, *23*, 413–33.

Ashton, D.H. (1975c). The root and shoot development of *Eucalyptus regnans* F. Muell. *Australian Journal of Botany*, *23*, 867–87.

Ashton, D.H. (1975d). The seasonal growth of *Eucalyptus regnans* F. Muell. *Australian Journal of Botany*, *23*, 239–52.

Ashton, D.H. (1976a). The development of even-aged stands in *Eucalyptus regnans* F. Muell. in central Victoria. *Australian Journal of Botany*, *24*, 397–414.

Ashton, D.H. (1976b). Phosphorus in forest ecosystems at Beenak, Victoria. *Journal of Ecology*, *64*, 171–86.

Ashton, D.H. (1979). Seed harvesting by ants in forests of *Eucalyptus regnans* F. Muell. in central Victoria. *Australian Journal of Ecology*, *4*, 265–77.

Ashton, D.H. & Frankenberg, J. (1976). Ecological studies of *Acmena smithii* (Poir). Merrill and Perry with special reference to Wilson's Promontory. *Australian Journal of Botany*, *24*, 453–87.

Attiwill, P.M. (1962). The effect of heat pre-treatment of soil on the growth of *E. obliqua* seedlings. In *Proceedings 3rd General Conference Institute of Foresters of Australia*. Melbourne: Institute of Foresters of Australia.

Beadle, N.C.W. (1962). Soil phosphate and the delimitation of plant communities in eastern Australia. *Ecology*, *43*, 281–8.

Beadle, N.C.W. & Costin, A.B. (1952). Ecological classification and nomenclature. *Proceedings of the Linnean Society of New South Wales*, *77*, 61–82.

Beetham, A.H. (1950). Aspects of forest practice in the regenerated areas of the Upper Latrobe Valley. Dip. For. thesis, Forests Commission of Victoria, Creswick.

Bray, J.R. & Gorham, E. (1964). Litter production in forests of the world. *Advances in Ecological Research*, *2*, 101–52.

Brookes, J.D. (1949). The relation of vegetation cover to water yield in Victorian mountain watersheds. M.Sc. thesis, University of Melbourne.

Christensen, P.E. (1971). Stimulation of seedfall in Karri. *Australian Forestry*, *35*, 182–90.

Christensen, P.E. (1974). Response of open-rooted Karri (*Eucalyptus diversicolor*) seedlings to nitrogen or phosphorus fertilizer. *Forests Department of Western Australia Research Paper No. 12.*

Committee of the South Gippsland Pioneers' Association (Eds). (1920). *Land of the Lyrebird*. Melbourne: Gordon & Gotch.

Connor, D.J., Legge, N.J. & Turner, N.C. (1977). Water relations of mountain ash (*Eucalyptus regnans* F. Muell.) forests. *Australian Journal of Plant Physiology*, *4*, 753–62.

Cremer, K.W. (1962). The effects of fire on eucalypts reserved for seeding. *Australian Forestry*, *26*, 129–54.

Cremer, K.W. (1965a). Effects of fire on seed shed from *Eucalyptus regnans*. *Australian Foresty*, *29*, 251–62.

Cremer, K.W. (1965b). Emergence of *Eucalyptus regnans* seedlings from buried seed. *Australian Forestry*, *29*, 119–24.

Cremer, K.W. (1975). Temperature and other climatic influences on shoot development and growth of *Eucalyptus regnans*. *Australian Journal of Botany*, *23*, 27–44.

Cremer, K.W. & Mount, A.B. (1965). Early stages of plant succession following the complete felling and burning of *Eucalyptus regnans* forest in the Florentine Valley, Tasmania. *Australian Journal of Botany*, *13*, 303–22.

Crocker, R.L. (1959). Past climatic fluctuations and their influence upon Australian vegetation. In *Biogeography and Ecology in Australia* (Monographiae Biologicae VIII), ed. A Keast, R.L. Crocker & C.S. Christian, pp. 283–90. The Hague: Junk.

Cunningham, T.M. (1960a). The natural regeneration of *Eucalyptus regnans*. *School of Forestry, Univ. of Melbourne, Bulletin No. 1.*

Cunningham, T.M. (1960b). Seed and seedling survival of *Eucalyptus regnans* and the natural regeneration of second-growth stands. *Appita*, *13*, 124–31.

Cunningham, T.M. & Cremer, K.W. (1965). Control of the understorey in wet eucalypt forest. *Australian Forestry*, *29*, 4–14.

Ellis, R.C. (1969). The respiration of the soil beneath some eucalypt forest stands as related to the productivity of the stands. *Australian Journal of Soil Research*, *7*, 349–58.

Ellis, R.C. (1971). Growth of eucalypt seedlings on four different soils, *Australian Forestry*, *35*, 107–18.

Evans, G., Cartwright, J.B. & White, N.H. (1967). The production of a phytotoxin, nectroline, by some root surface isolates of *Cylindrocarpon radicicola* wr. *Plant and Soil*, *26*, 253–60.

Florence, R.G. & Crocker, R.L. (1962). Analysis of blackbutt (*E. pilularis* Sm.) seedling growth in a blackbutt forest soil. *Ecology*, *43*, 670–9.

Floyd, A.G. (1966). Effect of fire upon weed seeds in the wet sclerophyll forests of northern New South Wales. *Australian Journal of Botany*, *14*, 243–56.

Floyd, A.G. (1976). Effect of burning on regeneration from seeds in wet sclerophyll forest. *Australian Forestry*, *39*, 210–20.

Frankcombe, D.W. (1966). The regeneration burn. *Appita*, *19*, 127–32.

Galbraith, A.V. (1937). *Mountain Ash* (Eucalyptus regnans F. Muell.). *A General Treatise on its Silviculture, Management and Utilization*. Melbourne: Government Printer.

Gardner, C.A. (1942). The vegetation of Western Australia with special reference to the climate and soils. *Journal of the Royal Society of Western Australia, 28,* 11–87.

Gilbert, J.M. (1959). Forest succession in the Florentine Valley, Tasmania. *Papers & Proceedings of the Royal Society of Tasmania, 93,* 129–51.

Gill, A.M. (1975). Fire and the Australian flora. *Australian Forestry, 38,* 4–25.

Gill, A.M. (1981). Patterns and processes in open-forests of *Eucalyptus* in southern Australia. In *Australian Vegetation,* ed. R.H. Groves, pp. 152–176. Cambridge University Press.

Gill, A.M. & Ashton, D.H. (1968). The role of bark type in relative tolerance to fire of three central Victorian eucalypts. *Australian Journal of Botany, 16,* 491–8.

Gloury, S.J. (1978). Aspects of fire resistance in three species of *Eucalyptus* L'Herit. with special reference to *E. camaldulensis* Denh. B.Sc. (Hons) thesis, University of Melbourne.

Grose, R.J. (1960). Effective seed supply for the natural regeneration of *Eucalyptus delegatensis* R.T. Baker, syn. *Eucalyptus gigantea* Hook.f. *Appita, 13,* 141–8.

Grose, R.J. (1963). The silviculture of *Eucalyptus delegatensis.* I. Germination and seed dormancy. *School of Forestry, Univ. of Melbourne, Bulletin No. 2.*

Hardy, A.D. (1935). Australia's great trees. *Victorian Naturalist, 51,* 231–41.

Hatch, A.B. (1960). Ash bed effects in Western Australian forest soils. *Forests Department of Western Australia Bulletin No. 64.*

Hooker, J.D. (1860). *The Botany (of) the Antarctic Voyage,* Part III. *Flora Tasmaniae,* Vol. 1. London: Lovell Reeve.

Hopkins, E.R. (1964). Water availability in mixed species *Eucalyptus* forest. Ph. D. thesis, University of Melbourne.

Howard, T.M. (1970). The ecology of *Nothofagus cunninghamii.* Ph.D. thesis. University of Melbourne.

Howard, T.M. (1973). Studies in the ecology of *Nothofagus cunninghamii.* I. Natural regeneration on the Mt Donna Buang massif, Victoria. *Australian Journal of Botany, 21,* 67–78.

Howard, T.M. (1981). Southern closed-forests. In *Australian Vegetation,* ed. R.H. Groves, pp. 102–120. Cambridge University Press.

Howard, T.M. & Ashton, D.H. (1973). The distribution of *Nothofagus cunninghamii* rainforest. *Proceedings of the Royal Society of Victoria, 86,* 47–76.

Institute of Foresters of Australia. (1976). Newsletter no. 17, p. 1.

Jackson, W.D. (1965). Vegetation. In *Atlas of Tasmania,* ed. J.L. Davies, pp. 30–5, Hobart: Lands & Survey Department.

Jackson, W.D. (1968). Fire, air, water and earth – an elemental ecology of Tasmania. *Proceedings of the Ecological Society of Australia, 3,* 9–16.

Jacobs, M.R. (1955). *Growth Habits of the Eucalypts.* Canberra: Commonwealth Government Printer.

Kemp, E. (1980). Evolution of the Australian biota in relation to fire. In *Fire and the Australian Biota,* ed. A.M. Gill, R.H. Groves & I.R. Noble, pp. 3–21. Canberra: Australian Academy of Science.

King, A.R. (1963). *The influence of colonisation on the forests and the prevalence of bush fires in Australia.* Melbourne: CSIRO, Australia, Division of Applied Chemistry (mimeographed report).

Langford, K.J. (1974). *Change in yield of water following a bushfire in a forest of Eucalyptus regnans.* Melbourne & Metropolitan Board of Works Departmental Report No. 31.

Leeper, G.W. (1970). Climates. In *The Australian Environment*, 4th edn (rev.), ed. G.W. Leeper, pp. 12–20. Melbourne: CSIRO, Australia & Melbourne University Press.

Loneragan, O.W. & Loneragan, J.F. (1964). Ashbed and nutrients in growth of karri seedlings. *Journal of the Royal Society of Western Australia, 47*, 74–80.

Luke, R.H. (1961). *Bush Fire Control in Australia*. Melbourne, Sydney & London: Hodder & Stoughton.

Luke, R.H. & McArthur, A.G. (1978). *Bushfires in Australia*. Canberra: Australian Government Publishing Service.

Lutz, H.J. & Chandler, R.F. Jr. (1947). *Forest Soils*. New York: John Wiley & Sons Inc.

McColl, J.B. (1966). Soil plant relationships in eucalypt forest. *Ecology, 50*, 355–62.

Mount, A.B. (1969). Eucalypt ecology as related to fire. *Proceedings 9th Tall Timbers Fire Ecology Conference*, pp. 75–108.

Mulvaney, D.J. (1969). *The Prehistory of the Aborigines*. London: Thames & Hudson.

Mutch, R.W. (1970). Wildland fires and ecosystems – a hypothesis. *Ecology, 51*, 1046–51.

Pryor, L.D. (1963). Ashbed response as a key to plantation establishment on poor sites. *Australian Forestry, 27*, 48–51.

Renbuss, M.A., Chilvers, G.A. & Pryor, L.D. (1972). Microbiology of an ashbed. *Proceedings of the Linnean Society of New South Wales, 97*, 302–310.

Select Committee, (1876). *Report on water reserves, Plenty Ranges*. Victoria Parliamentary Papers Sll.

Shirrefs, P.V. (1977). Interaction between *Eucalyptus regnans* F. Muell. and *Acacia dealbata* Link. B.Sc. (Hons.) thesis, University of Melbourne.

Singh, G., Kershaw, A.P. & Clark, R. (1980). Late Quaternary vegetation and fire history in Australia. In *Fire and the Australian Biota*, ed. A.M. Gill, R.H. Groves & I.R. Noble, pp. 23–54. Canberra: Australian Academy of Science.

Stretton, L.E.B. (1939). *Report to the Royal Commission to Enquire into the Causes and the Measures taken to Prevent the Bushfires of January, 1939 and to Protect Life and Property*. Melbourne: Government Printer.

Toumey, J.W. & Korstian, C.F. (1947). *Foundations of Silviculture upon an Ecological Basis*. New York: John Wiley & Sons Inc.

Underwood, R.J. (1973). *Natural fire production in the karri* (E. diversicolor F. Muell.) *forests*. Forests Department of Western Australia Research Paper No. 41.

Van Loon, A.P. (1969). *Prescribed burning in blackbutt forests*. Forests Commission of New South Wales Research Notes No. 23.

Vines, R.G. (1968). Heat transfer through the bark and the resistance of trees to fire. *Australian Journal of Botany, 16*, 499–514.

Vines, R.G. (1975). Bushfire research in CSIRO. *Search, 6*, 73–8.

Webb, L.J. (1959). A physiognomic classification of Australian rainforests. *Journal of Ecology, 47*, 551–70.

Webb, L.J. (1970). Fire environments in eastern Australia. In *Proceedings 2nd Fire Ecology Symposium, Monash University & Forests Commission of Victoria*.

Webb, L.J., Tracey, J.G., Williams, W.T. & Lance, G.N. (1969). Pattern of mineral return in leaf litter of 3 sub-tropical Australian forests. *Australian Forestry, 33*, 99–110.

Webb, R.N. (1966). The protection of eucalypt seed from fire by their capsules. B.Sc. (Hons.) thesis, University of Melbourne.

Williams, R.J. (1978). The comparative strategies of water use in selected xeromorphic and mesomorphic species at Lilly Pilly Gully, Wilson's Promontory, Victoria. B.Sc. (Hons.) thesis, University of Melbourne.

Willis, R.J. & Ashton, D.H. (1978). The possible role of soil lipids in the regeneration problem of *Eucalyptus regnans* F. Muell. In *Proceedings of the Open Forum on Ecological Research, Macquarie University, Bulletin of the Ecological Society of Australia, 8*, 11.

Zimmer, W.J. & Grose, R.J. (1958). Root systems and root/shoot ratios of some Victorian eucalypts. *Australian Forestry, 22*, 13–18.

7

Patterns and processes in open-forests of Eucalyptus in southern Australia

A. MALCOLM GILL

The two most species-rich genera of Australia, *Eucalyptus* and *Acacia*, are also the most prominent genera in Australia's woody plant communities. *Eucalyptus* was novel to the early students of the Australian flora but in the 200 year post-settlement history of the continent, approximately 400 species have been recognised (Pryor & Johnson, 1971). *Acacia* was not a new genus to science when settlement began near Sydney in 1788 but approximately 600 new species have been discovered subsequently (Burbidge, 1963).

Eucalyptus and *Acacia* may be regarded as genera typifying the Australian woody flora. Both are widespread; both have high levels of specific endemism; and both are conspicuous dominants of major plant formations. In more xeric areas, *Acacia* is often the community dominant (Chapter 9, Johnson & Burrows, 1981), whilst in more mesic areas *Eucalyptus* is dominant (Chapter 6, Ashton, 1981; Chapter 8, Gillison & Walker, 1981). The latter dominance reaches an extreme in the open-forests of *Eucalyptus* in southern Australia but *Acacia* retains a strong representation in the understorey of these forests.

In this chapter, these uniquely Australian forests will be described in relation to patterns of distribution and processes of community operation. First, however, I shall outline the development of the concept of 'open-forest' and define the forests I shall consider below.

The 'open-forest' concept

Diels (1906) began the difficult journey into the classification of Australia's eucalypt forests; the journey continues. Diels (1906) described the eucalypt forests of southwestern Australia as 'Sklerophyllen-Wald' (sclerophyll forest) apparently using the adjective 'sclerophyll' to define the leaf textures found in the understorey plants, rather than those of the eucalypt canopy (Specht, 1970; cf. Beadle & Costin, 1952; Williams 1955; Webb, 1959).

A quarter of a century after Diels, Prescott (1931) introduced the term 'wet sclerophyll forest' for the forests of better site qualities, particularly in Western Australia and Victoria (dominated by *Eucalyptus diversicolor* and *E. regnans* respectively). Another South Australian author, Wood (1937), later identified the two subdivisions of the sclerophyll forests as 'wet' and 'dry': the 'wet' forests occurred in areas with more than 100 cm annual rainfall and contained tree-ferns, whilst the 'dry' forests were in areas with less than 100 cm rainfall and contained many sclerophyllous shrubs.

Beadle & Costin (1952) reviewed the classification of Australian plant communities and although they did not attempt to delimit forest communities by rainfall they retained the use of the terms 'wet' and 'dry' to describe these forests: they noted the occurrence of mesomorphic shrubs in the 'wet' sclerophyll forest and xeromorphic shrubs in the 'dry'. In addition, they included a 'swamp' sclerophyll forest with an herbaceous understorey of helophytes. The latter term was not taken up generally and did not appear in the subsequent classification of Williams (1955).

A major change in nomenclature, along with refined definitions, came with Specht (1970) who prepared his new system in conjunction with a number of other Australian ecologists. 'Sclerophyll' forest became 'open-forest' which was defined as a community dominated by trees and which had a canopy cover of between 30 and 70%. Three major subcategories were then distinguished according to the height of the dominants: 'tall open-forest' was greater than 30 m tall and equivalent to the forest described as 'wet sclerophyll' by earlier authors; 'open-forest' was 10 to 30 m tall and equivalent both to the earlier 'dry sclerophyll forest' and to the layered open-forest of Williams (1955); 'low open-forest' was used to describe forests 5 to 10 m tall and largely composed of brigalow (*Acacia harpophylla*) as described in Chapter 9 (Johnson & Burrows, 1981).

In this chapter, the subcategory of forests called 'open-forests' by Specht (1970) will be considered but without his 'layered open-forest' phase. Basically, the forests considered will be the shrubby and grassy phases of the open-forest subformation, a grouping which roughly corresponds to the 'dry sclerophyll forests' of earlier authors. Confusion can occur in nomenclature because the formation-level 'open-forest' and the subformation 'open-forest', are indistinguishable by name. For the rest of this review 'open-forest' refers to the subformation *not* the formation, and does not include the 'layered open-forest' phase of the subformation, or the tall open-forest of the previous chapter.

Patterns of distribution in southern Australia
Broadly speaking, we can define three major zones of occurrence of the open-forests in southern Australia. One is in coastal and sub-coastal southeastern Australia from Adelaide to Brisbane; the second is in Tasmania; and the third is in the southwestern corner of Western Australia. In each zone, the wetter limits are marked by tall open-forests; the drier, more fertile limits are marked by grassy woodlands; the poorer-soil margins tend to abut shrubby woodlands; and sedges and other wetland plants may become prominent on poorly-drained margins. Below, three regions have been chosen for closer scrutiny. In each case the open-forest is associated with a variety of other communities and sufficient is known of the factors controlling distribution to warrant a brief review. The regions to be considered are: the Sydney Basin of New South Wales; the Eastern Highlands of Victoria; and the jarrah forest region in the vicinity of Perth, Western Australia.

Sydney Basin
The Sydney Basin is a geologically-defined region of Permian and Triassic shales and sandstones (Packham, 1962). In the Blue Mountains to the west of Sydney the topography is of extensive, gently undulating plateaux dissected by gorges and valleys with steep and colourful cliffs. The horizontally-bedded Hawkesbury sandstone often forms the plateau surface. It is underlain by the Narrabeen sandstones and shales but overlain by the Wianamatta shales. Together these three units comprise the Triassic sequence of the region. Study of the vegetation of these units led Beadle (1954, 1962) to suggest that the mosaic of vegetation formations and subformations in eastern Australia was controlled by soil phosphate levels, which, in turn, were a reflection of the phosphate status of the parent materials from which the soils developed.

Beadle (1962) found that where soil phosphate levels were very low in the Sydney region, such as on the Hawkesbury sandstones or on truncated lateritic profiles, 'xeromorphic woodlands and scrubs' occurred (Table 7.1). Where phosphorus levels were higher, however, as in complete lateritic profiles, 'dry sclerophyll' or 'open-forest' was found (Table 7.1). Even higher phosphate levels occurred on the Narrabeen shales and tall open-forests occurred there (Table 7.1) such as are described in the previous chapter. Rainforests were found on the most phosphorus-rich soils of the Narrabeen shales and sandstones (Table 7.1).

In passing, the reader's attention is drawn to the fact that Australian soils are extremely infertile. Phosphorus, particularly, occurs in very low amounts and the addition of superphosphate from the 1880s onwards has had an important effect on raising agricultural yields (Donald & Prescott, 1975). Both the relative stability and antiquity of landscapes and millions of years of leaching perhaps explain the inherent poverty of Australian soils (Donald & Prescott, 1975).

Beadle's (1962) results were for communities on soils of unimpeded drainage. In the region of his study, however, areas of impeded drainage occur as swamps (Davis, 1936) and these have their own distinct flora. Even within the open-forest, local seepages are found where ferns and sedges may be locally abundant.

Other variations in the vegetation of the region may be due to depth of soil. Davis (1936), for example, distinguished various stages of primary succession. Some of this variation is not only evident in large patches of seral communities but is also reflected in variations in understorey composition in the open-forest.

Thus, there are wide differences in the vegetation of the region associated with soil phosphate levels, soil drainage, and soil depth (moisture status?). Recently, a statistical survey of the causes of this wide variation was conducted by Burrough, Brown & Morris (1977) across an area of Hawkesbury sandstone about 110 km SSW of Sydney (between Fitzroy Falls and Barren Grounds (34°41′S; 150°44′E)). These communities range from sedgelands to heath to woodland to forest. Although underlain by the one geologic substratum (cf. Beadle, 1962), Burrough

Table 7.1. *Mean values for surface-soil phosphorus according to substrate and vegetation (data from Beadle, 1962): 'depauperate rainforest' is omitted*

	Hawkesbury sandstone	Laterite	Wianamatta shales	Narrabeen shales and sandstones
'Xeromorphic woodland/ scrub'	37	37	—	—
Open-forest	—	83	—	—
Tall open-forest	—	—	139	180–201
'Rainforest'	—	—	—	430

et al. (1977) noted that the distribution of heathlands and sedgelands appeared to be under the control of soil drainage as described by Pidgeon (1937, 1938, 1940) and Davis (1936, 1941), but they devoted most of their attention to plant communities of the well-drained and gentle middle to upper slopes of the landscape.

Across the survey region of Burrough *et al.* (1977), vegetation height and species composition were strongly correlated with mean annual rainfall, thereby suggesting that the availability of soil moisture was an important discriminant. Local variations in this vegetation pattern were interpreted in terms of local variations in effective rainfall or in soil drainage. It was not clear to them whether or not minor variations in vegetation were due to soil texture and porosity differences or to soil fertility. If the latter were the case, Burrough *et al.* (1977) thought that a combination of nutrients was probably involved rather than any particular one.

These studies in the Sydney Basin suggest that the open-forests of the region occur in a habitat with well-drained soils, moderate phosphorus content (for Australia) and moderate rainfall. Areas of poorer drainage may be occupied by sedgelands or heaths. Relatively rich soils with the most favourable soil-moisture conditions will have the largest number of rainforest species (Beadle, 1962), and the well-drained soils of the lowest fertility support 'xeromorphic woodland and scrub'.

Eastern Highlands of Victoria

As their name suggests, the Eastern Highlands are a geomorphic unit of the Victorian landscape. They consist of hills, mountains and, at the highest elevations, of plateaux remnants called 'high plains'. The Highlands are complex structurally and consist mainly of Palaeozoic sedimentary, igneous and metamorphic rocks (Hills, 1955).

In addition to the topographic and geologic complexity, the area has variable temperature and rainfall regimes. Temperatures drop with increasing altitude, which ranges from about 200 m to 2000 m. Rainfall varies from about 650 mm to more than 1500 mm and heavy snowfalls may occur in the higher country.

The environmental complexity of the region is reflected in the vegetation. To appreciate this complexity, some examples of the effect of aspect will be given later, but here the broad changes with altitude will be described. Because precipitation tends to increase with increasing altitude, the cline described tends to reflect the gradient in both precipitation and temperature. The example chosen is from Rowe (1967).

At the highest elevations are a variety of alpine and sub-alpine communities—fjaeldmark, herbfield, grassland, bog and woodland (see Chapter 16, Costin, 1981). Tall open-forests and woodlands occur in the next elevational bracket (750–1400 m), the woodlands tending to occur on the drier sites. The open-forests occur at lower elevations and give way to woodlands at still lower elevations. The open-forest here is associated with a 650 to 900 mm annual rainfall on well-drained soils.

Two examples of the importance of aspect are given by Ashton (1976*a*, *b*). The first is at the dry end of the open-forest spectrum and occurs just outside the Eastern Highlands at Mt Piper (37°12′S; 145°00′E) in central Victoria: rainfall is about 600 mm per year. The second example occurs at the wetter end of the open-forest spectrum and occurs at Beenak, 65 km east of Melbourne: annual rainfall is on average about 1400 mm per year. Below, each of these examples is considered briefly.

The study area at Mt Piper (Ashton, 1976*a*) is a conical hill rising about 230 m above the general surroundings. The form of this landscape feature is due to the indurated quartz capping of this generally sandstone structure. Soils are skeletal. The vegetation is open-forest and woodland. On deeper soils and protected aspects the open-forest is well developed, but on the rockier sites, exposed to greater insolation, the vegetation approaches woodland form. Whilst the contrast is not great between woodland and forest, the example suggests that aspect (insolation and moisture status) and soil depth affect the distribution of these vegetation types across what is really a continuum of types.

The second example is one in which contrasts are very strong. The Beenak study site (Ashton, 1976*b*) is a valley with two contrasted aspects with 15 to 20° slopes to the north and south. On small alluvial flats along the valley are closed-forests of *Nothofagus*, *Atherosperma* and *Acacia melanoxylon* up to 30 m in height, of the type described by Howard (1981) in Chapter 5. The south-facing aspect is dominated by tall open-forest up to 66 m in height and the north-facing aspect is covered by open-forest (although it reaches 45 m in height in places, which is greater than Specht (1970) considered for this type). Underlying the whole area is granite and the soils are generally deep. Apart from the differences in insolation and presumably moisture status of the soil, Ashton (1976*b*) found that the taller eucalypt forests occur on soils with higher fertility and that the shorter eucalypt forests are on phosphorus-deficient soils. These results, and similar results of Howard (1973), support Beadle's (1962) contention that soil phosphorus is an important variable associated with the delimitation of tall open-forests and open-forests as well as that of closed-forest and 'open-forest' (at the formation level).

Southwest Western Australia

This region has considerable interest since it is a vegetational and floristic outlier. The forest formations are widely separated from those in eastern Australia and the flora has 75% regional endemism (Wilson, 1973). Endemism is greatest in heathlands (Wilson, 1973) but is a feature of the forests as well (as we shall see later). Apart from forests and heathlands, the region supports areas of 'shrubby open-scrub' and 'grassy open-scrub' (Specht, 1970), previously known as 'mallee' and described in Chapter 10 (Parsons, 1981).

Physiographically (Mulcahy, 1973), the area is dominated by the Archaean shield which forms a low plateau delineated by the Darling scarp on the west and separated from the sea by a coastal plain of Quaternary sediments. To the south, the shield slopes gently to the sea and is covered with a discontinuous and thin veneer of Tertiary and Recent sediments. Areas of the shield are often mantled by laterites.

Tall open-forests of *E. diversicolor* occur in the far southwestern corner of the region where high rainfall (>750 mm per year) and substrates such as gneiss, granite, limestone, laterite and sand occur (Churchill, 1968). Similarly, a variety of soils support open-forests of *E. marginata* which continue into areas as far east as the 380 mm rainfall isohyet (Churchill, 1968) where the vegetation becomes a woodland. On the southern coastal plain, where soils are derived from limestone, an open-forest of *E. gomphocephala* occurs (Wilson, 1973).

Considerable local variation is obscured by regional descriptions which can give a false impression of the nature and complexity of landscape variation. To illustrate, some of this variety on the coastal plain just north of Perth is described. Hopkins (1960) found that *E. gomphocephala* forest occurred on shallow soils overlying limestone whilst mixed forests of *E. marginata* and *E. calophylla* were found where coffee rock or yellow sand was less than 2 m from the surface. *Eucalyptus* forests were absent from the deep sands and swamps also found in the area.

On the plateau, the forest canopy varies in composition along topographic, climatic and edaphic gradients (Havel, 1975a). *Eucalyptus marginata* is dominant on the nutritionally-poor gravels of lateritic origin but it does not occur in the loamy soils of relatively high fertility, either in the dry northeast or the moister southwest (Havel, 1975b). It is replaced by *E. wandoo* woodlands in the northeast, probably because of inadequate soil-moisture storage, and in the southwest by *E. diversicolor* (tall open-forest) which has a greater growth potential on soils of high fertility (Havel, 1975*b*). Many local patterns occur within the forest according to local variation in topographic and edaphic conditions and have been described for the northern part of the range of *E. marginata* by Havel (1975*a*, *b*).

Local and regional patterns of distribution of dominant species
In eastern Australia, eucalypt species (and sometimes the closely-related *Angophora*) of open-forests usually grow in association (Figure 7.1). This

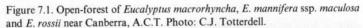

Figure 7.1. Open-forest of *Eucalyptus macrorhyncha*, *E. mannifera* ssp. *maculosa* and *E. rossii* near Canberra, A.C.T. Photo: C.J. Totterdell.

is borne out by vegetation surveys in which forest types may be known by the mixture of dominant eucalypt species, e.g. *E. pilularis–E. maculata* (Baur, 1962), *E. sieberana–E. gummifera–E. piperita* (Beadle & Costin, 1952), *E. fastigata–E. cypellocarpa* (Austin, 1978) and *E. obliqua–E. radiata* (Gill & Ashton, 1971). Although this is common it is not ubiquitous, e.g. *E. pauciflora* may form pure stands in forests abutting sub-alpine woodland of the same species. In Western Australia the pure stands seem to be common with the major types being dominated by *E. marginata* or *E. gomphocephala*. In New South Wales, results of surveys reported by Baur (1962, 1965) and Costin (1954) suggest that more than half the 'dry sclerophyll forests' required two or more eucalypt species to define them. Such data suggest high proportions for two species in many stands, whilst those designated by one eucalypt species only may have associates, but in smaller proportions.

Pryor (1959) emphasised the importance of breeding barriers between cohabiting species of *Eucalyptus* if the association is to persist. He found that genetic barriers occurred between subgeneric groups of the genus but that the great majority of species within any subgroup could interbreed freely. This suggested, then, that species in association should belong to different subgroups. Using data of Baur (1962, 1965) and Costin (1954) for species associations in 'dry sclerophyll forests', I found that whilst many associations involved species from different subgenera, about 40% had species associates from the same subgenus. In the latter category in Victoria is the widespread *E. obliqua–E. radiata* association (both *Monocalyptus*, Pryor & Johnson, 1971). Investigating this in the Hume Range (37°26′S; 145°08′E), Gill (1966) found that flowering of these species was separated in time, a mechanism which effectively prevented interbreeding. Observation of the same species association elsewhere in Victoria showed that flowering times could coincide but that successful cross pollination was rare or absent (D.H. Ashton, personal communication). A very rare putative hybrid was sterile (D.H. Ashton, personal communication). Thus Pryor's suggestion that barriers to breeding were necessary was upheld, but that these could be ecological as well as genetic was suggested. An alternative explanation of the association of species in the same subgenus is that interbreeding could occur but that any hybrid seedling formed would be unsuccessful in establishment. Evidence for such a possibility comes from the observation that hybrids typically occur where the habitat is disturbed (Pryor & Johnson, 1971), thereby implying that any habitat patchiness, favouring parental progeny, has been upset.

Another precondition for cohabitation is that tolerances of species overlap in such a way that the habitat is suited to both. That tolerances overlap is perhaps obvious but the way in which they overlap is not: an array of variables is involved. Part of this array of overlapping factors has been shown by the arrangement of species in relation to temperature and moisture (Morland, 1959; Rundle, 1977) or altitude and radiation (Austin, 1978) or nutrients and drainage classes (Gibbons & Downes, 1964) and is reflected in the many combinations that a group of species may form (Costin, 1954) in a particular region.

Demonstration of the site differences between cohabiting species which is evident from their regional distributions has proved to be very difficult experimentally. Thus, whilst Gill & Ashton (1971) found differences in the proportions of species in different height categories within forest stands in central Victoria, with *E. radiata* more 'tolerant' (in the forestry sense, Baker, 1950) than *E. obliqua*, experimental demonstration of this in the glasshouse was not achieved, although responses to shade, levels of mineral nutrition and drought were examined. The wide variability in seedling progeny, the difficulties of simulating the drastic natural selection of the field, and the fact that most regeneration appeared to be associated with post-fire environments, may explain this result. Other explanations could be that the factors examined were not critical, or that natural fine-grained variation in the field was not duplicated. Perhaps biotic rather than physical factors were more important in some cases.

Biotic explanations for cohabitation of eucalypts have been sought by Burdon & Chilvers (1974) and Morrow (1976). In both cases, leaf damage by invertebrate herbivores or fungi was found to be significant and host specificity was evident among certain groups of the damaging agents. Thus, theoretically, it was possible that compensatory host responses to waves of different host-specific predators could account for the cohabitations observed. Morrow (1976) found that in the absence of insects, one species, *E. stellulata*, had a higher growth rate than the dominant species, *E. pauciflora*. In the natural state, however, where insects were prevalent, *E. stellulata* had only reached half the height of *E. pauciflora* in 35 years. It appeared therefore that the greater intrinsic growth rate of *E. stellulata* in the sub-alpine environment of these experiments was balanced by the greater tolerance of *E. pauciflora* to insect attack. Indeed, the ability to cope with insect attack seemed to have led to the dominance of *E. pauciflora*.

Eucalypts are considered to have a remarkable sensitivity to habitat (Pryor, 1959) and this is often associated with ecotypic differentiation.

Both these observations suggest that eucalypts have high genetic flexibility in time and space. Although gene transport through seeds is for short distances only, e.g. approximately equal to tree height (Grose, 1960), that through pollen may take place over longer distances (Christensen, 1971). With compatible ecotypes (along a cline) or species (of the same taxonomic subgenus) pollen transfer and natural selection can provide the means for high genetic sensitivity to environmental change (e.g. Barber & Jackson, 1957).

National patterns in understorey floristics
If a comparison is made between the floristic lists for open-forests in Australia (from Havel's, 1975*a*, *b*, survey for Western Australia and personal communication; Specht & Perry, 1948, for South Australia; Ratkowsky & Ratkowsky, 1976, for Tasmania; and Gray & McKee, 1969, for the Australian Capital Territory) several points emerge.
   (i) There are numbers of genera common to both east and west;
   (ii) there are numerous disjunctions in species distributions;
   (iii) there are many species and genera endemic to southwest
      Western Australia, particularly.
These points are discussed briefly below.

Some genera which are common to both east and west, apart from the ubiquitous dominance of *Eucalyptus*, are shrubs of *Acacia, Daviesia, Hakea, Hibbertia, Leptospermum, Leucopogon* and *Pultenaea*. All these are large genera having more than 40 species per genus in Australia (Burbidge, 1963). Common herbaceous genera (all rhizomatous), viz. *Dianella, Lepidosperma* and *Lomandra*, and the ferns *Adiantum* and *Pteridium* are found on each of the above species lists. Green (1964), in his comparison of east and west, implicitly included vegetation formations other than forest in his comparisons. He suggested that there are 'several hundred' species common to east and west.

Green (1964) looked for large discontinuities (*c*. 1200 km) in species distributions across southern Australia. After eliminating many species on the basis of habitat (e.g. littoral niches) he demonstrated the phenomenon for 35 species. Noting that many of these species had very small seeds (e.g. orchids) he emphasised the possible role of long-distance dispersal in explaining the east-to-west disjunction.

Not only are there obvious species disjunctions between eastern and western Australia, but they occur in eastern Australia as well. The distribution of *Eucalyptus obliqua*, which is found across Bass Strait in Tasmania as well as on the mainland, is a good example of disjunction for a dominant species. Among understorey species are some *Acacia* species, such as *A. genistifolia, A verticillata* and *A. dealbata*, all of which occur in Tasmania as well as on the mainland (Gill & Ashton, 1968; Ratkowsky & Ratkowsky, 1976; Purdie, 1977*a*). Other, less obvious disjunctions than that imposed by Bass Strait occur, such as that posed by the Hunter River Valley in New South Wales and that between western Victoria and the Mt Lofty Ranges in South Australia but these have not been studied carefully from a floristic point of view. That they occur between populations of *E. obliqua* from open-forests has been documented by Gill (1966).

Wood & Williams (1960) pointed out that there are similarities in genera but differences between species in open-forests of three areas of eastern Australia, viz. South Australia, New South Wales and the Australian Capital Territory. They called these 'vicarious' species. Green (1964) suggested that the examples of Wood & Williams are better seen as species representative of common genera rather than as 'closely related allopatric species' (part of the definition of vicarious species of Cain, 1944). Unlike Wood & Williams' (1960) examples, Green's (1964) examples are for vicariism between southwestern and southeastern Australia and comprise 47 species pairs.

Not only may the differences and similarities between eastern and western Australia (or differences associated with disjunctions in eastern Australia) be seen in terms of disjunction in species distributions or vicariism, they may also be seen in the extent of local endemism. In comparisons of the species lists referred to at the start of this section, the isolation of the southwestern forests was reflected in the more than 20 endemic genera found there. Endemism to the local region at the generic level was absent or very low in the eastern States.

Disjunctions may be viewed as continuous populations broken by climatic, topographic or edaphic barriers bridged by long-distance dispersal of propagules, or be seen as insurmountable barriers to dispersal requiring a geological rather than a behavioural explanation for their presence. Green (1964), in noting the high proportion of minute-seeded orchids in his discontinuous species ranges between eastern and western Australia, could not dismiss the possibility that long-distance dispersal provided an explanation. However, for vicarious species, and the highly endemic genera of Western Australia, long periods of isolation are implied and geological explanations are required. Parsons (1969) suggested that lowered sea-levels in the Pleistocene, with the consequent exposure of coastal lowland now under the sea, may have provided the conditions whereby continuous populations of species could occur across southern Australia (and to Tasmania). Whether or not the time period that has elapsed since then is sufficient to account for the high endemism of the southwest flora remains unresolved.

Biological processes in open-forest

Nutrient cycling

Because Australian soils are so poor in plant nutrients, nutrient cycling is an important topic to a description of open-forests. Despite this, 'too little attention' has been given to it in forests (Attiwill, 1975). The importance of differences in soil phosphorus levels to vegetation distribution has already been emphasised together with the general poverty of this element in Australian soils. In agriculture, many other deficiencies occur, and the country has become renowned for its use of trace elements (Donald & Prescott, 1975). Whilst deficiency of trace elements has been marked, the only study of growth of native species in relation to them points to a possible role of a toxicity (manganese) in affecting eucalypt distribution (Winterhalder, 1963). Recently, however, some studies of cycling of trace elements in native forests have begun (Rogers & Westman, 1977; Westman & Rogers, 1977; Lee & Correll, 1978; O'Connell, Grove & Dimmock, 1978).

Emerging from studies of mineral composition of litter are correlations between certain elements. Lee & Correll (1978) found that concentrations of nitrogen, phosphorus, zinc, iron and copper were correlated: O'Connell *et al.* (1978), similarly, found that zinc levels were correlated with those of nitrogen, phosphorus and potassium. Such correlations may reflect functional differences in the plant, as suggested by Garten (1978), who found that nitrogen, phosphorus, iron, copper and sulphur were correlated and formed a set related to nucleic acid and protein metabolism. He also found that concentrations of calcium, magnesium and manganese were related and could be attributed to a 'structural–photosynthetic' set: Lee & Correll (1978) found such a set also in their analyses of eucalypt litter. Garten's (1978) correlations were found between elemental concentrations for 110 North American species of wide ecological amplitude, whilst those of the Australian studies were for litter concentrations within a single forest type.

Groupings of elements such as those indicated above may help in the construction of generalisations about the cycling of nutrients in the forest. Further research is necessary in this area before firm statements can be made, and this is particularly so for studies of litter decomposition.

The cycling of elements in the forest varies in rate and quantity according to the nature of the element and its carrier. Thus, elements such as potassium are particularly mobile and are leached from canopies by rain. Other elements may be carried in leaf or twig materials, or as animal excreta. Some elements may be circulated internally within the plant rather than externally through the soil. Below, these topics are reviewed briefly. Not considered here is the topic of nutrient uptake, an omission due to the lack of research in this area; an abundance of mycorrhizae in native forests is present, however, which may assist in nutrient uptake from their nutrient-poor milieu.

Leaching of plant canopies removes a great diversity of materials including both organic and inorganic substances (Tukey, 1970). In *Eucalyptus* forest, Guthrie, Attiwill & Leuning (1978) found foliage leaching of elements to be in the order Na > K > Ca > Mg, whilst Rowe & Hagel (1974) found that leaching from litter was in the order of K > (Ca, Mg) > Na. Among the anions, Rowe & Hagel (1974) found that leaching from litter was in the order of $HCO_3^- > SO_4^= > Cl^-$.

Amounts of materials leached will depend not only on the mobility of the element but also on the amount in the tissue being leached. Guthrie *et al.* (1978) found that, on an annual basis, leaching accounted for 66% Na, 56% K, 33% Mg and 14% Ca of the total present, with the balance being circulated through the litter. Of particular importance to the cycling of some elements, especially nitrogen and phosphorus, is the withdrawal of nutrients prior to leaf fall. This is an internal process of rapid recycling. For phosphorus this may amount to 70–80% in eucalypts (Ashton, 1976b; Attiwill, Guthrie & Leuning, 1978; O'Connell *et al.*, 1978). Similarly, O'Connell *et al.* (1978) reported that nitrogen withdrawal may amount to about 60–70%. However, in a forest with relatively abundant nitrogen, Westman & Rogers (1977) found that nitrogen withdrawal was inconsequential in eucalypt leaves. Perhaps the demand by young leaves affects the rate of withdrawal from older leaves?

Rapid movement of nutrients from the plant canopy to the soil may take place through insect frass and the excreta of vertebrates. Although the importance of some insect feeders has already been mentioned in another context, that of the jarrah leaf-miner, *Perthida glyphopa*, (Wallace, 1970) has not. Extensive leaf damage by such insects highlights the importance and role of insect feeders to nutrient cycling. The frass from such feeding is readily decomposed (Springett, 1978), so that, in addition to canopy nutrients being moved prematurely compared to those in an intact crown, the nutrients are also released to the soil more rapidly than for intact leaves.

About 2 to 4 t ha^{-1} litter falls from open-forest canopies each year (Lee & Correll, 1978) and is responsible for a large proportion of the movement of calcium and magnesium from the canopy (Guthrie *et al.*, 1978). Most of the material, especially the leaves, falls from the eucalypts in summer (e.g. Attiwill *et al.*, 1978). Because of the correlations noted earlier (Lee & Correll 1978), it might be expected that manganese would behave similarly to calcium and magnesium. Indeed, manganese concentrations parallel those of calcium and magnesium as the leaf matures (Westman & Rogers, 1977; O'Connell *et al.*, 1978), thereby supporting this idea.

The rate of return of the less mobile elements to the soil from litter may be estimated from decomposition rates. Gill (1964) estimated that decomposition of eucalypt leaves took from two to three years, and that for twigs took between four and eight years, but decomposition of large branches and logs takes many more years. Similar figures have been obtained for similar forests in other studies (Lee & Correll, 1978). Such rates of decay, and thereby of mineralisation, are slow compared with rates achieved by burning. However, burning has other consequences as well.

Fire releases elements from organic matter very rapidly but may volatilise elements such as nitrogen together with some phosphorus and sulphur. Some movement of nutrients may also occur in smoke and in local redistributions of ash by wind and water. If the fire is of high intensity and the crown is consumed, release of elements at the relatively high concentrations in the foliage occurs also. Death of trees may be associated with particular intensities, and stores of nutrients in trunk, branches and bark may then become subject to decay processes. The subject is a complicated one as the effects of fire on nutrient cycling depend not only on the characteristics of the plant species and the volumes of material present, but also on the intensities, frequencies and seasons of fire occurrence. No detailed work on these aspects of nutrient cycling in Australian forests has yet been conducted (Groves, 1977).

Nitrogen losses through volatilisation during fire may be restored through the activities of symbiotic nitrogen-fixing organisms. Nitrogen-fixing species are common in open-forests and include leguminous shrubs (such as *Acacia, Daviesia* and *Pultenaea*), non-leguminous shrubs (such as *Casuarina*) and cycads (such as *Macrozamia*). Halliday & Pate (1976) found that *Macrozamia riedlei* contributed about 19 kg ha^{-1} of nitrogen to a coastal community in Western Australia. Similar quantities could be expected where plants of the same genus occur in open-forests. *Macrozamia* plants are very fire-resistant and may be expected to fix nitrogen after fire occurs. Their input may vary according to any root damage during fire and according to population changes. For some *Acacia* species, at least, dramatic changes in population (and nitrogen fixation?) may occur according to the characteristics of fires to which they are subject (Shea, McCormick & Portlock, 1979).

Fire

Fires are a feature of open-forests. They occur in, and contribute to, one of the most severe fire situations in the world. In open-forest regions large fires may occur once every three to five years (Cheney, 1976). Fuels accumulate rapidly and can reach levels of 15 t ha^{-1} after 10 years, 22 t ha^{-1} after 20 years and 27 t ha^{-1} after 30 years (Wallace, 1966). These fuels may have very low moisture contents during summer and autumn when hot dry winds can also fan any fire that starts. Fires of high intensity under these conditions can be extremely dangerous to people in the area, as shown by the fires at Hobart, Tasmania in 1967 (Cheney, 1976).

The effects of fires on the vegetation vary according to sequences of fire events rather than to one event only. Particular sequences may be known as particular 'fire regimes' and consist of the variables of fire intensity, fire frequency and season of occurrence (Gill, 1975). Fire intensities can vary across three or four orders of magnitude, frequencies in open-forests may vary from once every five years to 50 or more years, and they can occur in particular sites in the open-forest zone at any time of year if ignited.

Although a single plant community may be affected by the same fire regime, the responses of individual plants and individual species may vary. Seedlings may be killed by a single fire but mature plants of the same species may survive many. Some species have individuals which are all killed by a single fire but the species persists because regeneration occurs from seed stored on the plant or in the soil. Even amongst species of the same genus in the same forest, variations occur: examples may be found in the genus *Acacia* where one species may have fire-sensitive individuals and another species have resistant individuals because of buds buried in the soil.

Most, if not all, the eucalypts of the open-forest have fire-resistant individuals (e.g. Gill & Ashton, 1968). They also have limited seed stores in the canopies of the trees which may be released in greater quantities than usual at the time of fire. Some shrub genera, such as *Hakea*, *Banksia* and *Casuarina*, may have species which store seed in the canopy and release it at the time of fire also. Some species, like the mistletoes, may have no storage in the canopy and none in the ground and rely on invasion from unburned areas for their persistence when individuals are killed by fire.

The validity of classical theories of secondary succession, including that after individual fires, is being increasingly questioned for Australian plant communities (Purdie & Slatyer, 1976; Noble & Slatyer, 1977). After fire in open-forest, Purdie (1977*a*, *b*) found that all species present before experimental fires were again present afterwards. She concluded that Egler's (1954) relay floristic model of succession was the most appropriate one to describe this situation (Purdie & Slatyer, 1976). The same conclusion could be drawn from the pioneering studies of Jarrett & Petrie (1929) in Victoria: these authors noted that vegetative regeneration was marked in their open-forests and that 'succession is confined usually to a sere of socies in the lower layers of the forest'. I take this to mean that changes in species abundances occur with time rather than there being changes in numbers of species present.

In South Australian studies, Cochrane, Burnard & Philpott (1962) identified four main 'seral stages' in open-forest. The first was typified by the presence of the liverwort *Marchantia polymorpha*, a lack of species dominance and low ground cover (a stage lasting 18 months); from two to five years after fire the composite *Ixodia achilleoides* was dominant; after six years, dominance changed to that of the slower-growing shrubs such as *Leptospermum myrsinoides*; whilst from seven to ten years, and persisting for at least 20 years, dominance of shrubs of *Acacia myrtifolia* and *Pultenaea daphnoides*, 1 to 2 m tall, was apparent. Variations around this pattern may be expected according to variations in the fire regimes but this has not been studied.

Growth and seed dispersal: selected aspects
Eucalypts have small seeds which germinate readily. Seedlings of the open-forest species soon develop a woody swelling in the axils of the cotyledons and first few leaves. These organs, known as 'lignotubers' (Kerr, 1925), are sites of regeneration after stresses, such as insect attack, shade, disease or fire, have been removed. As the plant grows and develops into a tree the lignotuber may be overgrown in some species but the presence of buds beneath the bark may aid survival after stress.

Shoot growth of intact trees occurs either as a unimodal peak in summer (Gill, 1964) or has a bimodal form with peaks in autumn and spring (Specht & Brouwer, 1975). Leaf fall, in southern areas at least, appears to be associated with shoot growth (Jacobs, 1955; Gill, 1964), although other factors such as drought (Pook, Costin & Moore, 1966), heavy flowering, insect attack and temperature changes (Jacobs, 1955) may affect the process. Whether or not shoot growth has a direct effect on litter fall or merely coincides with temperature (the factor of importance in leaf fall suggested by Attiwill *et al.*, 1978) is uncertain.

Inflorescence buds of most eucalypts form in leaf axils (Carr & Carr, 1959) as the leaves unfold but flowers may take two years to appear. The actual season of flowering may vary widely among the species although leafing phenology appears to be more uniform. Two years after flowering, capsule dehiscence begins and may continue over a two-year period. Seeds are heavily predated by ants (Jacobs, 1955).

Ants have been implicated in seed dispersal as well as harvesting. Berg (1975) indicated recently that many Australian species have seeds with 'elaiosomes' which are particularly attractive to ants. Many of the species with these structures are the shrubs of open-forests. Ants pick up the seeds by the elaiosome, transport it to the nest, remove the elaiosome and discard the seed either to the nest surface or within the nest itself. Diverse genera and families of plants are involved but most species are Australian endemics. Berg (1975) suggested that there may be interacting factors which have led to this abundance of seeds possessing elaiosomes and these factors may include fire and seed predation.

Utilisation

The open-forests of southern Australia are found close to the major centres of population. Although this is convenient for the transport of wood products it also results in demand for other uses, such as conservation of water and plant species, and for recreation.

When untouched eucalypt forests were first exploited, it was found that more than 50% of the volume of commercial species was so defective that it could not be used (Curtin, 1970). In the poorer forests where intensive management for wood production has been uneconomic, this situation has led to a tendency to exploit rather than to manage the resource. In better forests, clearfelling and intensive management have been possible because of the higher profit from wood products such as sawlogs. With recent emphasis on wood chips as a product, some foresters contend that harvesting for such materials will cause an upgrading of these forests for sawlog production (Ovington & Thistlethwaite, 1976). Eucalypts comprise 94% of the wood volume of native forests (Jacobs, 1970). In open-forests the more important species are *E. marginata* in Western Australia, and *E. obliqua*, *E. maculata* and *E. sieberi* in eastern Australia (Jacobs, 1970). All these species are used for sawlogs but in some areas they are also used for woodchips.

Woodchipping for export has become an important practice in New South Wales, Tasmania and Western Australia. In Western Australia, *E. marginata* and *E. calophylla* are the basis of the industry: in Tasmania, *E. viminalis*, *E. obliqua* and *E. amygdalina* are its backbone; whilst in New South Wales, *E. sieberi* and *E. gummifera* are the principal species (Heyligers, 1975). For further detail, the reader is referred to published reports of enquiries into the woodchip industry (Cromer *et al.*, 1975; Gilbert, 1976; Heyligers, 1975, 1977; Senate Standing Committee on Science and the Environment, 1977) and forest industries in general (Australian Conservation Foundation, 1974; 'FORWOOD', 1974; Routley & Routley, 1974).

Major problems in the management of open-forests concern the use of fire (Attiwill, 1975); the control of the root-rotting disease caused by *Phytophthora* species, especially in Western Australia (Ovington & Thistlethwaite, 1976); and the conservation of species (Attiwill, 1975). Both Attiwill (1975) and the Australian Conservation Foundation (1974) stress the need for an integrated view of the forest as a resource: wood production should be only one of a number of equally important goals.

From a species conservation point of view, many alliances of the open-forest are not conserved at all (Specht, Roe & Boughton, 1974) despite the observation of Tyndale-Biscoe & Calaby (1975) that the '*Eucalyptus* forests of southeastern Australia and Tasmania support a rich and varied fauna of mammals and birds and together form the single most important refuge for wildlife in Australia'. The high endemism of the southwestern Australian forests and the threat to their persistence due to *Phytophthora* and bauxite mining makes conservation an important and urgent need in that region where five alliances remain unconserved (Specht *et al.*, 1974).

Conclusions

Eucalyptus forests are an important and conspicuous element of the Australian landscape. Classification and nomenclature of these forests still poses problems which require resolution. In this chapter the sub-category of the formation-level 'open-forests', also known as 'open-forest', has been considered in its range of occurrence in southeastern Australia (including Tasmania) as well as in the southwest. These forests occur in areas of moderate rainfall and temperature where the phosphorus status of the soils is also moderate in the general spectrum of low values for Australian soils. The eucalypts dominating these forests often occur in associations and appear sensitive to slight changes in habitat. Distributions in some cases are disjunct as is the case with many understorey species. Disjunctions may be recognised at the levels of subspecies, species or genus and could be due to behavioural or historical-geological reasons. Because of the poor soils on which they grow, mineral cycling is an important topic and efficient withdrawal of phosphorus from leaves before leaf fall is a feature of these evergreen forests. Fire may strongly affect the cycling process but has been little studied in this context. Plants respond to single fires in different ways and, within a single community subject to a series of fires, many mechanisms for species persistence may be found. Fires tend to occur naturally in summer and autumn in litter dropped from the trees. Leaf fall from the eucalypts may be a response to leaf growth or other internally-controlled mechanisms but may be influenced as well by the external environment. The forests are used for wood products, recreation, and water supply as well as the conservation of a unique biota.

I would like to thank Dr J.J. Burdon and Mr J.J. Havel for their criticisms of the draft manuscript.

References

Ashton, D.H. (1976*a*). The vegetation of Mount Piper, central Victoria: A study of a continuum. *Journal of Ecology*, *64*, 463–83.

Ashton, D.H. (1976*b*). Phosphorus in forest ecosystems at Beenak, Victoria. *Journal of Ecology*, *64*, 171–86.

Ashton, D.H. (1981). Tall open-forests. In *Australian Vegetation*, ed. R.H. Groves, pp. 121–151. Cambridge University Press.

Attiwill, P.M. (1975). The eucalypt forest – resources, refuges, and research. *Australian Forestry*, *38*, 162–70.

Attiwill, P.M., Guthrie, H.B. & Leuning, R. (1978). Nutrient cycling in a *Eucalyptus obliqua* (L'Hérit.) forest. I. Litter production and nutrient return. *Australian Journal of Botany*, *26*, 79–91.

Austin, M.P. (1978). Vegetation. In *Land Use on the South Coast of New South Wales: A Study in Methods of Acquiring and Using Information to Analyse Regional Land Use Options*, ed. M.P. Austin & K.D. Cocks, pp. 44–67. Melbourne: CSIRO, Australia.

Australian Conservation Foundation (1974). Multiple use on forest land presently used for commercial wood production. *Search*, *5*, 438–43.

Baker, F.S. (1950). *Principles of Silviculture*. New York: McGraw Hill.

Barber, H.N. & Jackson, W.D. (1957). Natural selection in action in *Eucalyptus*. *Nature, 179*, 1267–9.

Baur, G.N. (1962). *Forest vegetation in north-eastern New South Wales*. Forests Commission of New South Wales Research Note Number 8.

Baur, G.N. (1965). *Forest types in New South Wales*. Forests Commission of New South Wales Research Note Number 17.

Beadle, N.C.W. (1954). Soil phosphate and the delimitation of plant communities in eastern Australia. *Ecology, 35*, 370–5.

Beadle, N.C.W. (1962). Soil phosphate and the delimitation of plant communities in eastern Australia. II. *Ecology, 43*, 281–8.

Beadle, N.C.W. & Costin, A.B. (1952). Ecological classification and nomenclature. *Proceedings of the Linnean Society of New South Wales, 77*, 61–82.

Berg, R.Y. (1975). Myrmecochorous plants in Australia and their dispersal by ants. *Australian Journal of Botany, 23*, 475–508.

Burbidge, N.T. (1963). *Dictionary of Australian Plant Genera*. Sydney: Angus & Robertson.

Burdon, J.J. & Chilvers, G.A. (1974). Fungal and insect parasites contributing to niche differentiation in mixed species stands of eucalypt saplings. *Australian Journal of Botany, 22*, 103–04.

Burrough, P.A., Brown, L. & Morris, E.C. (1977). Variations in vegetation and soil pattern across the Hawkesbury Sandstone plateau from Barren Grounds to Fitzroy Falls, New South Wales. *Australian Journal of Ecology, 2*, 137–59.

Cain, S.A. (1944). *Foundations of Plant Geography*. New York: Harper.

Carr, D.J. & Carr, S.G.M. (1959). Developmental morphology of the floral organs of *Eucalyptus*. I. The inflorescence. *Australian Journal of Botany, 7*, 109–41.

Cheney, N.P. (1976). Bushfire disasters in Australia, 1945–1975. *Australian Forestry, 39*, 245–68.

Christensen, P.E. (1971). The purple-crowned lorikeet and eucalypt pollination. *Australian Forestry, 35*, 263–70.

Churchill, D.M. (1968). The distribution and prehistory of *Eucalyptus diversicolor* F. Muell., *E. marginata* Donn. ex Sm. and *E. calophylla* R.Br. in relation to rainfall. *Australian Journal of Botany, 16*, 125–51.

Cochrane, G.R., Burnard, S. & Philpott, J.M. (1962). Land use and forest fires in the Mount Lofty Ranges, South Australia. *Australian Geographer, 8*, 143–60.

Costin, A.B. (1954). *A Study of the Ecosystems of the Monaro Region of New South Wales*. Sydney: Government Printer.

Costin, A.B. (1981). Alpine and sub-alpine vegetation. In *Australian Vegetation*, ed. R.H. Groves, pp. 361–376. Cambridge University Press.

Cromer, D.A.N., Eldershaw, V.J., Lamb, I.D., McArthur, A.G., Wesney, D. & Girdlestone, J.N. (1975). *Economic and Environmental Aspects of the Export Hardwood Woodchip Industry*, vols. I & II. Canberra: Australian Government Publishing Service.

Curtin, R.A. (1970). Increasing the productivity of eucalypt forests in New South Wales. *Australian Forestry 34*, 97–106.

Davis, C. (1936). Plant ecology of the Bulli district, Part I: Stratigraphy, physiography and climate; general distribution of plant communities and interpretation. *Proceedings of the Linnean Society of New South Wales, 61*, 285–97.

Davis, C. (1941). Plant ecology of the Bulli district. II. Plant communities of the plateau and scarp. *Proceedings of the Linnean Society of New South Wales, 66*, 1–19.

Diels, L. (1906). *Die Vegetation der Erde 7: Pflanzenwelt von West-Austration.* Leipzig: Engelmann.

Donald, C.M. & Prescott, J.A. (1975). Trace elements in Australian crops and pasture production, 1924–1974. In *Trace Elements in Soil–Plant–Animal Systems,* ed. D.J.D. Nicholas & A.R. Egan, pp. 7–39. London: Academic Press.

Egler, F.E. (1954). Vegetation science concepts. I. Initial floristic composition,a factor in old field vegetation development. *Vegetatio, 4,* 412–7.

'FORWOOD' (1974). *A series of reports prepared for the 'Forestry and Wood-Based Industries Development Conference', April 1974.* Canberra: Australian Government Publishing Service.

Garten, G.T. (1978). Multivariate perspectives on the ecology of plant mineral elemental composition. *American Naturalist, 112,* 533–44.

Gibbons, F.R. & Downes, R.G. (1964). *A study of the land in south-western Victoria.* Soil Conservation Authority of Victoria Technical Communication Number 3.

Gilbert, J.M. (ed.) (1976). *Woodchip Symposium Papers, 47th ANZAAS Congress.* Hobart: Tasmanian Forests Commission.

Gill, A.M. (1964). Soil-vegetation relationships near Kinglake West, Victoria. M.Sc. thesis, University of Melbourne.

Gill, A.M. (1966). The ecology of mixed species of *Eucalyptus* in central Victoria, Australia. Ph.D. thesis, University of Melbourne.

Gill, A.M. (1975). Fire and the Australian flora: a review. *Australian Forestry, 38,* 1–25.

Gill, A.M. & Ashton, D.H. (1968). The role of bark type in relative tolerance to fire of three central Victorian eucalypts. *Australian Journal of Botany, 16,* 491–8.

Gill, A.M. & Ashton, D.H. (1971). The vegetation and environment of a multi-aged eucalypt forest near Kinglake West, Victoria, Australia. *Proceedings of the Royal Society of Victoria, 84,* 159–72.

Gillison, A.N. & Walker, J. (1981). Woodlands. In *Australian Vegetation,* ed. R.H. Groves, pp. 177–197. Cambridge University Press.

Gray, M. & McKee, H.S. (1969). *A list of vascular plants occurring on Black Mountain and environs, Canberra, ACT.* CSIRO Australia, Division of Plant Industry Technical Paper Number 26.

Green, J.W. (1964). Discontinuous and presumed vicarious plant species in southern Australia. *Journal of the Royal Society of Western Australia, 47,* 25–32.

Grose, R.J. (1960). Effective seed supply for the natural regeneration of *Eucalyptus delegatensis* R.T. Baker, syn. *Eucalyptus gigantea* Hook. f. *Journal of the Australian Pulp & Paper Industry Technical Association, 13,* 141–7.

Groves, R.H. (1977). Fire and nutrients in the management of Australian vegetation. In *Proceedings of the Symposium on the Environmental Consequences of Fire and Fuel Management in Mediterranean Ecosystems,* ed. H.A. Mooney & C.E. Conrad, pp. 220–9, U.S. Forest Service General Technical Report WO–3.

Guthrie, H.B., Attiwill, P.M. & Leuning, R. (1978). Nutrient cycling in a *Eucalyptus obliqua* (L'Hérit.) forest. II. A study in a small catchment. *Australian Journal of Botany, 26,* 189–201.

Halliday, J. & Pate, J.S. (1976). Symbiotic nitrogen fixation by coralloid roots of the cycad *Macrozamia riedlei*: physiological characteristics and ecological significances. *Australian Journal of Plant Physiology, 3,* 349–58.

Havel, J.J. (1975a). *Site-vegetation mapping in the northern jarrah forest (Darling Range). 1. Definition of site-vegetation types.* Western Australian Forests Department Bulletin Number 86.

Havel, J.J. (1975b). *Site-vegetation mapping in the northern jarrah forest (Darling Range). 2. Location and mapping of site-vegetation types.* Western Australian Forests Department Bulletin Number 87.

Heyligers, P.C. (1975). Biological and ecological aspects related to the forestry operations of the export woodchip industry. In *Economic and Environmental Aspects of the Export Hardwood Woodchip Industry*, ed. D.A.N. Cromer *et al.*, Vol II, pp. 77–119. Canberra: Australian Government Publishing Service.

Heyligers, P.C. (1977). *The natural history of the Tasmanian, Manjimup and Eden-Bombala woodchip export concession areas.* Australian Government Department of Environment, Housing & Community Development Studies Bureau Report Number 22.

Hills, E.S. (1955). Physiography and geology. In *Introducing Victoria*, ed. G.W. Leeper, pp. 25–39. Melbourne: ANZAAS.

Hopkins, E.R. (1960). *The fertilizer factor in* Pinus pinaster *Ait. plantations on sandy soils of the Swan Coastal plain, Western Australia.* Western Australian Forests Department Bulletin Number 68.

Howard, T.M. (1973). Studies in the ecology of *Nothofagus cunninghamii* Oerst. 1. Natural regeneration on the Mt Donna Buang massif, Victoria. *Australian Journal of Botany*, *21*, 67–78.

Howard, T.M. (1981). Southern closed-forests. In *Australian Vegetation*, ed. R.H. Groves, pp. 102–120. Cambridge University Press.

Jacobs, M.R. (1955). *Growth Habits of the Eucalypts*. Canberra: Government Printer.

Jacobs, M.R. (1970). The forest as a crop. In *The Australian Environment*, 4th edn (rev.), ed. G.W. Leeper, pp. 120–30. Melbourne: CSIRO, Australia & Melbourne University Press.

Jarrett, P.H. & Petrie, A.H.K. (1929). The vegetation of Black's Spur region. II. Pyric succession. *Journal of Ecology*, *17*, 249–81.

Johnson, R.W. & Burrows, W.H. (1981). *Acacia* open-forests, woodlands and shrublands. In *Australian Vegetation*, ed. R.H. Groves, pp. 198–226. Cambridge University Press.

Kerr, L.R. (1925). The lignotubers of eucalypt seedlings. *Proceedings of the Royal Society of Victoria*, *27*, 79–97.

Lee, K.E. & Correll, R.L. (1978). Litter fall and its relationship to nutrient cycling in a South Australian dry sclerophyll forest. *Australian Journal of Ecology*, *3*, 243–52.

Morland, R.T. (1959). Erosion survey of the Hume catchment area. VI. Vegetation (cont.). *Journal of the Soil Conservation Service of New South Wales*, *15*, 176–86.

Morrow, P.A. (1976). The significance of phytophagous insects in the *Eucalyptus* forests of Australia. In *The Role of Arthropods in Forest Ecosystems*, ed. W.J. Mattson, pp. 19–29. New York: Springer.

Mulcahy, M.J. (1973). Landforms and soils of southwestern Australia. *Journal of the Royal Society of Western Australia*, *56*, 16–22.

Noble, I.R. & Slatyer, R.O. (1977). Post-fire succession of plants in Mediterranean ecosystems. In *Proceedings of the Symposium on the Environmental Consequences of Fire and Fuel Management in Mediterranean Ecosystems*, ed. H.A. Mooney & C.E. Conrad, pp. 27–36. U.S. Forest Service General Technical Report WO–3.

O'Connell, A.M., Grove, T.S. & Dimmock, G.M. (1978). Nutrients in the litter on jarrah forest soils. *Australian Journal of Ecology*, *3*, 253–60.

Ovington, J.D. & Thistlethwaite, R.J. (1976). The woodchip industry: environmental effects of cutting and regeneration practices. *Search*, *7*, 383–92.

Packham, G.H. (1962). An outline of the geology of New South Wales. In *A Goodly Heritage. ANZAAS Jubilee, Science in New South Wales*, ed. A.P.P. Elkin, pp. 24–35. Sydney: ANZAAS.

Parsons, R.F. (1969). Distribution and palaeogeography of two mallee species of *Eucalyptus* in southern Australia. *Australian Journal of Botany*, *17*, 323–30.

Parsons, R.F. (1981). *Eucalyptus* scrubs and shrublands. In *Australian Vegetation*, ed. R.H. Groves, pp. 227–252. Cambridge University Press.

Pidgeon, I.M. (1937). The ecology of the Central Coastal area of New South Wales. I. The environment and general features of the vegetation. *Proceedings of the Linnean Society of New South Wales*, *62*, 315–40.

Pidgeon, I.M. (1938). The ecology of the Central Coastal area of New South Wales. II. Plant succession on Hawkesbury Sandstone. *Proceedings of the Linnean Society of New South Wales*, *63*, 1–26.

Pidgeon, I.M. (1940). The ecology of the Central Coastal area of N.S.W. III. Types of primary succession. *Proceedings of the Linnean Society of New South Wales*, *65*, 221–49.

Pook, E.W., Costin, A.B. & Moore, C.W.E. (1966). Water stress in native vegetation during the drought of 1965. *Australian Journal of Botany*, *14*, 257–67.

Prescott, J.A. (1931). *The soils of Australia in relation to vegetation and climate*. Council for Scientific & Industrial Research, Australia, Bulletin Number 52.

Pryor, L.D. (1959). Species distribution and association in *Eucalyptus*. In *Biogeography and Ecology in Australia*, ed. A. Keast, R.L. Crocker & C.S. Christian, pp. 461–71 (Monographiae Biologicae VIII). The Hague: Junk.

Pryor, L.D. & Johnson, L.A.S. (1971). *A Classification of the Eucalypts*. Canberra: Australian National University.

Purdie, R.W. (1977a). Early stages of regeneration after burning in dry sclerophyll vegetation. I. Regeneration of the understorey by vegetative means. *Australian Journal of Botany 25*, 21–34.

Purdie, R.W. (1977b). Early stages of regeneration after burning in dry sclerophyll vegetation. II. Regeneration by seed germination. *Australian Journal of Botany*, *25*, 35–46.

Purdie, R.W. & Slatyer, R.O. (1976). Vegetation succession after fire in sclerophyll woodland communities in south-eastern Australia. *Australian Journal of Ecology*, *1*, 223–36.

Ratkowsky, D.A. & Ratkowsky, A.V. (1976). Changes in the abundance of the vascular plants of the Mount Wellington Range, Tasmania, following a severe fire. *Papers and Proceedings of the Royal Society of Tasmania*, *110*, 63–90.

Rogers, R.W. & Westman, W.E. (1977). Seasonal nutrient dynamics of litter in a sub-tropical eucalypt forest, North Stradbroke Island. *Australian Journal of Botany*, *25*, 47–58.

Routley, R. & Routley, V. (1974). *The Fight for the Forests, The Takeover of Australian Forests for Pines, Woodchips and Intensive Forestry*. Canberra: Australian National University Research School of Social Sciences.

Rowe, R.K. (1967). *A study of the land in the Victorian catchment of Lake Hume*. Soil Conservation Authority of Victoria Technical Communication Number 5.

Rowe, R.K. & Hagel, V. (1974). Leaching of plant nutrient ions from burned forest litter. *Australian Forestry*, *36*, 154–63.

Rundle, A.S. (1977). *A study of the land in the catchment of Lake Eildon*. Soil Conservation Authority of Victoria Technical Communication Number 11.

Senate Standing Committee on Science and the Environment (1977). *Woodchips and the Environment*. Canberra: Australian Government Publishing Service.

Shea, S.R., McCormick, J. & Portlock, C.C. (1979). The effect of fires on regeneration of leguminous species in the northern jarrah (*Eucalyptus marginata* Sm.) forest of Western Australia. *Australian Journal of Ecology*, *4*, 195–205.

Specht, R.L. (1970). Vegetation. In *The Australian Environment*, 4th edn (rev.), ed. G.W. Leeper, pp. 44–67. Melbourne: CSIRO, Australia & Melbourne University Press.

Specht, R.L. & Brouwer, Y.M. (1975). Seasonal shoot growth of *Eucalyptus* spp. in the Brisbane area of Queensland (with notes on shoot growth and litter fall in other areas of Australia). *Australian Journal of Botany*, *23*, 459–74.

Specht, R.L. & Perry, R.A. (1948). Plant ecology of part of the Mount Lofty Ranges. *Transactions of the Royal Society of South Australia*, *72*, 91–132.

Specht, R.L., Roe, E.M. & Boughton, V.H. (ed.) (1974). Conservation of major plant communities in Australia and Papua New Guinea. *Australian Journal of Botany, Supplementary Series* No. 7.

Springett, B.P. (1978). On the ecological role of insects in Australian eucalypt forests. *Australian Journal of Ecology*, *3*, 129–39.

Tukey, H.B. (1970). The leaching of substances from plants. *Annual Review of Plant Physiology*, *21*, 305–24.

Tyndale-Biscoe, C.H. & Calaby, J.H. (1975). Eucalypt forest as a refuge for wildlife. *Australian Forestry*, *38*, 117–33.

Wallace, M.M.H. (1966). Fire in the jarrah forest environment. *Journal of the Royal Society of Western Australia*, *49*, 33–44.

Wallace, M.M.H. (1970). The biology of the jarrah leaf miner, *Perthida glyphopa* (Lepidoptera). *Australian Journal of Zoology*, *186*, 91–104.

Webb, L.J. (1959). A physiognomic classification of Australian rainforests. *Journal of Ecology*, *47*, 551–70.

Westman, W.E. & Rogers, R.W. (1977). Nutrient stocks in a sub-tropical eucalypt forest, North Stradbroke Island. *Australian Journal of Ecology*, *2*, 447–60.

Williams, R.J. (1955). Vegetation regions. In *Atlas of Australian Resources*, 1st Series. Canberra: Department of National Development.

Wilson, P.G. (1973). The vegetation of Western Australia. In *Western Australian Year Book*, Vol. 12, pp. 55–62. Perth: Government Printer.

Winterhalder, E.K. (1963). Differential resistance of two species of *Eucalyptus* to toxic soil manganese levels. *Australian Journal of Science*, *25*, 363–4.

Wood, J.G. (1937). *Vegetation of South Australia*. Adelaide: Government Printer.

Wood, J.G. & Williams, R.J. (1960). Vegetation. In *The Australian Environment*, 3rd edn (rev.), pp. 67–84. Melbourne: CSIRO, Australia & Melbourne University Press.

8

Woodlands

A.N. GILLISON & J. WALKER

Australian vegetation is unique in many respects as a result of its long historical isolation from other parts of the Gondwanic platform, the low nutrient status of soils, severe fire climate, and low density of humans and browsing fauna. For woody vegetation these evolutionary aspects are perhaps nowhere better reflected than in woodlands. These woodlands cover approximately 25% or 1.94×10^6 km^2 of the continent, are also common on many of the offshore islands and extend into southern Papua New Guinea. The majority are important grazing lands (*sensu* Moore & Perry, 1970) as there are extensive grass or graminoid communities associated with them, and many thousands of square kilometres have been cleared since European settlement to make way for pasture and cultivation. Much of their present distribution reflects patterns of land use and most of the semi-intact woodlands are today under considerable grazing pressure.

Definition of a woodland

In this chapter a 'woodland' is regarded as a structural plant formation usually with a graminoid component, dominated by perennial woody plants over 2 m tall which do not have their crowns touching. In broad terms, 'woodland' is included within the open-forest and open-scrub formation classes of Fosberg (1970) and the IUCN/UNESCO units of woodlands and wooded savannas (IUCN, 1973).

In Australia, related vegetation types are included in Specht (1970) and Beard & Webb's (1974) classifications of 0.1–70% projected foliage cover classes of woody vegetation over 2 m tall (see also Williams, 1955; Gillison, 1970, 1976; Moore & Perry, 1970; Walker & Gillison, 1981).

Since most woodlands in Australia are at best semi-intact, they may bear little resemblance to their pre-European structure. We propose to describe woodlands as they exist today excluding only those subjected to recent gross disturbance. We therefore include areas grazed and burned for pastoral reasons, but exclude recently cleared or ring-barked areas.

Because of their structural similarity in the upper strata, the terms 'savanna' and 'woodland' are often interchangeable. This aspect is discussed more fully in a treatment of Australian savannas by Walker & Gillison (1981) where the woodland types described here are referred to as savanna woodland types, 'savanna' referring to the essentially graminoid (grasslike) component.

The widespread and conflicting use of the terms 'tree' and 'shrub' has confused the classification of Australian vegetation types: the commonly used definition of a shrub as a low multistemmed woody plant 'of smaller structure than a tree' (Carpenter, 1938), has little application in many parts of Australia where multistemmed (sympodial) woody plants 20 m tall (cf. an upper limit of 8 m in Specht, Roe & Boughton, 1974) and small single-stemmed (monopodial) woody plants 3 m tall are common. Further, some species exhibit both forms depending on the environment, e.g. *Eucalyptus microcarpa* and *Acacia aneura*.

For these reasons we emphasise that it is primarily the height and foliage cover of woody plants which characterise woodlands and we propose that these may be further described according to the occurrence of monopodial or sympodial forms. The terms are broadly analogous to those used in botanical nomenclature to describe inflorescence branching.

A descriptive framework for Australian woodlands

The main objective in establishing a framework for the description of Australian woodlands is to facilitate:

1. the recognition of various static and dynamic woodland types on a continental basis,
2. the correlation of their distributional patterns with major bioclimatic or environmental variables,
3. a systematic basis for the comparison of Australian woodlands with similar vegetation types in other countries.

Regardless of how they have been defined in the past, woodlands exhibit considerable variation along numerous environmental gradients. This inherent variability is further compounded by a variety of current and past land use practices. The variable nature of woodland vegetation has been partially responsible for a traditional classification with strong floristic connotations based on the dominant growth form or bark type for example: box, ironbark, bloodwood, paperbark, mallee etc. (Williams, 1955; Moore & Perry, 1970). The use of a classification system based on structural attributes (i.e. Beard & Webb, 1974; Specht *et al.*, 1974; Carnahan, 1976) has merely translated the traditional physiognomic woodland classification into a different format often with misleading interpretations, e.g. 'poplar box woodland' transgresses six of the structural categories described by Specht (1970) (see Figure 8.1).

In developing a descriptive framework for this volume our dilemma is to maintain compatibility with other chapters and at the same time avoid misleading the reader with a prescribed order which in our opinion does not exist.

Figure 8.1. A structuregram uses height and foliage projective cover to indicate the variation in woodland structure. The class boundaries shown are those of Specht *et al.* (1974), the values we prefer are shown for comparison.

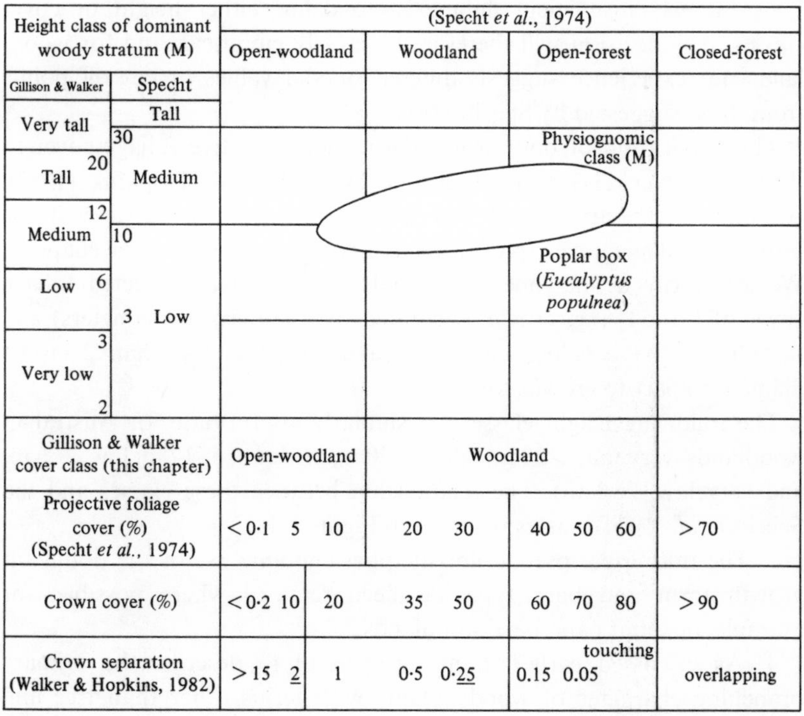

Height class of dominant woody stratum (M)	(Specht *et al.*, 1974)			
	Open-woodland	Woodland	Open-forest	Closed-forest
Gillison & Walker / Specht: Very tall / Tall — 30; Tall 20 / Medium; 12; Medium 10; Low 6; 3 Low; 3; Very low; 2			Physiognomic class (M) — Poplar box (*Eucalyptus populnea*)	
Gillison & Walker cover class (this chapter)	Open-woodland	Woodland		
Projective foliage cover (%) (Specht *et al.*, 1974)	<0·1 5 10	20 30	40 50 60	>70
Crown cover (%)	<0·2 10 20	35 50	60 70 80	>90
Crown separation (Walker & Hopkins, 1982)	>15 2 1	0·5 0·25	touching 0.15 0.05	overlapping

Woodland types

We employ four conceptual approaches to describe our woodland types.

1. The term woodland has already been defined here on a structural basis. However, our use of the height/cover values differs significantly from the discrete classes of Specht *et al.* (1974). We prefer to retain the continuum concept and for this reason we will describe the structure of each woodland type using an ordinative method developed by Walker & Gillison (1981) for classifying Australian savannas. Values for height and cover for sites within a woodland type are plotted as continuous variables on a height × cover table, and a line drawn around the range of values obtained (this is termed a structuregram). The resulting cloud allows the reader to see at a glance the full known structural range of a recognised woodland type. In the example given in Figure 8.1 the widespread *E. populnea* woodland type has cover values which vary from 6 to 60% and height values from 8 to 20 m. By this means we retain the use of an otherwise artificial tabular system as a framework for the interpretation and display of real data.

2. For descriptive purposes it is necessary to subdivide height and cover continua into categories which reflect tallness and openness. Unfortunately insufficient data exist to define either height or cover classes. Any decision will therefore be largely empirical, and for woodlands our experience suggests different modal values or cut off points from those suggested by Specht (1981).

The application of both crown cover and projective foliage cover to delineate cover classes is discussed by Walker & Hopkins (1982). Although we prefer classes based on crown separation, we will use projective foliage cover (pfc) to remain compatible with other chapters. We arbitrarily define 'open-woodlands' as having a projected foliage cover of 0.1–5% (i.e. crowns separated by 2–30 crown diameters) and 'woodlands' have values between 5 and 60% (i.e. less than 2 crown diameters apart to crowns touching).

The following height classes are similarly appropriate for Australian woodlands: very tall, > 20 m; tall, 12–20 m; medium, 6–12 m; low, 3–6 m; and very low, 2–3 m). The relationship between these classes and the Specht *et al.* (1974) classes is shown in Figure 8.1.

3. The traditional pseudo-floristic/physiognomic use of the dominant growth form and bark type has been retained where possible, for example, box and paperbark woodlands.

4. As discussed earlier, it appears useful to describe the primary branching character of woody plants with terms other than tree and

shrub; we suggest monopodial and sympodial. For example, we distinguish between monopodial box-barked (*E. microcarpa*) woodlands of central eastern Australia and the sympodial (mallee) form of the same box-species in harsher environments to the south and west. For mulga (*A. aneura*), there is a rapid monopodial to sympodial change on a steep mesic to arid continental gradient.

To simplify the classification of Australian woodlands we have ignored a primary division based on seasonal leaf fall (cf. Fosberg, 1970) as the majority of woodlands are evergreen. The only partly or wholly deciduous woodlands that occur in northern Australia are dealt with as an 'anomalous' type.

The attributes which combine to give a classification of woodland type are shown in Table 8.1.

Woodland distribution and bioclimate

To broadly describe the distribution of Australian woodlands we use the observation that a marked correlation exists at a continental scale between the structural characteristics of woodland vegetation and bioclimate as defined by indices which reflect plant growth responses to varying combinations of water, light and temperature. We consider edaphic, geomorphological and other specific factors known to be

Table 8.1. *Classification of Australian woodland types*

Woodland type	Structural class	Cover class	(i) woodland 10–60% pfc
			(ii) open woodland 0.1–10% pfc
		Height class	(i) very tall > 20 m
			(ii) tall 12–20 m
			(iii) medium 6–12 m
			(iv) low 3–6 m
			(v) very low 2–3 m
	Physiognomic class	Stem class	(i) monopodial (M)
			(ii) sympodial (S)
		Other (bark, leaves etc.)	

Example:
Low (M) paperbark woodland
(*Melaleuca viridiflora*/*M. nervosa*)

correlated with particular woodland types as subsidiary to these primary bioclimatic factors.

In Walker & Gillison (1981), a set of bioclimatic provinces for the Australian region were recognised on the basis of plant growth indices together with a 'seasonality' coefficient which expressed variability of annual available water (CV%). The general bioclimatic models have been described by Nix (1976) and the specific indices used were calculated by H.A. Nix using several unpublished routines. Further detail is available in Walker & Gillison (1981).

The distribution of the bioclimatic provinces is shown in Figure 8.2 and the derivation of the coding in Table 8.2. Each province is given a code according to its region, sub-region and seasonality type, e.g. province 3 is coded as AIIIS which indicates it belongs to the megatherm region (A), in sub-region III and is seasonal (S).

The plant growth index value chosen as a bioclimatic boundary was

Figure 8.2. Distribution of bioclimatic provinces in Australia (see Table 2 for details).

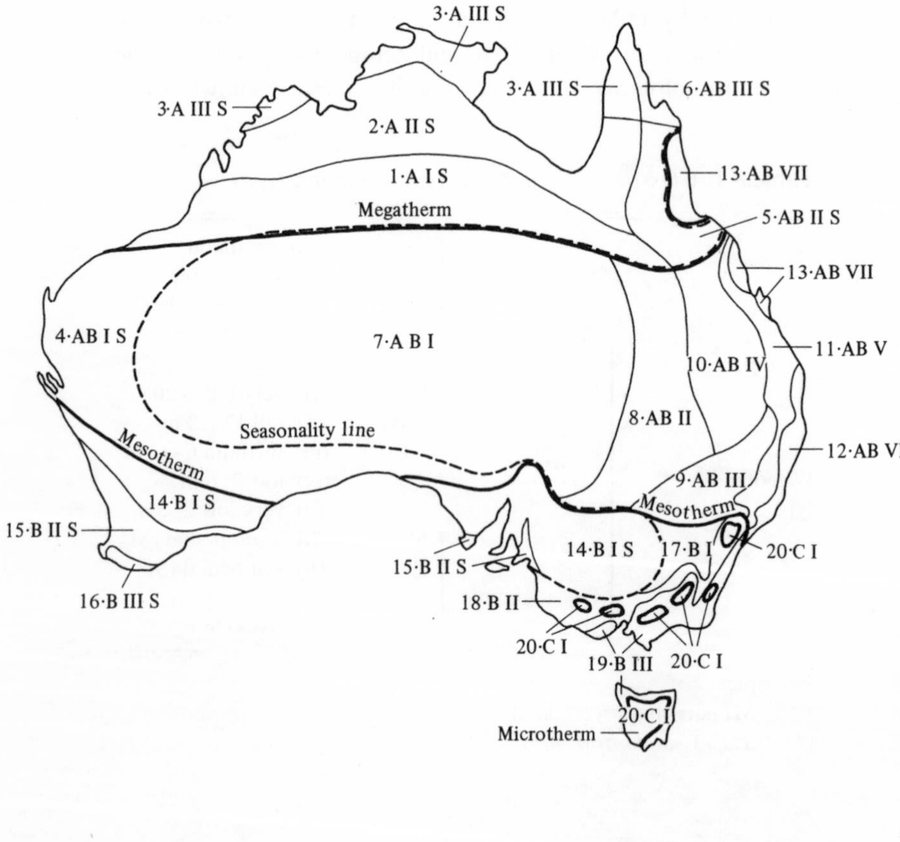

Table 8.2. *Bioclimatic provinces for Australia. See Walker &*
Gillison (1981) for further details

Province No.	Code	Bioclimatic region	Bioclimatic sub-region	Thermal optima°C
1	AIS	Megatherm	I	28
2	AIIS		II	28
3	AIIS		III	28
4	ABIS	Megatherm–	I	28/19
5	ABIIS	Mesotherm	II	28/19
6	ABIIS		III	18/19
7	ABI	Megatherm–	I	28/19
8	ABII	Mesotherm	II	28/19
10	ABIII		III	28/19
11	ABV		V	28/19
12	ABVI		VI	28/19
13	ABVII		VII	28/19
14	BIS	Mesotherm	I	19
15	BIIS		II	9
16	BIIIS		III	19
17	BI	Mesotherm	I	19
18	BII		II	19
19	BIII		III	19
20	CI	Microtherm	I	12

selected in some cases on the basis of a coincidence with known major
vegetation boundaries, e.g. desert–open woodland boundary in northern
Australia. To avoid circularity we emphasise that the bioclimatic provinces are for descriptive purposes only to indicate the important plant
growth conditions experienced by major woodland types.

Broad floristics of Australian woodlands
Woodlands range in species richness of perennial flowering plants from as
low as three species ha^{-1} in some low open-woodlands in the interior to
80 species ha^{-1} in sub-coastal woodlands (Gillison, unpublished data).
Their floristic composition is unique in that the majority of dominant
woody plants are composed of sclerophyllous taxa from within the
Myrtaceae (*Eucalyptus*) and Fabaceae (*Acacia*). Of the 530 eucalypt taxa
currently described, approximately 80% occur in woodlands. Most of
these eucalypts comprise the relatively small 'sub-genera' *Blakella*,
Corymbia, Eudesmia, Idiogenes and the largest 'sub-genus' *Symphyo-*
myrtus (after Pryor & Johnson, 1971). The 'sub-genus' *Monocal-*

yptus is more or less restricted to the more dense forest formations of the coastal and upland regions of the east, although some species occur in highland woodland formations (e.g. *E. niphophila, E. pauciflora, E. stellulata*).

Like *Eucalyptus*, most *Acacia* spp. are included in woodland formations. *Acacia* tends to replace *Eucalyptus* in northern areas where annual rainfall is <600 mm, except where there is extra run-on water (see Pryor, 1959 quoted in Pedley, 1978).

Apart from the floristically and structurally dominant *Acacia* and *Eucalyptus*, the following subsidiary woody perennial taxa characterise woodlands: Anacardiaceae (*Buchanania*), Bombacaceae (*Adansonia, Bombax*), Casuarinaceae (*Casuarina*), Cochlospermaceae (*Cochlospermum*), Combretaceae (*Terminalia*), Cycadaceae (*Bowenia, Cycas, Macrozamia*), Euphorbiaceae (*Excoecaria, Petalostigma*), Fabaceae (*Albizia, Bauhinia, Cassia, Erythrophleum*), Malvaceae (*Abelmoschus, Hibiscus*), Myrtaceae (*Melaleuca*), Palmae (*Livistona*), Proteaceae (*Banksia, Grevillea, Hakea, Persoonia*), Rhamnaceae (*Alphitonia, Ventilago*), Rubiaceae (*Gardenia*), Rutaceae (*Atalaya, Eremocitrus, Eremophila, Flindersia, Geijera*), Sapindaceae (*Dodonaea, Heterodendrum*), Sapotaceae (*Planchonia*), Sterculiaceae (*Brachychiton*), and Xanthorrhoeaceae (*Xanthorrhoea*). Many of these genera intergrade into woodland systems of Papua New Guinea and it is almost certain that many would have occupied the formerly exposed Sahul shelf, a former Pleistocene land bridge linking northern Australia and the present island of Papua New Guinea (Chapter 2, Walker & Singh, 1981). Other than this, the continent with greatest floristic affinity in woodland taxa is Africa, where convergent structural development is evident, particularly in the Fabaceae (*Acacia*) and in the Combretaceae and Sterculiaceae. It is of interest that in the Bombacaceae, the baobab *Adansonia*, a characteristically African genus, occurs in the far northwest of Australia and may indicate an extant Gondwanic link.

Adaptive traits in Australian woodlands
Although floristic assemblages in Australian woodlands tend to be discrete, certain elements (e.g. *Alphitonia, Dolichandrone, Flindersia, Livistona, Ventilago*) indicate possible links with a closed seasonal woody vegetation formerly more extensive and rather like the present-day 'monsoon' forest in northern Australia which has presumably fluctuated with the arid cycles (the most recent of which took place probably between 10 000 and 15 000 years ago). It is likely that the climatic 'norm'

for the Quaternary was considerably wetter than at present so that woodlands now exist in a relatively arid phase in a condition that has been further modified by man in more recent times through changes in fire regimes and the introduction of exotic grazing and browsing animals. Present woodlands therefore reflect a confounded aspect of these elements, both past and present, but there are some overall aspects that strongly suggest adaptation to environmental pressures, of which fire and water stress are paramount.

Many of the life forms and leaf features indicate a xerophilous trend. Sclerophylly is well developed (i.e. thickened cuticles, increased glaucousness, rolled margins, dense indumentum, high stomate density, high specific gravity and volatile oils, and so on). Woodland eucalypts are commonly pendulous, isobilateral sclerophylls with a high oil content, and there is a unique development of phyllodineous acacias. (Convergence to phyllode-like leaves is also evident in the geographically widespread *Melaleuca*). These features, together with solid leaf types, such as *Hakea*, *Callitris*, and the phyllocladous *Casuarina* characterise many Australian woodland plants. Reduction in leaf size is evident with increased water stress and in the more seasonal regions there is a pronounced development in deciduousness of which some of the woodlands dominated by *Bauhinia* and *Terminalia* in northern Australia are prime examples. Increases in the amount of volatile oils (Rutaceae, Myrtaceae) and laticiferous tissue (Moraceae, Euphorbiaceae) are also linked with increasing water stress.

Resistance to a variety of fire regimes is well developed, especially in *Eucalyptus*, where below-ground lignotuberous and sometimes rhizomatous organs or root suckers allow for rapid vegetative regeneration after fire (Jacobs, 1955; Carrodus & Blake, 1970; Lacey, 1974). In some eucalypt species, above-ground epicormic shoots and specific bark types also assist in recovery from fire. Many other woodland genera have below- and above-ground adaptations which ensure adequate fire survival (Gill, 1980).

The acacias as a whole appear better equipped to withstand extremes of water stress rather than fire, unlike the eucalypts. Mulga (*Acacia aneura*) is capable of withstanding a pressure of −120 bars (Slatyer, 1962) and brigalow (*A. harpophylla*) can withstand −50 bars (Connor & Tunstall, 1968). Australian *Acacias* are distinguished from their African counterparts by the very low incidence of thorns, which may indicate an historical lack of association with an African-type browsing 'savanna' fauna.

Description of selected woodland types
In this volume, a number of vegetation types described by other authors
fall within our definition of woodland (e.g. *Acacia* woodlands in Chapter
9, Johnson & Burrows, 1981; mallee in Chapter 10, Parsons, 1981; sub-
alpine woodland in Chapter 16, Costin, 1981; and open-forests in
Chapter 7, Gill, 1981). We have avoided duplication in selecting some of
the remaining common woodland types.

Very tall woodland
Physiognomic class: (M) woollybutt–stringy bark (*E. miniata, E.
tetrodonta*)
Provinces 2, 3. As indicated in the accompanying diagram (Figure
8.3), the type structure is variable and changes from short regenerative
'pole' woodland with monopodial stems (Figure 8.4) to tall woodland
that becomes increasingly dense as one moves into Province No. 3.

Figure 8.3. Structuregrams of woodland types described in this chapter. *a.* Very
tall woodland, (M) woolly butt – stringy bark (*Eucalyptus miniata, E. tetrodonta*). *b.* Tall woodland, (M) grey box (*E. moluccana*). *c.* Medium height
woodland, (M) poplar box (*E. populnea*). *d.* Low woodland, (M) paperbark
(*Melaleuca viridiflora, M. nervosa*). *e.* Very low woodland, (M) silverleaf box and
snappy gum (*E. pruinosa, E. brevifolia*). *f.* Low (deciduous) woodland, (M)
deciduous *Bauhinia, Terminalia.*

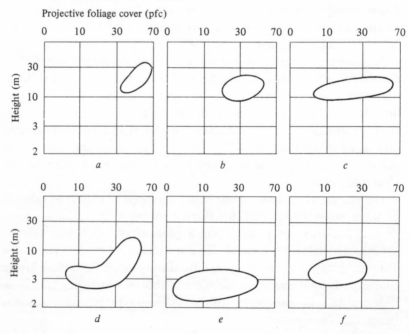

Understorey tall shrubs (*Acacia* spp., *Gardenia megasperma*, *Grevillea heliosperma*, *Grevillea pteridiifolia*) have a crown cover of 20%, and graminoids occur often with a proliferation of small shrubs, *Acacia* spp., *Bossiaea bossiaeoides* and *Petalostigma* spp. This association is usually found on deeply-weathered red sandy soils where presumably it is able to tap water at depth. Leaf size varies from microphyll to notophyll (*sensu* Webb, 1959) for *E. miniata* and *E. tetrodonta*. In this relatively low nutrient environment with annual ground fires a conspicuous divergence of life forms is evident in both eucalypts. In *E. miniata*, the one species provides two life forms; a tall seed-producing tree out of fire reach, and a low rosette propagated by vegetative shoots at or below ground-level in response to fire. *Eucalyptus tetrodonta* is capable of producing extensive root suckers and to date is the only documented case within *Eucalyptus*. This facultative characteristic is evident in many other plants in this bioclimatic Province (see also Lacey, 1974) and may represent an evolutionary response to a relatively high fire frequency and low nutrient status. The remarkable capacity of plants in this area to produce underground storage and reproductive tissue is similar in some ways to

Figure 8.4. Regenerating very tall woodland of woollybutt (*Eucalyptus miniata*) in northern Australia. Note monopodial form. Photo: A.N. Gillison.

the geoxylic suffrutices described by White (1977) for parts of central Africa.

On Melville Island the understoreys consist of tall shrubs of *Gardenia*, *Grevillea* and *Owenia* which are replaced in mesic sites by *Acacia*, *Alphitonia*, *Cycas* and *Phyllanthus*. In the west of Province 3 the palms *Cycas media* and *Livistona humilis* are common understorey elements.

Tall woodland
Physiognomic class: (M) grey box (*E. moluccana*)

Provinces 11, 12, (5). The grey box woodlands form a closely-knit taxonomic assemblage comprised of a number of species but mostly of *E. microcarpa* and *E. moluccana*. The former tends to occupy the semi-arid regions of eastern Australia and the latter occurs in the wetter zones to the east. As indicated in the structuregram (Figure 8.3), *E. moluccana* extends into tall open-forest and sometimes closed-forest categories but is commonly tall woodland. The best form of this woodland commonly occurs as pure stands 17 m tall with 50% foliage cover in Province 11, as ssp. *queenslandica*. In the more seasonal Province 5 a shorter, more open woodland of ssp. *crassifolia* is common (14 m tall with 45% foliage cover).

Grey box woodlands become more dense to the south (Province 12) with ssp. *pedicellata* and ssp. *moluccana*. Grey box is commonly contiguous with ironbark communities that tend to be upslope (*E. crebra*, *E. drepanophylla*, *E. sideroxylon*) or with blue gums (*E. tereticornis*) downslope on flats. In Province 12 it is commonly associated with spotted gum (*E. maculata*). In the northern Provinces 5, 11 and 13 soils tend to be agronomically poor solodics, whereas in the southern Province 12 grey box woodlands are common on brown earths of high agronomic value and this has led to the elimination of many fine stands in favour of cultivated crops and pastures.

Understorey components and strata are both variable. In the north (Provinces 5, 11, 13) mid to tall grasses are common (*Dichanthium* spp., *Heteropogon contortus*, *Imperata cylindrica* and *Themeda australis*) and shrubs of *Acacia* spp. are plentiful. In Province 12 the woodlands have an understorey shrub layer of *Acacia*, *Melaleuca* and *Pultenaea* spp., with graminoids *Dianella*, *Lomandra*, and short tussock grasses *Poa* and *Imperata*.

Associated with its broad geographic range is a series of genetically-fixed growth forms that correspond closely to climate changes. Gillison (1976) found that when grown in a controlled environment, seedlings from northern Provinces were monopodial, and large-leaved, with small lignotubers, whereas seedlings from the southern Province 12 were shorter, hemi-sympodial, with large lignotubers and relatively small leaves. More recently, Gillison, Lacey & Bennett (1980) found at a remote southern locality at Bungonia, southwest of Sydney, on low-nutrient soils, that *E. moluccana* ssp. *moluccana* is capable of reproducing vegetatively by means of rhizo-stolons. This feature, combined with the overall capacity of grey box to coppice following stem damage is indicative of the range of survival modes in the genus *Eucalyptus*.

Other tall woodlands in eastern Australia include the ironbarks, *E. crebra*, *E. drepanophylla*, *E. melanophloia* and *E. sideroxylon* and the bloodwoods *E. intermedia* and *E. polycarpa*.

In far northern Australia in Provinces 2 and 3, the woollybutt–stringy bark associations also occur as tall woodlands.

Woodlands of medium height
Physiognomic class: (M) poplar box (*E. populnea*)

Provinces 8, 9, 10, 11 (17). In Australia, woodlands of medium height are the most widespread. In the example given, the geographic range is over 10° of latitude and, as with grey box, the poplar box woodlands appear to be similarly differentiated in growth form relative to climate, ranging from monopodial forms in the north (Province 13) to hemi-sympodial forms in the south (Province 9).

Poplar box is typical in form of many other eucalypt box-barked species (so-called because of the bark resemblance to turkey box, *Buxus sempervirens*) that form woodlands of medium height, in particular, *E. microcarpa*, *E. microneura*, *E. normantonensis* and *E. tectifica*. The structure of poplar box communities varies with soil type, the red and brown solodics being the most important. The community usually has a simple, two-layered, tree–grass structure, on heavy clays in Province 10, often associated with *E. orgadophila*, whereas on red earths, extensive shrub associates are common, particularly where grazing has been intensive (the semi-arid shrub woodlands of Moore & Perry, 1970).

Non-eucalypt low trees associated with poplar box on red earths are *Acacia excelsa*, *A. oswaldii*, *Albizia basaltica*, *Bauhinia carronii*, *Flindersia dissosperma*, *Geijera parviflora*, *Heterodendrum oleifolium* and *Ventilago viminalis*. On sandy soils a number of other *Acacia* spp. exist, often with *Callitris columellaris*, and on heavier clays, isolated pockets of *Casuarina cristata* and *C. luehmannii* are common. The eucalypts *E. cambageana*, *E. orgadophila* and *E. thozetiana* are also common on these heavy soil types together with brigalow (*A. harpophylla*).

The grasses *Aristida contorta* and *A. jerichoensis* are common throughout poplar box woodlands, whilst in the northern areas *A. inaequiglumis* and *A. pruinosa* are also very common. Other grass species in the central and southern Provinces are *Bothriochloa decipiens*, *Chloris acicularis* and *C. truncata*, with *Neurachne mitchelliana* and *Stipa variabilis*.

Other tree species that make up medium-height woodlands are too numerous to list here. In northern Australia, the bloodwoods *Eucalyptus dichromophloia*, *E. nesophila*, *E. polycarpa* and *E. porrecta* are typical of this woodland together with the ironbarks *E. crebra*, *E. drepanophylla*, *E. jensenii*, *E. melanophloia* and *E. shirleyi*. The gumbarked eucalypts *E. alba*, *E. brevifolia*, *E. confertiflora* and *E. papuana* are also conspicuous elements. In the central eastern and southern provinces west of the Great Dividing Range, box and ironbark woodlands are common. Apart from poplar box, *E. microcarpa*, *E. odorata* and *E. pilligaensis* are notable dominants in southern box woodlands, although in Province 14 these tend to give way to mallee formations. In Western Australia *E. ptychocarpa* and *E. salmonophloia* are typical of a number of eucalypt species that form medium to low woodlands on sandy soils.

Low woodland
Physiognomic class: (M) paperbark (*Melaleuca viridiflora* and *M. nervosa* complex (Figure 8.5)

Provinces 2, 5, (11). Low woodlands are most common in the semi-arid, highly seasonal bioclimatic Provinces in northern Australia, but also occur along the eastern fringe of Queensland, especially in Province 5, in swampy (low-nutrient) localities. A variety of eucalypt species may occur in low woodlands (*E. brevifolia*, *E. dichromophloia*, *E. jensenii*, *E. papuana*, *E. pruinosa*) but one of the most unique and physiognomically significant is paperbark woodland. This is characteristically a woodland type dominated by monopodial species of *Melaleuca*, of which the two most common are *M. nervosa* and *M. viridiflora*. The unique feature of this woodland is the laminated paper-thin sections of bark and often mesophyll-sized coriaceous phyllode-like leaves. The most widespread paperbark woodlands are in the floodplain areas of the Gulf of Carpentaria, although they occur throughout the megatherm seasonal region of both Australia and Papua New Guinea. Paperbarks, or 'tea-trees' as they are sometimes known, may be dwarfed or twisted, usually about 3–7 m tall but may assume 'pole' forest proportions in backing swamps, often up to 18–20 m tall.

The *M. viridiflora* community is usually two-layered with a short to mid-height (10–50 cm) ground layer of graminoids. Although both *M. viridiflora* and *M. nervosa* occur mainly on solodic plains, they may also occur on quartzite ridges or black soil flats, associated variously with the bloodwoods *E. dichromophloia* and *E. polycarpa* and particularly deciduous species of *Terminalia* (*T. aridicola*, *T. oblongata*, *T. platyphylla*) and the deciduous *Bauhinia cunninghamii*. *Melaleuca tamarascina* sometimes occurs on solodic flats either as a co-dominant with *M. viridiflora* or else as a pure stand. Other species of *Melaleuca* (*M. argentea*, *M. cajuputi* and *M. leucodendron sens. lat.*) are scattered throughout the region usually in mixed stands with other woody genera (*Terminalia*, *Petalostigma*, *Alphitonia* and eucalypts).

The low-nutrient site conditions that support paperbark communities usually carry only very mild fires as a consequence of a low fuel load. Because of their unique bark, the melaleucas are also fire-tolerant. Together with the capacity of paperbarks to establish profusely from seed, these factors combine to make the communities highly effective colonisers and competitors in woody plant communities. This capability has created an endemic 'woody weed' problem in parts of northern

Figure 8.5. Low woodland of the paperbark *Melaleuca nervosa* in Province 2, northern Australia. Photo: A.N. Gillison.

Australia and an exotic 'woody weed' problem in Florida in the United States of America.

Low woodlands also occur in the sub-alpine regions of Australia where the relatively harsh environment is associated with a reduction in tree height. Such woodlands consist commonly of the snow-gums *E. niphophila* and *E. pauciflora* described in Chapter 16 (Costin, 1981).

Very low woodland

Physiognomic class: (M) silver-leaf box and snappy gum (*E. pruinosa–E. brevifolia*)

Provinces 2, 3. These woodlands are characterised by dwarf, essentially monopodial trees 2–4 m tall that vary locally from 2 to 60% cover. Crowns are usually dense and rounded and the communities are typically two-layered tree–graminoid associations. Such woodlands occur in harsh environments and exist mainly in Province 2, which is characterised by periods of extreme water stress.

In such an environment that usually includes low-nutrient, shallow soils, dwarf woodland trees tend to be evergreen and relatively fire- and drought-resistant. Trees that occupy this type commonly range into low woodland. In Province 2, dwarf woodland eucalypts are of two leaf types; either small (microphyll) and shiny (*E. brevifolia* and *E. jensenii*) or large (approaching mesophyll size) and glaucous as in *E. pruinosa*.

The distribution of snappy gum (*E. brevifolia*) dwarf woodland ranges from the Ord River in northwestern Australia, through to the Leichhardt River that flows into the Gulf of Carpentaria and commonly occurs on sandstone or quartzite ridges. Silver-leaf box (*E. pruinosa*) tends to favour heavier soils under these extreme seasonal conditions and occupies much the same geographic range.

The ironbark *E. jensenii* is restricted more to northwestern Australia on shallow stony slopes and ridges (Figure 8.6). Grass species commonly associated with this woodland type are the hummock 'spinifex' grasses (*Triodia*, *Plectrachne*) and low grasses *Eragrostis* and *Chrysopogon*. These woodlands represent the poorer pasture associations in northern Australia.

Low (deciduous) woodland

Physiognomic class: (M) deciduous *Bauhinia*, *Terminalia*

Provinces 1, 2, 3. Throughout the megatherm seasonal region, run-on basins with seasonally-inundated cracking clays are characterised by a unique vegetation type. It is a deciduous variant of the low evergreen

Figure 8.6. Very low woodland of ironbark (*Eucalyptus jensenii*) in the Kimberley region of northern Australia. Photo: A.N. Gillison.

Figure. 8.7. Low deciduous *Bauhinia*, *Terminalia* woodland in the Gulf country of northern Australia. Photo: N.H. Speck.

eucalypt woodlands. Here the non-eucalypt genera *Bauhinia* (Papilionaceae) and *Terminalia* (Combretaceae) predominate (Figure 8.7).

Stands vary floristically from mixtures to pure stands of each genus, sometimes associated with semi-deciduous eucalypts (*E. alba*, *E. foelscheana* and *E. bleeseri*). During the post-wet season, these localities support a simple two-layered community of trees and mid-height grasses with the tree crowns dense and green. Under conditions of increasing water-stress towards the latter part of the year, the grasses are usually fired and the trees become either totally or partially deciduous.

A peculiar phenomenon associated with these deciduous woodlands is the growth flush and precocious flowering that takes place about two months before the onset of the monsoonal rains. This feature has not yet been fully explained, but is presumably associated with the highly predictable rainy season where a plant in an otherwise deciduous community can be assumed to have a competitive advantage if it is in full or partial flush before the soil-water store is replenished.

Bauhinia cunninghamii, *Terminalia aridicola*, *T. ferdinandiana*, *T. oblongata* and *T. platyphylla* are among the most common species. Other non-eucalypts, particularly in sub-coastal areas are *Dolichandrone*, *Erythrophleum*, *Ventilago* and *Excoecaria*. Associated grass genera are *Aristida*, *Astrebla*, *Chrysopogon*, *Eragrostis*, *Sorghum* and *Themeda*.

The type is not restricted to northern Australia and is widespread in the megatherm region of Papua where genera of Malesian affinity (*Albizia*, *Antidesma*, *Bombax*, *Cordia*, *Desmodium* and *Timonius*) co-exist with eucalypts and terminalias. Detailed descriptions for deciduous woodlands in Australia are generally lacking, although there are limited descriptions in Specht (1958), Perry & Lazarides (1964), Speck (1965), and Story (1970).

Conclusions

In this descriptive and limited treatment of Australian woodlands we have emphasised the variable nature and unique physiognomic character of Australian woodlands. For reasons of brevity, important dynamic aspects of woodlands have not been discussed. The effects of disturbance on woodland systems have received considerable attention in the literature particularly in relation to grazing by domestic animals. Demographic studies have been carried out but detailed results are as yet unpublished.

Herbage and shrub responses to woodland thinning have been documented by Walker, Moore & Robertson (1972), and Beale (1973), and responses to fire have been reviewed by Leigh & Noble (1980), Stocker & Mott (1980) and Lacey, Walker & Noble (1981). An important aspect of woodland dynamics is the control of unwanted shrubs for grazing purposes (Moore & Walker, 1972). Woody plant responses and problems associated with overgrazing in the poplar box (*E. populnea*) woodlands are outlined in Tunstall *et al.* (1981).

Our understanding of woodland dynamics is far from complete and more work in this area should be carried out in the future. We hope that the descriptive framework outlined will provide a basis for such research.

References

Beale, I.F. (1973). Tree density effects on yields of herbage and tree components in south west Queensland mulga (*Acacia aneura* F. Muell.) scrub. *Tropical Grasslands*, 7, 135–42.

Beard, J.S. & Webb, M. (1974). *Vegetation Survey of Western Australia: Great Sandy Desert—Part 1. Aims, Objectives and Methods*. Perth: University of Western Australia Press.

Carnahan, J.A. (1976). Natural vegetation. In *Atlas of Australian Resources*, Second Series. Canberra: Department of National Resources.

Carpenter, J. (1938). *An Ecological Glossary*. New York: Hafner (1962 reprint).

Carrodus, B.B. & Blake, T.J. (1970). Studies on the lignotubers of *Eucalyptus obliqua* L'Herit. I. The nature of the lignotuber. *New Phytologist*, 69, 1069–72.

Connor, D.J. & Tunstall, B.R. (1968). Tissue water relations for brigalow and mulga. *Australian Journal of Botany*, 16, 487–90.

Costin, A.B. (1981). Alpine and sub-alpine vegetation. In *Australian Vegetation*, ed. R.H. Groves, pp. 361–376. Cambridge University Press.

Fosberg, F.R. (1970). A classification of vegetation for general purposes. In *Guide to the Check Sheet for IBP Areas* (IBP Handbook No. 4), ed. G.F. Peterken, Appendix 1, pp. 73–120. Oxford & Edinburgh: Blackwell Scientific Publications.

Gill, A.M. (1980). Fire adaptive traits of vascular plants. In *Fire regimes and ecosystem properties*, ed. H.A. Mooney, J.M. Bonnicksen, N.L. Christensen, J.E. Lotan & W.A. Reiners, (in press). Washington, DC: USDA Forest Science General Technical Report.

Gill, A.M. (1981). Patterns and processes in open-forests of *Eucalyptus* in southern Australia. In *Australian Vegetation*, ed. R.H. Groves, pp. 152–176. Cambridge University Press.

Gillison, A.N. (1970). Dynamics of biologically-induced grassland-forest transitions in Papua New Guinea. M.Sc. thesis, Australian National University.

Gillison, A.N. (1976). Taxonomy and autecology of the Grey Box-*Eucalyptus moluccana* Roxb. (*s. lat.*). Ph.D. thesis, Australian National University.

Gillison, A.N., Lacey, C.J. & Bennett, R.H. (1980). Rhizo-stolons in *Eucalyptus*. *Australian Journal of Botany*, 28, 229–304.

IUCN (1973). *A Working System for Classification of World Vegetation*. IUCN Occasional Paper No. 6. Morges: International Union for Conservation of Nature and Natural Resources.

Jacobs, M.R. (1955). *Growth Habits of the Eucalypts*. Canberra: Government Printer.

Johnson, R.W. & Burrows, W.H. (1981). *Acacia* open-forests, woodlands and shrublands. In *Australian Vegetation*, ed. R.H. Groves, pp. 198–226. Cambridge University Press.

Lacey, C.J. (1974). Rhizomes in tropical eucalypts and their role in recovery from fire damage. *Australian Journal of Botany*, *22*, 29–38.

Lacey, C.J., Walker, J. & Noble, I.R. (1981). Fire in Australian savannas. In *Proceedings of SCOPE Workshop on Dynamic Changes in Savanna Ecosystems, Pretoria, S. Africa 1979*, ed. B.J. Huntley & B.H. Walker, (in press). Amsterdam: Elsevier Scientific Publishing Company.

Leigh, J.H. & Noble, J.C. (1980). The role of fire in the management of rangelands in Australia. In *Fire and the Australian Biota*, ed. A.M. Gill, R.H. Groves & I.R. Noble, pp. 471–495. Canberra: Australian Academy of Science.

Moore, R.M. & Walker, J. (1972). *Eucalyptus populnea* shrub woodlands. Control of regenerating trees and shrubs. *Australian Journal of Experimental Agriculture & Animal Husbandry*, *12*, 437–40.

Moore, R.M. & Perry, R.A. (1970). Vegetation. In *Australian Grasslands*, ed. R.M. Moore, pp. 59–73. Canberra: Australian National University Press.

Nix, H.A. (1976). Environmental control of breeding, part-breeding dispersal and migration of birds in the Australian region. In *Proceedings of the 16th International Ornithological Congress*, pp. 272–305. Canberra: Australian Academy of Science.

Parsons, R.F. (1981). Eucalyptus scrubs and shrublands. In *Australian Vegetation*, ed. R.H. Groves, pp. 227–252. Cambridge University Press.

Pedley, L. (1978). A revision of *Acacia* Mill. In Queensland. *Austrobaileya*, *1*, 75–337.

Perry, R.A. & Lazarides, M. (1964). Vegetation of the Leichhardt-Gilbert area. CSIRO, Australia, Land Research Series No. 11, pp. 152–91.

Pryor, L.D. & Johnson, L.A.S. (1971). *A Classification of the Eucalypts*. Canberra: Australian National University.

Slatyer, R.O. (1962). Internal water balance of *Acacia aneura* F. Muell. in relation to environmental conditions. *Arid Zone Research*, *16*, 137–46.

Specht, R.L. (1958). The climate, geology, soils and plant ecology of the northern portion of Arnhem Land. In *Records of the American–Australian Scientific Expedition to Arnhem Land*, Vol. 3, *Botany and Plant Ecology*, ed. R.L. Specht & C.P. Mountford, pp. 333–414. Melbourne: University Press.

Specht, R.L. (1970). Vegetation. In *The Australian Environment*, (4th edn rev.), ed. G.W. Leeper, pp. 46–67. Melbourne: CSIRO, Australia & Melbourne University Press.

Specht, R.L. (1981). The use of foliage projective cover. In *Vegetation Classification in the Australian Region*, ed. A.N. Gillison & D.J. Anderson, (in press). Canberra: CSIRO, Australia & Australian National Univeristy Press.

Specht, R.L., Roe, E.M. & Boughton, V.H. (1974). Conservation of major plant communities in Australia and Papua New Guinea. *Australian Journal of Botany, Supplementary Series*, No. 7.

Speck, N.H. (1965). Vegetation and pastures of the Tipperary area. CSIRO, Australia, Land Research Series No. 13, pp. 81–98.

Stocker, G.C. & Mott, J.J. (1980). Fire in the tropical forests and woodlands of northern Australia. In *Fire and the Australian Biota*, ed. A.M. Gill, R.H. Groves & I.R. Noble, pp. 425–439. Canberra: Australian Academy of Science.

Story, R. (1970). Vegetation of the Mitchell-Normanby area. CSIRO, Australia, Land Research Series No. 26, pp. 75–88.

Tunstall, B.R., Torssell, B.W.R., Walker, J., Moore, R.M., Robertson, J.A. & Goodwin, W.F. (1981). Vegetation changes in a poplar box woodland. Effects of tree killing and domestic livestock. *Australian Rangelands Journals*, (in press).

Walker, J. & Gillison, A.N. (1981). Australian savannas. In *Proceedings of SCOPE workshop on Dynamic Changes in Savanna Ecosystems, Pretoria, S. Africa 1979*, ed. B.J. Huntley & B.H. Walker. Amsterdam: Elsevier Scientific Publishing Company, (in press).

Walker, J. & Hopkins, M. (1981). Description and classification of Australian vegetation. In *Australian Soil and Land Survey Handbook*, vol. 1, ed. R.C. McDonald, R.F. Isbell & J.G. Speight, (in press). Standing Committee on Soil Conservation.

Walker, J., Moore, R.M. & Robertson, J.A. (1972). Herbage response ot tree and shrub thinning in *Eucalyptus populnea* shrub woodlands. *Australian Journal of Agricultural Research*, *23*, 405–10.

Walker, D. & Singh, G. (1981). Vegetation history. In *Australian Vegetation*, ed. R.H. Groves, pp. 26–43. Cambridge University Press.

Webb, L.J. (1959). A physiognomic classification of Australian rainforests. *Journal of Ecology*, *47*, 551–70.

White, F. (1977). The underground forests of Africa. In *Tropical Botany*, ed. D.J. Mabberley & Chang Kiaw Lan, pp. 57–71. *The Gardens Bulletin Singapore* Vol. XXIX.

Williams, R.J. (1955). Vegetation regions. In *Atlas of Australia*, 1st Series. Canberra: Department of National Development.

9

Acacia *open-forests, woodlands and shrublands*

R.W. JOHNSON & W.H. BURROWS

Acacia communities, together with the eucalypts, dominate the Australian landscape. Whilst *Eucalyptus* open-forests and woodlands are characteristic of semi-arid and more mesic coastal Australia, *Acacia* shrublands and low woodlands largely replace *Eucalyptus* communities in drier regions of the continent. They become dominant in southern areas when the annual rainfall is winter-incident and less than 250 mm. In northern Australia they predominate in areas receiving less than 350 mm. Within the central and western arid zone, particularly in the northwest and in drier desert areas, *Acacia* often forms only a sparse shrub layer over *Triodia* grassland. In these communities scattered mallee eucalypts may occur, though *Eucalyptus* is more frequently restricted to moister habitats along streams and around depressions. Structurally, the *Acacia* communities of central Australia are either shrublands or open-shrublands, although some of the important species, such as mulga (*A. aneura*), form woodlands on more favourable sites.

Acacia communities, however, also extend into more mesic areas in northeastern Australia, where they usually form single dominant open-forests and woodlands interspersed with areas of grassland and semi-arid eucalypt woodlands. In central Queensland, open-forests of brigalow (*A. harpophylla*) reach almost to the coast where they receive more than 900 mm a year. Much less significant areas of *Acacia* open-forests and woodlands are found in temperate southern Australia.

Even in *Eucalyptus* open-forests and woodlands, species of *Acacia* may be common understorey trees and shrubs, particularly on sandy and texture-contrast soils. A few species, such as *A. fasciculifera*, *A. hylonoma* and *A. bakeri*, occur in rainforests, usually where soils are of low fertility. Tussock grasslands, such as the Mitchell grass (*Astrebla*) and blue grass

(*Dichanthium*) downs of Queensland, become associated with scattered trees and shrubs of a number of *Acacia* species.

Stebbins (1972) pointed out that shrubs do not constitute either a taxonomic or an evolutionary category. The same taxonomic unit may exist as a tree in one place and as a shrub in another. Also it is often difficult to differentiate low trees from tall shrubs (Moore, 1973). These comments are particularly relevant to many of the dominant *Acacia* spp. which exist as tall trees in wetter and stunted shrubs in drier parts of their range.

These dominant *Acacia* spp., like most other *Acacia* spp. in Australia, are phyllodineous and lack spines. This provides the vegetation with a character vastly different from the *Acacia* communities of the Afro-Asian and American continents.

Acacia aneura and *A. harpophylla*, the most widespread species of the shrublands and the open-forests and woodlands respectively, show a number of adaptations which enable them to survive and flourish under low and variable rainfall and high evaporation. Both possess phyllodes which exhibit extreme resistance to desiccation (Connor & Tunstall, 1968). Despite adverse water conditions associated with soils of heavy texture, which are often very saline, *A. harpophylla* is able to grow and form forests of high standing biomass (Moore, Russell & Coaldrake, 1967).

Acacia aneura has terete to slightly flattened phyllodes which are vertically rather than horizontally aligned, thereby aiding water redistribution and minimising heat absorption. The phyllodes have a hairy, scurfy and resinous covering (Francis, 1925) and are sclerophyllous and drought resistant by means of dormancy. The plants aestivate when drought occurs and resume growth within four days after water again becomes available (Slatyer, 1961). They also have the capacity to channel water down their phyllodes and stems so that rainfall is concentrated at the base of the trunk (Slatyer, 1965; Pressland, 1973). By contrast, the amount of rainfall reaching the soil surface as stem flow is usually small in *A. harpophylla* communities, although it is thought that leaf drip is important in water redistribution (Russell, Moore & Coaldrake, 1967).

Spatial patterning is also one of the characteristics of the adaptation of these species to their environment. In central and western Australia and the more arid parts of Queensland and New South Wales, *A. aneura* frequently occurs in groves that receive run-off water from the sparsely vegetated intergrove areas. The groves are usually positioned along the contour in a discontinuous fashion. About 50% of the soils supporting *A.*

harpophylla possess gilgai microrelief. Here, trees are confined mainly to elevated areas so that the depressions, which prevent general run-off, accumulate surface moisture locally yet grow little perennial vegetation.

Acacia communities occupy land mainly used for pastoral production. Semi-arid and arid conditions combined with highly unreliable rainfall impose constraints on productivity and make management difficult. This is further exacerbated by the presence of a standing biomass which is predominantly of little use to man or his animals, though some species such as *A. aneura* are important drought reserves and are lopped to feed starving sheep. The *Acacia* dominants, in addition, greatly limit the abundance and growth of edible grasses and forbs. In higher rainfall areas *A. harpophylla* communities have been cleared for cultivation and sown pasture, whilst thinning of stands of *A. aneura* in the wetter parts of its range improves native pastures. In the most arid areas these lands remain unoccupied except for localised use in abnormally wet seasons.

Descriptive accounts of *Acacia* communities covering much of northern Australia have appeared in numerous CSIRO Land Research Series reports. These, together with Queensland Department of Primary Industries' Land Use Study reports and the explanatory notes accompanying the vegetation survey of Western Australia by Beard, have provided much of the background information for this chapter and these sources are gratefully acknowledged.

Acacia open-forests and woodlands of northeastern Australia

A number of species of *Acacia* of similar growth-type form the almost unispecific canopy layer of open-forests and woodlands in semi-arid regions of northeastern Australia. They occupy two distinct situations in the landscape. One is the shallow, mainly coarse-textured, infertile acid soils on scarp retreats and tablelands. Such soils represent the upper catenary sequence of the dissected Tertiary lateritic land surface. The other situation is the fertile, deep, usually alkaline clay soils on undulating to lowland plains associated mainly with the erosional-depositional mid-to lower-slopes of the catenary sequence.

Acacia spp. from different sections of the genus favour these different situations. The most successful species on the Tertiary lateritic surface and exposed underlying sandstones are members of the Section Juliflorae, which have flowers in spikes, whilst the deep clays are dominated by members of the Microneurae group of the Section Plurinerves. The latter group has flowers in heads and phyllodes with crowded longitudinal nerves.

The most widespread species on laterites and sandstone are lancewood (*A. shirleyi*) and bendee (*A. catenulata*) and on the heavy clay soils *A. harpophylla* and gidgee (*A. cambagei*) are characteristic. On very rare occasions members of different groups form narrow ecotones but usually they are separated by *Eucalyptus* woodlands which occupy intermediate habitats.

Acacia communities are found in areas receiving up to 900 mm rainfall a year, but they are most extensive in rainfall zones receiving less than 750 mm a year. In undisturbed situations in more mesic areas they form open-forest communities, but with decreasing rainfall the canopy becomes more open and the trees become shorter to form low woodlands. Some species, such as *A. cambagei* and *A. catenulata*, penetrate the central drier zone of *Acacia* shrublands where they also form shrubby communities. In contrast *A. aneura*, the most abundant *Acacia* of the arid shrublands, forms extensive open-forests and woodlands in less arid areas of its range.

Although *Acacia* spp. form an almost unispecific upper canopy layer, a few species of *Eucalyptus* enter the canopy or occur as emergents. They are rarely common, however, and are usually found in marginal situations. *Casuarina cristata* alone forms mixed stands with *A. harpophylla* over wide areas of southern Queensland. Occasionally ecotonal areas of mixed *Acacia* spp. of the same group may be found. Even in these zones different species of *Acacia* generally prefer rather distinct habitats.

Whilst the canopy species of each group are from distinct sections of *Acacia*, the associated species also tend to be from taxonomically distinct groups. In mesic areas, woody understorey species in lowland clay situations have affinities with the Australian rainforest element, whilst on shallow soils on tablelands affinities lie with the Eremean flora (Chapter 1, Doing, 1981). The grassy ground flora also indicates different origins, with the Chlorideae and Sporoboleae relatively more prominent in the former group and the Aristideae and Andropogoneae in the latter.

On each of these edaphic-physiographic situations, species within each group tend to form a replacement series controlled by moisture availability. *Acacia harpophylla* is replaced, with increasing aridity, by *A. cambagei* on fine-textured soils and a similar but less marked transition occurs on shallow acid soils with *A. shirleyi* being replaced by *A. petraea* (Pedley, 1978).

The present distribution of the major species is controlled primarily by climate and secondarily by soil type within climatic zones, but past history has also had an influence. Cycles of mesic and arid conditions have left isolated post-climaxes of *Acacia* communities in favourable sites well outside their normal climatic range. *Acacia harpophylla*, a species unsuited to spread by long-range dispersal, occurs in widely separated areas receiving as little as 300 mm annual rainfall. Pedley (1973) recorded refugia of *A. aneura* communities which dominate the arid interior in areas where rainfall exceeds 500 mm.

Acacia *open-forests on shallow coarse-textured acid soils*
The most widespread associations are dominated by *A. shirleyi* and *A. catenulata*, although a number of other *Acacia* spp. of the Section Juliflorae (such as *A. sparsiflora*, *A. burrowii*, *A. petraea* and *A. rhodoxylon*) form structurally similar communities.

All the species are associated with the eroded residual Tertiary lateritic surfaces and sandstones, often lateritised, which have been exposed following removal of the Tertiary surface. Soils are usually very acid to acid, occasionally neutral and are of three major types.

1. Lateritic Red and Yellow Earths. These are generally shallow soils overlying massive or concretionary laterite. They usually occupy gently sloping situations often on the crests of eroded tablelands.
2. Skeletal Soils. They are usually uniform coarse-textured soils developed on lateritised sandstone. They are mainly sands or sandy loams, occasionally light clays, and occur on moderately to steeply dissected slopes often on scarp retreats.
3. Lateritic Podsolic Soils. These are generally shallow soils developed on quartzose sandstones capped by laterite. They are less widespread than the previous type and occur on less steep slopes.

The soils are generally infertile with low levels of available nitrogen, phosphorus and carbon; calcium and magnesium may be limiting for growth in some soils.

Lancewood (Acacia shirleyi). Lancewood is the name given to a group of species each of which form unispecific canopy layers of slender, closely-packed trees on residual tablelands and scarps. Of these, *A. shirleyi* is the most widespread. It occupies these habitats in an arc stretching from northwest of the Darling Downs (31°03′S; 145°45′E) in Queensland to the base of Cape York Peninsula and across the Barkly Tableland into the Northern Territory. Outliers occur as far west as the Victoria River catchment. It is most widespread in the 500 to 750 mm rainfall belt. In wetter areas it forms open-forests up to 15 m in height. In drier parts of its range, low and low-open woodlands predominate.

A wide variety of *Eucalyptus* spp. occur as canopy associates and as emergents although they are rarely abundant. The most prevalent species are members of the bloodwood group, including *E. trachyphloia, E. polycarpa, E. dichromophloia* and *E. citriodora*, and of the ironbark group, including *E. drepanophylla, E. decorticans, E. melanophloia* and *E. crebra*. Some members of the box group, such as *E. exserta* and *E. thozetiana*, also occur.

Most communities lack a well defined understorey tree or shrub stratum, although scattered small trees and shrubs may occur. Species composition varies with latitude and moisture. In tropical areas *Macropteranthes kekwickii, Petalostigma banksii* and *Erythrophleum chlorostachys* may be found. In semi-arid Queensland *Erythroxylum australe, Petalostigma pubescens* and, particularly in areas of deeper soil, *Alphitonia excelsa* occur. Ground vegetation is very sparse and comprises mainly wiry grasses of low quality for grazing. *Aristida* spp. are prominent with *A. pruinosa* frequent in tropical and *A. caput-medusae* in subtropical communities. Other widespread grasses include *Cymbopogon, Schizachyrium*, and *Cleistochloa subjuncea. Heteropogon contortus* and *Themeda australis* may occur in wetter areas; in more xeric areas are found *Triodia* and *Plectrachne pungens*.

Bendee (Acacia catenulata). This species forms open-forest communities which resemble, structurally and floristically, the lancewood communities. The trees are slender and of even height, up to 13 m in mesic areas, with a mid-storey tree and shrub layer which is usually very sparse but may be conspicuous. It extends on to slightly deeper and more loamy soils than *A. shirleyi* and in these situations the woody understorey layer becomes more prominent. It grows in the 150 to 700 mm rainfall belt, occurring in isolated pockets from south of Charters Towers to the western Darling Downs and the Grey Range, west of Thargomindah.

Associated canopy trees are not common and are mainly other *Acacia* spp., such as *A. shirleyi*, and in drier areas, *A. petraea, A. aneura* and *A. aprepta*, together with *Eucalyptus* spp. similar to those occurring in the lancewood communities. *Eucalyptus exserta* is a widespread associate and because the soils may be deeper and more loamy than in the latter community, the boxes such as *E. populnea* and *E. cambageana* are more prominent.

Although the mid-storey layer is usually very sparse, small-leaved myrtaceous and rutaceous shrubs such as *Lysicarpus angustifolius*, *Micromyrtus* sp., *Phebalium glandulosum*, *Thryptomene hexandra* and *Baeckia jucunda* may form a heath-like understorey. *Eremophila mitchellii* and, in drier western areas, *E. latrobei* and *Cassia* spp. are also found.

The ground flora is sparse, being composed mainly of arid scrub grasses such as *Aristida caput-medusae* and other *Aristida* spp., *Thyridolepis mitchelliana*, *Digitaria* spp., and *Paspalidium* spp. The fern *Cheilanthes sieberi* is a frequent associate. *Sida* spp. are among the most abundant forbs and in drier areas *Bassia* spp. become more prominent.

Acacia open-forests and woodlands on deep fine-textured alkaline soils
Acacia harpophylla and *A. cambagei* dominate the most widely occurring communities in this group. Blackwood (*A. argyrodendron*), boree (*A. tephrina*) and Georgina gidgee (*A. georginae*) communities are less widespread but because all five species occupy similar habitats, pairs of species may form mixed communities where their ranges overlap. A number of other species, all of the Microneurae group of the Section Plurinerves, form structurally and floristically similar communities which are important locally. These include yarran (*A. omalophylla* and *A. melvillei*), womal (*A. maranoensis*) and bowyakka (*A. microsperma*).

In contrast to the previous group these species are usually associated with the mid- to lower-catenary land surfaces within the Tertiary weathered zone. They are most abundant on soils developed on argillaceous sedimentary and basic volcanic rocks exposed following removal of the Tertiary surface and on lateritised Tertiary sediments. They also occur on more recent fine-textured alluvial soils. The soils are usually alkaline, occasionally neutral or even acid, with carbonate concretions and often gypsum at depth, although many of the soils become strongly acid at depth. They can be conveniently divided into 3 main types.

1. Clay Soils. These are usually deep, grey to red-brown in colour and are by far, the most widespread type. Three major subgroups occur depending on the origin of the parent material. Deep cracking clays, often gilgaied, are heavy and often saline soils, developed on pre-Tertiary sediments which have been altered through the lateritisation of the Tertiary land surface or are located on transported Tertiary-weathered zone material. Gilgais vary from incipient to large depressions to 2 m in depth, the latter usually associated with the deeper soil profiles (Isbell, 1962). The second subgroup are the sedentary clay soils which

are of moderate depth and occur on argillaceous Permian, Jurassic and Cretaceous sedimentary rocks and basic volcanics which were exposed following the complete removal of the Tertiary land surface. They are mainly medium- to heavy-clays and gilgais are rare. Alluvial clay soils are the third subgroup and are calcareous cracking and non-cracking clays to clay loams occupying the flood plains of major rivers and streams. Gilgais are infrequent.

2. Texture-contrast Soils. These are usually deep soils with a sandy or loamy surface, less than 40 cm thick, on a strongly alkaline clay subsoil. Where these occur the communities often contain various *Eucalyptus* spp.

3. Red and Yellow Earths. The soil is invariably loamy at the surface, becoming clayey with depth. They are usually alluvial or aeolian in origin and occur mainly in the more arid areas.

The soils, particularly the clays, are relatively fertile though the available phosphate content is low, except in the alluvial clays and the sedentary soils derived from basalt (Isbell, 1962).

Brigalow (Acacia harpophylla). These communities represent the most mesic extreme of this group of associations, generally being found on clay soils on gently undulating lowlands and plains. They extend from about Charters Towers in the north to south of the Queensland–New South Wales border, occupying an area of more than 60 000 km^2. *Acacia harpophylla* assumes prominence usually as an emergent over a semi-evergreen vine thicket in coastal and sub-coastal areas receiving up to 900 mm rainfall a year. The most extensive and richest development of the brigalow communities occurs in the 750 to 500 mm annual rainfall belt in areas receiving two or preferably four months of effective winter rainfall (Farmer, Everist & Moule, 1947). In this region, *A. harpophylla* dominates open-forest communities usually 10 to 20 m high (Figure 9.1), but with decreasing annual rainfall the trees become lower in stature and less dense. In areas receiving less than 500 mm rainfall, woodlands and low open-woodlands are the most common structural form. In these drier areas *A. harpophylla* communities are found in pockets receiving surface run-on moisture. These outliers extend as far west as the Grey Range, west of Quilpie, where annual rainfall is less than 350 mm.

Throughout its range, *A. harpophylla* commonly occurs as a shrub, often forming dense shrublands. In most circumstances these are the result of major disturbance, such as unsuccessful clearing of open-forest communities or fire (Johnson, 1964).

In northern areas, *A. harpophylla* typically occurs as the unispecific canopy of a low-layered open-forest. In sub-tropical areas *Casuarina cristata* is frequently co-dominant and the canopy varies from all *C. cristata* to all *A. harpophylla*. A number of *Eucalyptus* spp. may also be present as canopy trees and emergents. There is less variety of species than in the former group on the shallow acid soils. Members of the box group are the most prominent. The eucalypts also often occur in broad ecotonal communities and their individual presence reflects the habitat. North of Taroom and Injune, *E. cambageana* forms mixed communities with brigalow on red-brown texture-contrast soils, with *Eremophila mitchellii* and *Carissa ovata* frequent as mid-storey species. On alluvial clay soils along flood plains of major rivers, *Eucalyptus microtheca* is a common associate with *Terminalia oblongata* the characteristic understorey tree in tropical areas. *Eucalyptus populnea* forms mixed communities on texture-contrast soils, particularly in sub-tropical areas with *Eremophila mitchellii* the characteristic mid-storey species. Other associated *Eucalyptus* spp. include *E. thozetiana*, *E. microcarpa*, *E. orgadophila*, *E. brownii*, *E. crebra* and *E. ochrophloia*. In addition to other *Acacia* and *Eucalyptus* spp., the most common canopy associates are

Figure 9.1. Open-forest of *Acacia harpophylla* in southeastern Queensland. Photo: R.W. Johnson.

Brachychiton spp., *Lysiphyllum* spp., *Cadellia pentastylis*, and *Ventilago viminalis*.

A number of floristic associations based on the composition of the mid-storey low tree and shrub layer can be recognised (Johnson, 1964), their distribution depending on habitat and climatic factors. On uniform clay soils in sub-tropical areas, *Geijera parviflora* is the characteristic understorey species; in tropical areas *Terminalia oblongata* is predominant. *Eremophila mitchellii* characterises the understorey on texture-contrast soils, although in drier western areas it replaces both *T. oblongata* and *G. parviflora* on more clayey soils. Other frequent mid-storey species include *Myoporum deserti*, *Heterodendrum diversifolium*, *Carissa ovata*, *Eremocitrus glauca* and *Rhagodia spinescens*. Lianes such as *Capparis lasiantha*, *Cissus opaca*, *Jasminum racemosum* and various asclepiads are widespread although most individuals rarely climb into the canopy layer.

Grasses are characteristic of the ground layer, but as the canopy of the mid-storey and upper layers closes, forbs become relatively more prominent. *Paspalidium caespitosum* is usually predominant with *P. gracile*, *Chloris* spp. and *Enteropogon acicularis* also abundant, whilst *Sporobolus* spp., mainly *S. caroli* and *S. scabridus*, *Leptochloa* spp. and *Eriochloa* spp. are characteristic species. On texture-contrast soils, *Aristida ramosa*, *Cymbopogon refractus* and *Calyptochloa gracillima* are prominent. The sedges *Cyperus gracilis*, and on heavy clay soils, *C. bifax* are often present. The most abundant forbs are *Abutilon oxycarpum*, *Rhagodia nutans*, *Bassia tetracuspis*, *Brunoniella australis*, *Enchylaena tomentosa* and various *Sida* spp.

Gidgee (Acacia cambagei). This *Acacia* replaces *A. harpophylla* as rainfall decreases to below 550 mm a year. The eastern extent of its range lies between Charters Towers and Clermont where it forms open-forest communities with a canopy up to 15 m high. In the interface, *A. harpophylla* extends along run-on sites in drier areas with *A. cambagei* occupying the drier elevated sites in wetter areas. With increasing aridity the canopy becomes lower and more open, and west of the Great Dividing Range most communities are woodlands (Figure 9.2) or shrublands.

Tongues of *A. cambagei* communities extend westward around the margin of the arid centre. In the northern tongue, *A. cambagei* occupies light clay and loamy soils, but as annual rainfall approaches 250 mm it is most often found on red earths and calcareous desert loams in low-lying

sites. Scattered stands occur in the southeastern part of the Barkly
Tableland extending into the Northern Territory. In the south it extends
into northwestern New South Wales and across far northern South
Australia.

The open-forest communities are similar, structurally and floristically,
to the adjacent *A. harpophylla* communities. The suite of *Eucalyptus*
species, which occur as emergents or co-dominants in the canopy layer, is
similar to that occurring in the brigalow communities. *Eucalyptus po-*
pulnea, *E. cambageana* and *E. microtheca* occupy their characteristic
habitats. Because *A. cambagei* is found in more arid habitats than *A.*
harpophylla, other eucalypts become prominent. *Eucalyptus terminalis*, in
tropical areas, and *E. ochrophloia* in alluvial plains in southwestern
Queensland occur in ecotonal communities with *A. cambagei*.

In tropical areas *Terminalia oblongata* and *Eremophila mitchellii* are
the characteristic mid-storey species with *Geijera parviflora* replacing *T.*
oblongata in southern areas of the mesic fringe. Because *A. cambagei*
occupies a drier environment than *A. harpophylla* the more xeric
E. mitchellii tends to be the most characteristic mid-storey species.
Other mid-storey shrubs characteristic of *A. harpophylla* open-forests

Figure 9.2. Woodland of *Acacia cambagei* in southeastern Queensland.
Photo: R.W. Johnson.

such as *Myoporum deserti, Carissa ovata, Rhagodia spinescens* and *Heterodendrum diversifolium* extend into *A. cambagei* open-forests. *Cassia* spp. are frequent components of the mid-storey shrub layer with *C. nemophila* common in mesic areas and varying mixtures of other species in more arid zones. In southern areas which receive predominantly winter rainfall *Atriplex nummularia* is the most abundant understorey shrub.

The ground layer in the more mesic areas is predominantly grassy with *Paspalidium* spp., *Enteropogon acicularis* and *Sporobolus* spp. the most frequent species and *Enchylaena tomentosa, Abutilon* spp. and *Rhagodia nutans* widespread forbs. With increasing aridity *Paspalidium* spp. become infrequent and species of *Enneapogon* and *Eragrostis* increase in importance. In the shrublands, *Dactyloctenium radulans* becomes more prominent whilst *Bassia* spp., *Atriplex* spp. and other chenopodiaceous forbs become the major components of the ground stratum. Hence the gradient from open-forest to shrubland is paralleled by gradual replacement of grass by forbs.

In central-western Queensland, *A. cambagei* woodlands are closely associated with the Mitchell grass (*Astrebla* spp.) downs on heavy cracking clays. They often fringe or form mosaics with the tussock grassland and species of the grasslands become associated in the ground layer.

Blackwood (Acacia argyrodendron). *Acacia argyrodendron* forms an almost unispecific canopy of open-forest communities which lie roughly in the overlap of the more mesic *A. harpophylla* and more xeric *A. cambagei* communities to the south of Charters Towers in the basins of the Cape, Suttor and Belyando Rivers. Outliers occur southwards in an arc along the western slopes of the Great Dividing Range and into the Isaacs River basin as far south as about 22°30'. Average rainfall varies from 475 to 655 mm. These communities occupy lowlands covered in Tertiary weathered sediments.

Associated species are similar to those in adjoining *A. harpophylla* open-forests with *Eremophila mitchellii, Terminalia oblongata* and *Carissa ovata* occurring in the woody understorey and *Eucalyptus cambageana, E. microtheca* and *E. brownii* in particular habitats occurring as emergents. *Paspalidium caespitosum* is the most abundant herb; *Sporobolus caroli, Brunoniella australis* and *Dipteracanthus corynothecus* are moderately frequent.

Boree (Acacia tephrina). Boree forms woodland and open-woodland communities in central-western Queensland from Hughenden to Tambo in the 425 to 550 mm annual rainfall belt. In the more arid areas, communities are reduced to open-shrublands. At their northern limit scattered stands are found eastward to the coast at 20°S.

Acacia tephrina occurs frequently on cracking clay soils and is often found in association with *A. cambagei*. Communities of *A. tephrina* and *A. tephrina–A. cambagei* often occur on the fringes of and form mosaics with *Astrebla* spp. grasslands. *Lysiphyllum carronii*, *Heterodendrum oleifolium*, *Atalaya hemiglauca* and *Flindersia maculosa* may be found scattered through the canopy layer particularly in more mesic areas. *Eremophila mitchellii*, *E. maculata* and in more western areas *E. dalyana* form an often very sparse low shrub layer. *Astrebla*, *Sporobolus* and many of the grass species abundant in the adjacent *A. cambagei* communities contribute to an open ground-layer. With increasing aridity the typical trend in forbs towards dominance of chenopodiaceous species, such as *Atriplex*, commonly *A. lindleyi* and *A. spongiosa*, and *Bassia* spp. occurs.

Georgina gidgee (Acacia georginae). Skirting the shrublands of the central arid area in the tropical 200 to 250 mm annual rainfall belt, *A. georginae* forms woodlands and open-woodlands up to 8 m tall from the Georgina River basin in western-central Queensland through the Barkly Tableland region to northeast of Alice Springs. It is best developed on medium to heavy grey and brown calcareous clays, although it also occurs on lighter soils including somewhat acid red earths.

On heavy clays it forms a very open canopy over *Astrebla* spp. and other grasses and forbs characteristic of the Mitchell grass downs, whilst on better drained, lighter soils, a sparse mid-storey shrub layer with *Cassia* spp. and *Eremophila* spp. becomes evident. *Enneapogon* spp. are prominent in the short grass ground layer in these situations.

Utilisation of Acacia *Open-Forests and Woodlands of Northeastern Australia*

The relatively fertile clay soils supporting *Acacia* open-forests and woodlands provide a sound base for a productive agricultural and pastoral industry. Because of the density of standing trees and shrubs, productivity of the undisturbed communities is low and it has generally been necessary to modify severely or replace the existing vegetation. Clearing has occurred over large areas, particularly in higher rainfall zones, by ring-barking and in more recent decades by pulling trees down with a heavy chain dragged between high-powered bulldozers.

Because of the high potential agricultural and pastoral productivity of lands supporting *A. harpophylla* open-forest, the forests have been replaced by crops and pastures of introduced species such as buffel grass (*Cenchrus ciliaris*), Rhodes grass (*Chloris gayana*) and green panic (*Panicum maximum* var. *trichoglume*) (Johnson, 1964). Few large areas of undisturbed vegetation remain. *Acacia argyrodendron* open-forests have also been converted to sown pastures. Carrying capacity may be increased initially ten to twentyfold. However, continued production is partly dependent on management to control regrowth of *A. harpophylla* and associated woody species. Both *A. harpophylla* and *A. argyrodendron* sucker from horizontal roots and other species such as *Eremophila mitchellii*, *Terminalia oblongata* and various *Eucalyptus* spp. regenerate from butts and lignotubers. Ploughing, chemical control, fire and severe grazing with sheep have been used to bring regrowth of *A. harpophylla* under control.

Acacia cambagei open-forests have also been replaced by sown pastures (Purcell, 1964). Clearing has occurred mainly along the wetter eastern fringe where more prevalent understorey trees and shrubs limit the growth of native grasses and where conditions are more favourable for establishment of sown pastures. Regrowth from *Eremophila mitchellii* is often a problem after clearing (Beeston & Webb, 1977).

In contrast, *Acacia* open-forests on the shallow acid soils have been left largely undisturbed. *Acacia shirleyi* is used locally to provide rails for fencing but soil fertility is too low to warrant expenditure on clearing. Limited clearing of *A. catenulata* open-forests has occurred on deeper red and yellow earths where pulling, without burning, and sowing buffel grass (*Cenchrus ciliaris*) is recommended (Tiller, 1971).

Acacia **open-forests and woodlands of southern Australia**
In areas to the south and southeast of the central Australian *Acacia* shrublands, most open-forests and woodlands are dominated by *Eucalyptus* spp. However, north of the Great Australian Bight as far west as Spencer Gulf, *Eucalyptus* open-forests and woodlands are infrequent and there are large areas of woodlands and open-woodlands characterised by the trees western myall (*Acacia papyrocarpa*), *Casuarina cristata* and *Myoporum platycarpum*.

Apart from *A. papyrocarpa*, myall (*A. pendula*) is the only other species of *Acacia* which forms woodland communities over considerable areas in southern Australia. Both species of *Acacia* are members of the Microneurae group of the Plurinerves and occur on similar soils and topography as related species in northeastern Australia.

Western Myall (Acacia papyrocarpa *(syn.* A. sowdenii)). The most extensive *Acacia* woodland formation excluding the *A. aneura* woodlands and those of northeastern Australia is that characterised by *A. papyrocarpa.* It forms woodlands to 10 m in height in the 200 to 250 mm rainfall belt to the northwest of Port Augusta and Whyalla in South Australia (Specht, 1972) and extends into Western Australia as a fringe skirting the Nullarbor Plain (Beard, 1975). However, in western South Australia and Western Australia the canopy is lower and very open and western myall often occurs merely as very scattered trees in a semi-succulent shrubland.

A. papyrocarpa is found mainly on shallow uniform loam to clay-loam soils or red and brown earths, both of which are calcareous and similar to those occupied by *A. cambagei* and *A. georginae* in the drier parts of their range. In woodland communities, *Myoporum platycarpum* is a common canopy associate. *Casuarina cristata* may also occur. A well defined semi-succulent mid-storey shrub layer is characteristic with *Maireana sedifolia* predominant and *Atriplex vesicaria* and *A. stipitata* commonly present. Tall shrubs associated with the *Acacia* woodlands of northeastern Australia such as *Heterodendrum oleifolium* and *Cassia* spp., such as *C. sturtii* and *C. phyllodinea*, extend into these southern communities. Low chenopodiaceous shrubs and forbs and, in favourable seasons a wide variety of annuals, occupy the ground layer with *Enneapogon* and temperate grasses such as *Danthonia* and *Stipa.*

Myall (A. pendula). *Acacia pendula* forms woodland and open-woodland communities, rarely widespread but locally important, from Clermont in central Queensland to the Riverina (34°30′S; 145°00′E) area of southern New South Wales. They are or formerly were best developed in New South Wales, southwest of Hay, and on the northwestern plains in the Macquarie District (Beadle, 1948; Moore, 1953) as well as on the western Darling Downs in Queensland. Annual rainfall varies from 375 to 550 mm with communities occupying the wetter end of the range in Queensland and the drier end in southern areas. They occupy similar habitats to the *Acacia* communities of the lowland clays of northeastern Australia and represent the southern attenuation of this complex.

In northern areas, *A. pendula* woodlands are mainly restricted to clay soils. Occasionally *Eucalyptus microtheca* and *E. populnea* occur as emergents and other canopy trees include *Heterodendrum diversifolium* and *Acacia stenophylla.* No well defined understorey occurs but scattered *Eremophila maculata* is a common associate. The ground flora resembles that of the *Astrebla* spp. and *Dichanthium sericeum* grasslands. In southern areas, *Atriplex nummularia* may form a conspicuous low shrub layer with other associated shrubs such as *Rhagodia spinescens* and *Enchylaena tomentosa*. Temperate grasses, such as *Danthonia* and *Stipa*, and *Chloris truncata* are prominent in the herbaceous layer. Communities also occur on better drained red earth soils in southern areas.

Utilisation of Acacia *Open-Forests and Woodlands of Southern Australia*
Acacia papyrocarpa woodlands are used for sheep and wool production from native pastures and the timber is cut locally for fence posts and also for firewood (Lange & Purdie, 1976). Present usage coupled with the infrequency of germination events threatens the survival of existing communities. Similarly, *A. pendula* woodlands have been largely replaced by grasslands in southern areas as a result of felling associated with grazing by sheep (Moore, 1953).

Acacia low woodlands and shrublands of central and western Australia
Acacia shrublands are the dominant woody vegetation of Australia's semi-arid and arid interior. The most characteristic species is mulga (*Acacia aneura*). Mulga communities together with mixed mulga–hummock grass communities and mulga mid-height grass communities occupy about 1 500 000 km² or about 20% of the total area of the continent (Everist, 1971). *Acacia aneura* is by far the most ubiquitous *Acacia* within this range but it may be replaced by other localised dominants such as *A. brachystachya*, *A. stowardii*, *A. kempeana* and the *A. linophylla–A. ramulosa* complex, largely in response to topographic and edaphic factors. These communities show relationships with the open-forests of the shallow coarse-textured soils of northeastern Australia with the major dominants also being members of the Section Juliflorae.

Two types of *Acacia* low woodlands and shrublands can be recognised: one with tussock grasses and forbs characterising the ground layer, the other with hummock grasses predominant.

Acacia *with tussock grasses*
Mulga (Acacia aneura). *Acacia aneura* associations extend in a discontinuous belt from near the Western Australian coast, across the southern edge of the central deserts, to western New South Wales and southwest Queensland, with another substantial occurrence in the north of the central arid area in the Northern Territory. *Acacia aneura* is

adapted to environments where the soil water regime is almost always limiting for growth, but where there is some possibility of recharge at all seasons. *Acacia aneura* is mostly found in areas receiving from 200 to 500 mm mean annual rainfall, but it is conspicuously absent from the semi-arid regions with a regular summer or winter drought (Nix & Austin, 1973).

In Central Australia, Perry & Lazarides (1962) consider that climate has been the primary factor in selection of species and life forms and is the controlling factor influencing the structure and density of the communities. However, soil characteristics are of major importance in determining floristics and distribution of communities within the area.

The most extensive occurrence of *A. aneura* is on plains and sand plains which often receive some run-on water from adjacent hills and low ranges. Nevertheless a great diversity of habitat can be identified from dissected residual soils on ridges to desert sandhills and solonized brown soils, red earths and texture-contrast soils of the flats and plains.

Acacia aneura is commonly found on red earth soils throughout most of its Australian distribution. These soils are light textured with a hard coherent subsoil strongly impregnated with ferruginous compounds. It rarely occurs in calcareous habitats, especially in Queensland and the Northern Territory, although it can be found in areas with neutral to alkaline subsoils, mainly in the south and west of the continent. In the latter habitats, *A. aneura* is usually sparser and more stunted. *A. aneura* rarely occurs on clay soils of heavy texture. Although there is some variation evident in the tolerance to soil acidity levels, a general feature of all soils supporting *A. aneura* is that they have a very low available phosphorus content (< 20 ppm measured by acid extraction).

The variable nature of the *A. aneura* phyllode has contributed to taxonomic problems. At least three morphological phyllode variants can be distinguished: 'broad', 'long-narrow' and 'short-narrow' (Pedley, 1973). This variation could be related to polyploidy within the species, diploid and tetraploid races of *A. aneura* having been identified (I. DeLacy, personal communication). Pedley (1973) suggested that the complex pattern of variation in growth form and phyllode dimensions of *A. aneura* is possibly due to the retreat of the species to refugia during arid periods at the end of the Tertiary, followed by a recent expansion of its range. Specht (1972) considered that calcifuge and calcicole forms of the complex would well repay investigations.

Acacia aneura communities (Figure 9.3) vary widely in density and composition. Highest *A. aneura* densities are achieved in the eastern mesic extremity where up to 8000 stems ha^{-1} have been recorded. Over most of its range, densities of the order of 100 to 300 stems ha^{-1} would be more common. In more xeric habitats, *A. aneura* is quite sparse.

It occurs as a tree (10–15 m tall) in the mesic areas, but only exists as a stunted low shrub (2–3 m tall) in very xeric habitats or where it occurs on very shallow or calcareous soils. *Eucalyptus* spp. occur comparatively infrequently in *A. aneura* communities.

Whatever the habit of the dominant *A. aneura*, multi-layering characterises almost all these communities. The physiognomic complexity of *A. aneura* associations decreases along a gradient from more favourable to harsh environments (Boyland, 1973). A low, rarely continuous, shrub stratum is often present. Associated shrubs may be either sclerophyllous or with hairy semi-succulent leaves (Specht, 1972). The most commonly occurring genus is *Eremophila* with over 100 species represented in the *A. aneura* areas of Western Australia (Speck, 1963) and 19 species recorded in *A. aneura* areas of Queensland (Boyland, 1974). Since many of the *Eremophila* species have distinctive habitat preferences they are often used in community identification. Widespread among the shrub layer genera are *Cassia*, *Dodonaea* and *Maireana*, the last being more common

Figure 9.3. Tall shrubland of *Acacia aneura* in southern Queensland. Photo: W.H. Burrows.

on saline soils and rare in Queensland. In general, *A. aneura* associations of Western Australia are more shrubby than those in the east.

A herbaceous layer of perennial or seasonal forbs and grasses is usually well developed, although it may be sparse. A feature of mulga lands is the floristic independence of ground-storey and upperstorey communities. Although mulga communities are relatively poor, floristically, there is considerable variation in ground layer grass and forb composition, more so on a north–south rather than an east–west axis. Perry & Lazarides (1962) recognised 18 distinct ground-storey communities associated with *A. aneura* in the Northern Territory. Speck (1963) identified some 75 communities, most of which contained *A. aneura*, in the Wiluna–Meekatharra area of Western Australia. Additional distinctive community types are found in the remaining States where *A. aneura* occurs, although those of Queensland and the Northern Territory have much in common.

Acacia aneura is frequently associated with short-grass communities in which *Eragrostis eriopoda* and *Monochather paradoxa* are characteristic perennial tussock grasses. Genera of forbs are predominantly from the families Asteraceae, Chenopodiaceae, Amaranthaceae and Malvaceae. There is a notable paucity of leguminous forbs in mulga shrubland.

In southwestern Queensland, communities dominated by *Acacia aneura* are characteristic of the transported detritus and dissected residuals of the laterite and silcrete land surface. Mulga is often associated in less arid areas with *Eucalyptus populnea* with which it may be dominant, co-dominant or form an understorey tree stratum. On its eastern margin, *A. aneura* grades into *E. populnea*–dominated communities (see previous chapter, Gillison & Walker, 1981). These communities are usually found on plains of deep sandy red earths and loamy texture-contrast soils and are equally extensive in New South Wales and in Queensland. Other associated upperstorey species can include *E. intertexta* and *E. melanophloia*. *Callitris columellaris* frequently occurs as a tree, mainly in New South Wales. A shrubby understorey is usually present although it is more prominent in southern areas. In drier environments, *E. populnea* is replaced by *E. terminalis* and less frequently *E. papuana*.

Cassia spp., such as *C. artemisioides* and *C. nemophila*, and *Eremophila* spp., such as *E. gilesii*, *E. bowmanii* and *E. mitchellii*, particularly where *Eucalyptus populnea* occurs, are characteristic understorey shrubs. *Themeda australis* and *Dichanthium sericeum* are important grasses on the mesic fringe but *Eragrostis eriopoda* is the most widespread and abundant species. Other widespread species include

Aristida contorta and *A. jerichoensis, Monachather paradoxa, Thyridolepis mitchelliana* and *Enneapogon*. Temperate species such as *Stipa variabilis* are more abundant in New South Wales. *Bassia, Sida, Ptilotus* and *Euphorbia* are among the most frequent forbs with the ferns *Cheilanthes sieberi* and *C. tenuifolia* abundant locally.

On dune fields, *Acacia aneura* often forms open shrublands with *Atalaya hemiglauca*. Scattered trees of *Hakea leucoptera* and *Grevillea juncifolia* may be present and other *Acacia* such as *A. tetragonophylla, A. ligulata, A. murrayana* and *A. calcicola* are frequently associated where the habitat is suitable. *A. tetragonophylla* is found on shallow soils whilst *A. ligulata* and *A. murrayana* are frequent on the extended flanks of dunes. *A. calcicola* occurs with mulga on low eroded dunes. *Cassia desolata, Dodonaea angustissima, Eremophila duttonii* and *E. sturtii* occur as low shrubs and *Aristida contorta, Enneapogon, Eragrostis* and *Eriachne* are the most frequent grasses. These dune communities also extend into northwestern New South Wales.

Mulga communities extend westward into northern South Australia and into the Alice Springs region of the Northern Territory. In northern arid South Australia, *Acacia brachystachya* is frequently associated with *A. aneura* forming tall shrublands on sands, earths and duplex soils with a sandy and loamy surface. A low shrub layer is usually present although it is usually not dense. *Cassia* spp., with *C. nemophila* frequent on the deeper sands, and *Eremophila* spp., such as *E. latrobei* and *E. glabra*, are found in the understorey shrub layer. On shallow soils, *C. sturtii* and *Acacia kempeana* become more frequent. *Atriplex vesicaria* and *Maireana sedifolia*, except on the deepest sands, are widespread whilst *Aristida contorta* and *Enneapogon* are frequent herbs. On deeper sands, *Monochather paradoxa* and *Eragrostis eriopoda* occur.

Mulga associations in the Northern Territory are best developed on plains adjacent to mountains and hills where they are extensive on coarse- to medium-textured red earth soils (Perry & Lazarides, 1962). Other *Acacia*, such as *A. brachystachya*, are frequent associates. *Hakea* and *Acacia estrophiolata* may occur as scattered trees and *Cassia* and *Eremophila*, particularly *E. gilesii*, may form a sparse to medium-dense low shrub layer. *Eragrostis eriopoda* is the most abundant species in the ground layer of most communities and is usually associated with *Thyridolepis mitchelliana*. Short grasses, such as *Aristida contorta* and *Enneapogon*, and forbs, such as *Helipterum*, particularly *H. floribundum, Ptilotus helipteroides* and *P. atripliciplius*, grow between the perennial grass tussocks or in some communities form the ground layer.

In Western Australia *Acacia aneura* low woodlands and tall shrublands dominate the area south of the Tropic of Capricorn, extending as far south as about 30°S. They are best developed on plains covered with red earth and loamy soils over siliceous hardpans, between coastal areas on the west and southwest and the deserts on the east. Other *Acacia* are frequently present, the specific composition of the communities depending on habitat. *A. pruinocarpa* is widespread although it does not extend as far south as *A. aneura*. *A. ramulosa* and *A. ligulata* are more frequent in sandy habitats, *A. grasbyi* on granite or in creek beds, *A. sclerosperma* on calcrete and *A. victoriae* on salty flats. Towards the northwest, *A. aneura* and *A. xiphophylla* communities merge.

Where the surface is sandy, mallees such as *Eucalyptus kingsmillii* and *E. oleosa*, may be found. Other associated trees and large shrubs include *Brachychiton gregorii*, *Heterodendrum oleifolium*, *Hakea suberea* and *Canthium latifolium*. *Cassia* spp., such as *C. helmsii*, *C. sturtii*, *C. artemisioides*, *C. nemophila*, *C. chatelainiana* and *C. desolata* and *Eremophila* are prominent in the small shrub layer. Species of *Eremophila* are useful indicators of habitat type with *E. foliosissima* and *E. margarethae* frequent on deep loams, *E. fraseri* more prominent on stony ground and *E. granitica* on hills. Other frequent species are *E. leucophylla*, *E. cuneifolia* and *E. platycalyx*.

About the Tropic where rainfall has a larger summer component, annual grasses are important in the ground layer; forbs are favoured by winter rainfall. Frequent ephemerals include *Ptilotus*, *Helipterum*, *Goodenia*, *Schoenia cassiniana* and *Waitzia acuminata* whilst the most characteristic grasses are *Monochather paradoxa* and *Eragrostis eriopoda* with *Aristida*, *Eriachne*, *Enneapogon* and *Eragrostis*.

Throughout the Western Australian mulga zone, *Acacia aneura* forms shrublands on stony hills of granite and gneiss and on lateritic scarps and breakaways. *Acacia quadrimarginea* and *A. grasbyi* are frequent associates in western areas and *A. linophylla*, *A. ramulosa* and *A. brachystachya* may also be present. *Cassia*, *Eremophila* spp. (particularly *E. latrobei*) and *Ptilotus obovatus* are among the more widespread lower shrubs. *Acacia aneura* shrublands are also found in saline areas where they often occupy sandy patches. In these situations *A. linophylla* and *A. ramulosa* are abundant and *A. sclerosperma* and *Hakea preissii* frequent associates. *Eremophila oldfieldii* becomes more prominent and halophytes such as *Maireana pyramidata*, *Atriplex* and *Bassia* more abundant among the forbs and grasses. Similar communities are associated with greenstone hills.

On the lateritic plains of the Great Victoria Desert and extending into the south of the Great Sandy Desert, Beard (1968) described a formation known as mulga parkland which is a mosaic of mulga scrub and *Hakea–Acacia* scrub steppe. The former occurs on loamy soil in draws and depressions and the latter occupies the rises of hard ironstone.

In the area separating the Great Victoria Desert from the succulent shrublands of southern Australia, *A. aneura* forms a low tree or shrubby layer over semi-succulent and succulent halophytic shrubs and forbs. The soils are saline red sands and loams with a siliceous hardpan. Characteristic associates are *Casuarina cristata* and *Myoporum platycarpum*. It merges with *A. papyrocarpa* open-woodland which fringes the Nullarbor Plain, although the mulga communities are usually found on drier and sandier soils. *Eucalyptus* spp., such as *E. oleosa*, are locally present. The understorey may be predominantly *Maireana sedifolia*, particularly on calcareous soils, or *Atriplex* spp., such as *A. vesicaria*, particularly in saline areas. In South Australia, *A. brachystachya* becomes frequent. Similar communities with an understorey of semi-succulent shrubs are associated with the *Atriplex* and *Maireana* shrublands of northeastern South Australia and northwestern New South Wales.

Within its distribution range *Acacia aneura* is sometimes replaced by other shrubby *Acacia* species such as *A. stowardii, A. brachystachya, A. kempeana, A. linophylla* and *A. ramulosa*. This replacement is usually related to topographic and soil features.

Bastard mulga (Acacia stowardii). This occurs as a sparse low open-shrubland 1 to 2 m high. It is confined mainly to shallow red earths and red lithosols on dissected plains and low hills within the mulga lands of eastern Australia, although communities extend into Western Australia. *Acacia stowardii* occasionally forms pure stands or it may be co-dominant with *A. aneura* or *Eucalyptus exserta*. The lower shrub layer is usually well developed. In addition to species of *Dodonaea, Eremophila, Acacia* and *Canthium*, shrubs from genera not normally associated with the surrounding *A. aneura* communities, such as *Phebalium glandulosum* and *Westringia rigida*, also occur. The ground cover in these communities is usually quite sparse. *Eriachne pulchella* is prominent amongst the grasses that are present.

Turpentine mulga (Acacia brachystachya). This species occurs as a low to tall open-shrubland 1 to 3 m in height. It is usually associated with *A.*

aneura and occurs on shallow red earths in Queensland and on sand dunes in South Australia. There may be a well-defined lower shrub layer of *A. stowardii, Canthium latifolium, Cassia* spp. and *Dodonaea* spp. The ground cover is usually very sparse with grasses in the genera *Aristida, Eragrostis* and *Eriachne* and some forbs such as *Maireana, Ptilotus* and *Sida* being most frequent.

Witchetty bush (Acacia kempeana). *Acacia kempeana* is a very frequent shrub of arid areas, particularly in the Northern Territory and north-western South Australia. It occurs infrequently in Queensland. It is a shrub 2 to 3 m high with many stems from the base. Perry & Lazarides (1962) claimed that *A. kempeana* has a lower water requirement than *A. aneura* which leads to *A. kempeana* occupying the more droughty habitats. However, *A. kempeana* also appears almost exclusively on the more calcareous soils, whereas *A. aneura* occurs mainly on acid to neutral soils. Associated ground-layer species with *A. kempeana* are mainly short grasses and forbs. The commonest include *Aristida contorta, Enneapogon* spp. and *Helipterum floribundum.*

Sandhill mulgas (Acacia linophylla, A. ramulosa). This community occurs mainly in South Australia, occupying the area between the mallee in the south and the *A. aneura* communities of the drier north. *Acacia linophylla* and *A. ramulosa* form a tall shrubland formation 3 to 5 m high which is mainly found on deep, red, non-calcareous sand dunes. In more southerly regions *Casuarina cristata* is a common associate in the upper layer; in the north *Callitris columellaris* may partly or wholly replace the sand mulgas. *Cassia nemophila* is a common undershrub. Grasses in the genera *Aristida* (particularly *A. browniana*) and *Enneapogon* are well represented and annual forbs in the family Asteraceae are prevalent.

In Western Australia south of the Tropic and west of the Great Victoria Desert, *A. ramulosa* and *A. linophylla* are also prominent species in tall shrublands on sandy rises in claypans and on sand dunes in sand plains. Other *Acacia* spp. including *A. murrayana, A. acuminata, A. burkittii* and *A. sclerosperma* are frequent. Associated shrubs include *Eremophila* spp., such as *E. leucophylla* and *E. clarkei, Grevillea* spp., such as *G. eriostachya, G. stenobotrya* and *G. stenostachya, Melaleuca uncinata* and *Thryptomene johnsonii*. In deeper sand, mallees such as *Eucalyptus leptopoda* and *E. oldfieldii* may be conspicuous. Under favourable seasonal conditions ephemerals are abundant, especially composites such as *Myriocephalus guerinae, Brachycome ciliocarpa,*

Helichrysum davenportii and *Podolepis auriculata*, and *Ptilotus*. Grasses such as *Monochather paradoxa*, *Eragrostis* and *Eriachne helmsii* also occur. In saline situations, halophytes are more abundant.

Snakewood (Acacia xiphophylla). In the Ashburton region of Western Australia which receives rainfall of summer and bimodal incidence, *Acacia xiphophylla* forms extensive shrublands. They occur mainly on alluvial plains with clay soils carrying surface stone. *Acacia victoriae* and *A. tetragonophylla* are frequent associates; in southern areas *A. sclerosperma* becomes more prominent. Scattered small shrubs of *Eremophila cuneifolia* and *Cassia* may occur and the ground vegetation is usually sparse. Ephemerals such as *Ptilotus* and tussock grasses are present but often halophytes such as *Maireana, Bassia* and *Atriplex* may be more abundant.

Other Acacia *shrublands*. In a brief survey of *Acacia* shrublands it is not possible to cover all the local variants that occur. For instance, *A. victoriae* and *A. tetragonophylla* are common understorey shrubs in the drier regions of areas dominated by *A. aneura*. Occasionally both of the former species may be dominant in their own right, particularly in situations with deeper soils and adjacent to water courses. *Acacia victoriae* is usually found on heavier, often saline, soils but it is also frequent on shallow stony soils.

In western areas of Western Australia south of the Tropic, *A. sclerosperma* with the same two species forms shrublands which are usually associated with limestone and calcrete. *Hakea preissii* may be prominent. In more mesic areas in southwestern Western Australia on the fringe of the *Acacia* shrublands, tall shrublands dominated by *Acacia* spp., such as *A. resinomarginea*, occur. They are characterised by a low heath-like layer of sclerophyllous shrubs.

Acacia *spp. over hummock grasses*
Throughout Central Australia, hummock grasses (*Triodia, Plectrachne*) are important constituents of the vegetation of sand plains, dune fields and many rocky hills (Chapter 13, Groves & Williams, 1981). True hummock grasslands are however rare and there is usually a sparse cover of shrubs and low trees. Individual species of hummock grasses vary in importance according to geographical location and habitat. *T. pungens* is dominant north of the Tropic whilst *T. basedowii* replaces *T. pungens* in southern areas. *Plectrachne schinzii* predominates on deeper and coarser

sands. In sub-tropical areas *A. aneura* is the most common *Acacia* found in the overstorey but in northern areas other *Acacia* become prominent. These communities are more abundant in western parts of the continent and in sub-tropical rather than in temperate areas.

Kanji (Acacia pyrifolia). To the north of the Tropic in summer rainfall areas, *Acacia pyrifolia* shrublands with hummock grasses occupy large areas of hard alkaline red soils on granite plains and basaltic rises in the Pilbara Region. Frequently associated shrubs are *Hakea suberea* and *Grevillea pyramidalis* and *A. ancistrocarpa* and *A. tetragonophylla* may also be present. *Triodia pungens* is the most abundant hummock grass although on more stony ground on foot-slopes of hills *T. wiseana* becomes prominent, whilst in more southerly areas *T. basedowii* replaces *T. pungens*. Small understorey shrubs of *Cassia* and *Acacia* spp., such as *A. bivenosa* and *A. translucens*, may be present. On heavier soils *A. xiphophylla* becomes dominant.

Acacia ancistrocarpa. To the west of the Great Sandy Desert, *Acacia ancistrocarpa* shrublands with hummock grasses occur on shallow earthy sands over laterite. *Acacia monticola* is a characteristic species with *Grevillea wickhamii*, *G. eriostachya*, *G. refracta* and *Hakea suberea* frequent associates. *Triodia pungens* is more abundant in northern areas and *T. basedowii* in southern areas whilst on deeper sands *Plectrachne schinzii* replaces *Triodia*. On sandplains and between sandhills the above association forms a mosaic with *A. coriacea–Hakea suberea* shrubland with hummock grasses. The latter usually occupies more low-lying sandy areas.

To the south of the Tropic, *A. aneura* becomes prominent but in the sandy interdunal areas north of the Great Victoria Desert, *A. aneura* is rare and similar *A. ancistrocarpa* communities occur.

Mixed Mallee–Acacia. In the deserts south of the Great Sandy Desert in the mulga region, *Acacia aneura* is not common in sandy interdunal areas. Various *Acacia* spp., such as *A. helmsiana*, *A. pachyacra*, *A. grasbyi* and *A. linophylla*, form mixed shrublands with mallees such as *Eucalyptus gamophylla* and *E. kingsmillii* and *Triodia basedowii* is the dominant hummock grass. *Plectrachne schinzii* is often present on deeper sands. *Hakea*, *Eremophila leucophylla* and *Grevillea* spp. including *G. juncifolia* and *G. eriostachya* are frequent associates with *Newcastelia cephalantha*, *Dicrastylis exsuccosa* and *Thryptomene maisonneuvii*.

Similar mixed communities occur on sandplains and dunefields in regions of the Northern Territory. Trees and shrubs are sparse. Numerous *Acacia* spp. including *A. maitlandii, A. murrayana, A. kempeana, A. dictyophleba, A. coriacea* and *A. tenuissima* occur in the shrublayer with the mallees *E. gamophylla* and *E. pachyphylla*. Lower shrubs of *Cassia, Grevillea* spp. (including *G. juncifolia*), *Eremophila* spp. (including *E. longifolia* and *E. latrobei*), *Hakea* and *Keraudrinia integrifolia* are widespread. In addition to *Triodia pungens* in northern areas and *T. basedowii* in the south, there is a sparse cover of grasses and forbs.

The communities extend into Queensland where mallees are absent and *Eucalyptus terminalis* and *E. papuana* become more prominent. In southern Queensland and the arid areas of northern New South Wales and South Australia, *A. aneura* becomes more prominent in these communities. *Hakea, Grevillea, Acacia, Cassia* and *Eremophila* are still characteristic genera although the specific composition varies with location. *Triodia basedowii* is the most abundant hummock grass. Grasses such as *Aristida, Enneapogon, Eragrostis* and *Eriachne* together with forbs such as *Helipterum* and *Ptilotus* occur between the hummocks.

Mulga–mallee–hummock grassland communities are common in the west, mainly south of the Tropic of Capricorn in the Wiluna–Warburton area. They are found on sandplains with or without dunes and in the latter case are often restricted to the inter-dunal areas. *Triodia basedowii* is the common hummock grass. *Acacia aneura* is often associated with *A. pruinocarpa, Hakea suberea* and *Grevillea juncifolia* and the most widespread mallees are *Eucalyptus kingsmillii* and *E. gamophylla*. Occasionally the tree *Eucalyptus gongylocarpa* occurs as an emergent. Smaller shrubs such as *Eremophila leucophylla* and *Alyogyne pinoniana* and forbs such as *Helipterum stipitatum* and *Ptilotus polystachyus* are frequent associates.

Utilisation of Acacia *shrublands*

Large areas of *Acacia* shrublands are unoccupied or not in commercial use, largely because of aridity, lack of surface water or the unsatisfactory nature of the understorey species (e.g. *Triodia* hummock grassland) as pasture plants. In the areas used commercially, beef cattle raising is almost exclusively practised in the northwest of Western Australia, the north of South Australia, the far west of Queensland and in the Northern Territory. In the southern portion of *Acacia* shrublands, and in eastern Australia, sheep raising, along with some cattle, is predominant.

Acacia aneura shrublands are by far the most important associations utilised commercially. *Acacia aneura* is Australia's premier fodder shrub, not because it is the most nutritious, but because it is palatable, abundant and widespread (Everist, 1971). During recurrent and cyclical droughts *A. aneura* phyllodes confer stability on an otherwise fragile grazing system by virtue of stock browsing the phyllodes within reach, or particularly in eastern Australia, trees and tall shrubs being felled to provide stock access. In *A. aneura*–tussock grassland communities, sheep have been observed to select up to 70% of *A. aneura* phyllodes in their diet when grass is dry and dormant (Beale, 1975). Methods of using *A. aneura* phyllodes as stock feed were discussed by Everist (1949).

Since *Acacia aneura* is such a useful 'drought' reserve there has been a tendency for it to become depleted, especially in the marginal pastoral zones of southern and western Australia. Legislative controls over the cutting of *A. aneura* scrub have been instigated in Western Australia. Hall, Specht & Eardley (1964) and Preece (1971) reported on the infrequency of *A. aneura* regeneration in the south of the continent. However, at the more mesic end of its range, regeneration of *A. aneura* is adequate (Burrows, 1973).

In areas of Queensland receiving greater than 450 mm mean annual rainfall *A. aneura* densities are so high that different means of thinning stands to promote growth of understorey pasture species have been studied (Beale, 1973; Pressland, 1975). Both authors reported an inverse relationship between shrub density and pasture yield. Artificial thinning of *A. aneura* shrublands becomes questionable in the more arid areas as disturbance may quickly lead to a disclimax of unpalatable shrubs, such as species of *Eremophila* and *Dodonaea*, or at worst a denuded landscape from which shrubs and grasses are both absent.

The potential for replacement of *Acacia* shrublands with more productive pasture species is limited. It is likely, however, that some 40 000 km² of *A. aneura* shrublands will be converted to pasture based on the introduced species, *Cenchrus ciliaris*, within the next 30 to 50 years.

A broader perspective on the distribution, use and potential of *Acacia aneura* shrublands as a vegetation resource in Australia's semi-arid and arid regions can be found in the proceedings of a symposium published in *Tropical Grasslands* (1973), Volume 7 (1). *Acacia* shrublands other than *A. aneura* can be expected to be of little commercial importance and should retain their floristic integrity for the foreseeable future.

References

Beadle, N.C.W. (1948). *The Vegetation and Pastures of Western New South Wales*. Sydney: Government Printer.

Beale, I.F. (1973). Tree density effects on yields of herbage and tree components in south west Queensland mulga (*Acacia aneura* F. Muell.) scrub. *Tropical Grasslands, 7,* 135–42.

Beale, I. F. (1975). Forage intake and digestion by sheep in the mulga zone of Queensland, Australia. Ph.D. thesis, Colorado State University.

Beard, J.S. (1968). Drought effects in the Gibson Desert. *Journal and Proceedings of the Royal Society of Western Australia, 51,* 39–50.

Beard, J.S. (1975). *Vegetation Survey of Western Australia: Nullarbor.* Perth: University of Western Australia Press.

Beeston, G.R. & Webb, A.A. (1977). *The ecology and control of* Eremophila mitchellii. Queensland Department of Primary Industries, Botany Branch Technical Bulletin No. 2.

Boyland, D.E. (1973). Vegetation of the mulga lands with special reference to south-western Queensland. *Tropical Grasslands, 7,* 35–42.

Boyland, D.E. (1974). Vegetation. In *Western Arid Region Land Use Study*, Part I, pp. 47–74. Queensland Department of Primary Industries, Division of Land Utilization Technical Bulletin No. 12.

Burrows, W.H. (1973). Regeneration and spatial patterns of *Acacia aneura* in southwest Queensland. *Tropical Grasslands, 7,* 57–68.

Connor, D.J. & Tunstall, B.R. (1968). Tissue water relations for brigalow and mulga. *Australian Journal of Botany, 16,* 487–90.

Doing, H. (1981). Phytogeography of the Australian Floristic Kingdom. In *Australian Vegetation*, ed. R.H. Groves, pp. 3–25. Cambridge University Press.

Everist, S.L. (1949). Mulga (*Acacia aneura* F. Muell.) in Queensland. *Queensland Journal of Agricultural Science, 6,* 87–139.

Everist, S.L. (1971). Continental aspects of shrub distribution, utilisation and potentials: Australia. In *Wildland Shrubs–Their Biology and Utilisation*, pp. 16–25. US Forest Service General Technical Report INT-1.

Farmer, J.N., Everist, S.L. & Moule, G.R. (1947). Studies in the environment of Queensland. I. The climatology of semi-arid pastoral areas. *Queensland Journal of Agricultural Science, 4,* 21–59.

Francis, W.D. (1925). Observations on the plants of Charleville: characteristics of the western flora. *Queensland Agricultural Journal 24,* 598–602.

Gillison, A.N. & Walker, J. (1981). Woodlands. In *Australian Vegetation*, ed. R.H. Groves, pp. 177–197. Cambridge University Press.

Groves, R.H. & Williams, O.B. Natural grasslands. In *Australian Vegetation*, ed. R.H. Groves, pp. 293–316. Cambridge University Press.

Hall, E.A.A., Specht, R.L. & Eardley, C.M. (1964). Regeneration of the vegetation on Koonamore vegetation reserve, 1926–1962. *Australian Journal of Botany, 12,* 205–64.

Isbell, R.F. (1962). *Soils and vegetation of the brigalow lands, eastern Australia.* CSIRO, Australia, Division of Soils, Soil & Land Use Series No. 43.

Johnson, R.W. (1964). *Ecology and Control of Brigalow in Queensland.* Brisbane: Queensland Department of Primary Industries.

Lange, R. & Purdie, R. (1976). Western myall (*Acacia sowdenii*), its survival prospects and management needs. *Australian Rangelands Journal, 1,* 64–9.

Moore, A.W., Russell, J.S. & Coaldrake, J.E. (1967). Dry matter and nutrient content of a subtropical semiarid forest of *Acacia harpophylla* F. Muell. (brigalow). *Australian Journal of Botany*, *15*, 11–24.

Moore, C.W.E. (1953). The vegetation of the south-eastern Riverina, New South Wales. I. The climax communities. *Australian Journal of Botany*, *1*, 485–547.

Moore, R.M. (1973). Australian arid shrublands. In *Arid Shrublands*, ed. D.N. Hyder, pp. 6–11. Denver: Society for Range Management.

Nix, H.A. & Austin, M.P. (1973). Mulga: a bioclimatic analysis. *Tropical Grasslands*, *7*, 9–22.

Pedley, L. (1973). Taxonomy of the *Acacia aneura* complex. *Tropical Grasslands*, *7*, 3–8.

Pedley, L. (1978). A revision of *Acacia* Mill. in Queensland. *Austrobaileya*, *1*, 75–337.

Perry, R.A. & Lazarides, M. (1962). Vegetation of the Alice Springs area. CSIRO, Australia, Land Research Series No. 6, pp. 208–36.

Preece, P.B. (1971). Contributions to the biology of mulga. II. Germination. *Australian Journal of Botany*, *19*, 39–49.

Pressland, A.J. (1973). Rainfall partitioning by an arid woodland (*Acacia aneura* F. Muell.) in south western Queensland. *Australian Journal of Botany*, *21*, 235–45.

Pressland, A.J. (1975). Productivity and management of mulga in south-western Queensland in relation to tree structure and density. *Australian Journal of Botany*, *23*, 965–76.

Purcell, D.L. (1964). Gidyea to grass in the central west. *Queensland Agricultural Journal*, *90*, 548–58.

Russell, J.S., Moore, A.W. & Coaldrake, J.E. (1967). Relationships between subtropical and semiarid forest of *Acacia harpophylla* (brigalow), micro-relief, and chemical properties of associated gilgai soil. *Australian Journal of Botany*, *15*, 481–98.

Slatyer, R.O. (1961). *Principles and problems of plant production in arid regions.* CSIRO, Australia, Division of Land Research Regional Survey Technical Memorandum No. 61/22.

Slatyer, R.O. (1965). Measurements of precipitation interception by an arid plant community (*Acacia aneura* F. Muell.). *Arid Zone Research*, *25*, 181–92.

Specht, R.L. (1972). *The Vegetation of South Australia*, 2nd edn. Adelaide: Government Printer.

Speck, N.H. (1963). Vegetation of the Wiluna-Meekatharra area. CSIRO, Australia, Land Research Series No. 7, pp. 143–61.

Stebbins, G.L. (1972). Evolution and diversity of arid-land shrubs. In *Wildland Shrubs–Their Biology and Utilization*, pp. 111–20. US Forest Service General Technical Report INT–1.

Tiller, A.B. (1971). Is bendee country worth improving? *Queensland Agricultural Journal*, *97*, 258–61.

10

Eucalyptus *scrubs and shrublands*

R.F. PARSONS

Eucalyptus scrubs and shrublands are defined in this chapter as any vegetation where the tallest stratum is made up of eucalypt shrubs 2 to 8 m high. For scrubs, the dominant shrubs are denser than in shrublands and have a projective foliage cover greater than 30% (Specht, 1970; Carnahan, 1976). By far the major portion of these vegetation types is dominated by eucalypts having many stems arising from a large, underground, woody swelling composed of stem tissue called a lignotuber (syn. 'burl'). Eucalypts with this growth habit are commonly called mallees (an Aboriginal word). This term is also widely used to describe the plant communities and regions where these plants predominate and will be so used here for conciseness.

Mallee communities extensive enough to be mapped at a scale of 1 : 6 000 000 (Carnahan, 1976) extend across Australia from Western Australia (longitude 117°E) to New South Wales (longitude 146°E) with a latitudinal range of from 22° to 37°S. The great majority are located, however, between 30° and 36°S (Carnahan, 1976).

In this main area of mallee occurrence, climate is broadly of mediterranean-type with predominantly winter rainfall. Significant falls of summer rain can occur, however, and it is possible that there is a generally higher proportion of summer rainfall than in otherwise comparable climates in other countries (Rowan & Downes, 1963; Specht, 1969, 1973). Nevertheless, the summer rainfall is highly erratic and summer droughts are characteristic (Leeper, 1970).

Using Köppen's classification, the wetter mallee areas are 'Csb' climates and most of the drier ones are 'BSfk' with smaller areas of 'BWk' and 'BWh' (*Atlas of Australian Resources*, 1973; Climate). Although the widespread BSfk climates have uniform rainfall by Köppen's definition, the ratio of May–October to November–April median rainfall at Australian sites exceeds 1.3 (*Atlas of Australian Resources*, 1973; Climate) and these climates are regarded as 'modified mediterranean' (Leeper, 1970). Coastal mallee areas, such as at Ceduna, South Australia, can have frosts (screen minima 2°C or less) on average for 18 nights per year and heavy frosts (0°C or less) for five nights per year. For inland stations, e.g. Kalgoorlie in Western Australia, the figures rise to twenty-seven and seven respectively (*Atlas of Australian Resources*, 1973: Temperature).

Virtually all mallee mapped at the 1 : 6 000 000 scale by Carnahan (1976) is located between the 178 mm and the 762 mm isohyets for mean annual rainfall. It is most common and often predominant between the 254 mm and 457 mm isohyets. At their upper rainfall limit, mallee communities are usually replaced by woodlands dominated by single-stemmed eucalypts (Chapter 8, Gillison & Walker, 1981), and at their lower rainfall limit by *Acacia* shrublands (Chapter 9, Johnson & Burrows, 1981), *Myoporum* woodlands and other arid communities lacking eucalypts. On going north from the major mallee areas, the proportion of rain falling in summer increases. North of latitude 22°S, at sites with a definite peak in summer rainfall, mallee eucalypts are much less common. Sites with soils and annual rainfall levels suitable for mallee instead usually carry various single-stemmed sub-tropical eucalypts, such as *Eucalyptus dichromophloia* (Beard, 1974). Overall, mallee can be regarded as the most arid of the eucalypt-dominated communities of temperate Australia. Two exceptions to this are the mallee areas of *E. odontocarpa* mallee found between 18 and 22°S in Western Australia, Northern Territory and Queensland, including areas with a very marked peak in summer rainfall (Chippendale, 1963; Hall & Brooker, 1974*b*), and *E. normantonensis* which has a latitudinal and climatic range similar to *E. odontocarpa* (Hall & Brooker, 1974*a*).

Eucalypts with a mallee growth form can also be found in areas wetter than those already considered and where single-stemmed eucalypts are predominant. This occurs in a variety of unfavourable habitats, either by eucalypts which are normally single-stemmed assuming a mallee habit (e.g. *E. baxteri*) or by the occurrence of distinct wet-country mallees of restricted distribution specific to such habitats. Examples of the latter include the sub-alpine *E. kybeanensis* on exposed sites in southern Australia at altitudes up to 1600 m (Hall & Brooker, 1973), *E. rupicola* from sandstone cliff faces in the Blue Mountains of New South Wales (Kleinig & Brooker, 1974) and *E. codonocarpa* from sites with very infertile soils on rhyolite in Queensland and a mean annual rainfall of 1520 mm (Jones, 1964).

From these examples and others, it is clear that the mallee growth-habit can occur in response to a variety of stress conditions. The wet-country mallee communities will not be considered further in this chapter, but a brief account of some of them is given in Chapter 16 (Costin, 1981); for the main areas of mallee vegetation considered in this chapter, the major stress involved is shortage of water.

The mallee growth habit

Mallees are usually 3 to 9 m tall, but can exceptionally reach heights up to 18 m (e.g. some stands of *E. diversifolia* on Kangaroo Island, South Australia). The lignotubers (which occur in most single-stemmed species of *Eucalyptus* as well) arise as swellings in the axils of the cotyledons and first few leaves. They become large, woody, convoluted swellings often 0.3–0.6 m in diameter and sometimes up to 1.5 m. The largest recorded is 10 m across, which carried 301 living stems, in *E. gummifera* (Mullette, 1978).

Lignotubers have the same anatomical characteristics as normal stems but with greatly contorted xylem elements (Chattaway, 1958; Bamber & Mullette, 1978). Although Carrodus & Blake (1970) found no differences in starch contents between some lignotubers and stems, lignotuber wood can have almost twice the proportion of storage tissue as stem wood and thus a larger potential for starch storage (Bamber & Mullette, 1978). Lignotubers also contain a very large number of concealed dormant buds (Carrodus & Blake, 1970).

The frequent fires which occur in mallee areas rarely damage the largely-buried lignotubers. Usually all aerial stems and leaves are killed and new shoots are produced from the dormant buds in the lignotuber. Mallee lignotubers may carry up to 70 shoots six months after fire and this can diminish to about 20–30 seven years later and to less than ten by 100 years (Holland, 1969c).

The stems of mallees usually branch sparingly and bear leaves only at the end of the branches, thereby forming a narrow band of leaves along the tops of their crowns. The canopy resulting from many plants is often very even and horizontal, giving typical mallee communities a very distinctive appearance (Figure 10.1).

Many mallee species occur occasionally as single-stemmed trees. Conversely, many eucalypt species which are usually single-stemmed, occur in multi-stemmed form under adverse site conditions or after destruction of the main stem by fires, termites or felling. The multi-stemmed character of mallee eucalypts is under partial genetic control (Mullette, 1976).

Variation in the extent to which the multi-stemmed character is expressed means that it can be difficult to classify any given species as a mallee or non-mallee. This problem is compounded by the existence in dry areas of Western Australia of shrubby species similar to mallees but with the lignotuber absent or poorly developed. These are called 'marlocks' (Burbidge, 1952; Chippendale, 1973). Despite these problems in delimiting mallees, rough estimates of the numbers of species of mallee eucalypts are provided below.

Evolutionary history

Very little is known about the evolutionary history of mallee vegetation and the few general points that can be made refer almost entirely to the mallee eucalypts themselves. At present, there is a total of about 515 described and generally accepted species of eucalypts (Brooker & Blaxell, 1978; Kelly, 1978). Of these, approximately 108 are mallee eucalypts,

Figure 10.1. Mature stand of mallee, about 8 m tall, dominated by *Eucalyptus viridis*, *E. dumosa* and *E. calycogona* at Kiata Lowan Sanctuary, northwestern Victoria. Photo: T. Pescott.

excluding 'wet country' species (Blakely, 1955; Chippendale, 1973). These 108 species are scattered through at least three different subgenera (Pryor & Johnson, 1971). From this and their floral morphology, it is assumed that the mallee habit is a secondary development in *Eucalyptus* (Burbidge, 1952).

Of the 108 species, 71 occur only in Western Australia, 16 occur only east of Western Australia and 21 are shared by both areas. The much higher species richness of southwest Western Australia than of south-eastern Australia is true of many other plant groups as well. The causes for this striking feature of Australian plant geography (see also Chapter 1, Doing, 1981) are still not well understood. In the present case, Burbidge (1960) suggested that, in Western Australia, many mallee eucalypts arose relatively recently from the older, more stable, 'relic' flora of the wetter parts of southwest Western Australia in response to increasing climatic dryness. The mallee areas of southwest Western Australia are seen as the 'primary centre' for speciation of mallee eucalypts from which a number of species migrated to eastern Australia (Burbidge, 1960).

General ecology
Much of this section has to centre around the ecology of the dominant *Eucalyptus* species as very little detailed community ecology has been done and little is known of the relationship of the understorey species to habitat factors.

The 'core areas' of mallee
In the rainfall zone from 380 to 250 mm, mallee is predominant on a range of calcareous soils; it is this combination which traditionally has been regarded in Australia as constituting the 'typical' mallee country to be dealt with in this section.

Eastern Australia. The major areas are on Quaternary aeolian deposits (see e.g. Lawrence, 1966). Soil texture is very variable and topsoil textures range from sand to clay. Although the soils are predominantly calcareous, some of the deep sands like the Berrook sands can have siliceous surface horizons (Rowan & Downes, 1963). Nevertheless, these are mapped as brown sands with calcareous horizons rather than as leached sands (Northcote, 1960). Whilst topsoil pH for these brown sands is around neutral, for the other main soil groups it is usually strongly alkaline with average values higher than pH 8 (Rowan & Downes, 1963).

The ecologically best-known area is northwestern Victoria (Rowan & Downes, 1963; Connor, 1966; Parsons & Rowan, 1968; Noy-Meir, 1971, 1974) and adjoining parts of South Australia and New South Wales (Noy-Meir, 1971, 1974); accordingly this area will be emphasised here (Table 10.1). The same general relationships seem to apply in the other main areas of mallee in eastern Australia (Smith, 1963; Specht, 1972).

The available literature frequently refers to the marked differences in mallee eucalypt size between communities. This is reflected in the use of various terms like 'big mallee', 'small mallee' etc. (Zimmer, 1937; Rowan & Downes, 1963; Noy-Meir, 1971). These have never been standardised or quantified. Partly after Rowan & Downes (1963), the following will be used here: big mallee usually has three to four stems per plant, stem diameters more than 15 cm at maturity, and height over 6 m; mallee has many thinner stems per plant and height about 3.5–6 m; small mallee is like mallee but with height less than 3.5 m at maturity.

In the region being discussed, soil texture is usually the most important single factor affecting the distribution of native plants (Noy-Meir, 1974). This is related both to increasing levels of macronutrient with increasing clay content (Parsons & Rowan, 1968) and to the 'inverse texture effect' (Rowan & Downes, 1963; Noy-Meir, 1974), whereby soil water supply to plants decreases with increasing clay content. This inverse texture effect is a result of, firstly, the small depth of penetration of rainfall into the soils with higher clay contents, with subsequent increases in evaporation losses, and secondly, the larger amount of rainfall needed to bring the clayier soils from air-dry condition up to the 'available water' range of soil water potential, so that light showers can make water available on dry sandy soils but not on dry clayey ones (Rowan & Downes, 1963; Noy-Meir, 1974). These effects will be discussed in more detail later.

The small mallee, mallee and big mallee categories mentioned above show some obvious general relationships to soil texture, with size of eucalypt plants usually increasing with clay content (Table 10.1). In Victoria, big mallee can occur on clays in the wetter part of the area, but in the driest parts (annual rainfall less than 280 mm), grassland often occupies these soils (Rowan & Downes, 1963). Big mallee is particularly prominent in these dry areas, however, on fertile sandy loams to clay loams. It seems likely that, on these soils, height of mallee eucalypts increases with decreasing rainfall in the area and that this change can

Table 10.1. *Main mallee communities in northwestern Victoria and adjoining areas; annual rainfall 230 to 380 mm*

Main eucalypt species	Eucalypt growth form	Understorey type	Characteristic species	Predominant topsoil texture	Main references
E. incrassata E. foecunda	Small mallee	Sclerophyllous shrubs	E. incrassata Callitris verrucosa Aotus ericoides	Sand	Parsons & Rowan (1968); Noy-Meir (1971).
E. socialis E. dumosa	Small mallee	Hummock grasses (*Triodia*)	E. socialis Triodia spp. Bassia parviflora	Sand[a]	Parsons & Rowan (1968); Noy-Meir (1971).
E. oleosa E. gracilis E. dumosa	Mallee	Mixed shrubs	Acacia colletioides Eremophila glabra Cassia nemophila	Probably loamy sand to sandy loam	Parsons & Rowan (1968); Noy-Meir (1971).
E. oleosa E. gracilis E. dumosa	Mallee, big mallee	Semi-succulent shrubs (chenopods etc.)	E. oleosa E. gracilis Bassia diacantha	Sandy loam to clay loam	Parsons & Rowan (1968); Noy-Meir (1971).
E. calycogona E. dumosa	Mallee, big mallee	Not known	Not known	Clay	Parsons & Rowan (1968); Litchfield (1956).
E. behriana[b]	Mallee, big mallee	Chenopods, low shrub *Acacia*	Not known	Clay	Connor (1966); Litchfield (1956).

[a] These sands are usually more fertile than those carrying *E. incrassata*.
[b] This species mostly occurs where annual rainfall exceeds 330 mm.

occur even when soil fertility stays approximately constant. One possible explanation is that in the wetter, fertile areas, water supply is sufficient to allow ample regeneration, thereby causing high densities of mature eucalypts so that competition between trees restricts the maximum height attained. In contrast, in the drier, fertile areas, lower rainfall may restrict regeneration, thereby causing lower densities of mature eucalypts and thus allowing individual plants to attain greater heights than in dense stands. Data on density and other parameters are badly needed to confirm or reject this hypothesis. Certainly, general observation suggests that densities of mature eucalypts are usually lower in big mallee than in mallee (Rowan, 1971).

It has been suggested that low eucalypt density in big mallee on clay plains in wetter areas may be related to inherent subsoil salinity (Rowan, 1971).

Regarding overall floristics in the mallee communities of the area, the most marked discontinuity is between the *Eucalyptus incrassata* type (Table 10.1) with its species-rich, sclerophyllous understorey (*Hibbertia*, *Aotus*, *Leptospermum* etc.) of 'Southern Temperate' affinities and all the other types, in which semi-succulent shrubs, especially chenopods, are prominent (*Bassia*, *Maireana*, *Zygophyllum* etc.). This floristic series is said to have 'semi-arid, Eremaean' affinities (Noy-Meir, 1971).

From a correlation of floristic data with habitat factors using multiple regression techniques (including both mallee and non-mallee communities in the area), Noy-Meir (1974) suggested that the floristics are determined mainly by variables related to soil texture. The strongest correlations were with depths of soil wetting by various typical falls of rain (calculated as the ratio of amounts of rain to soil-water capacity) and with topsoil water capacity. Other variables contributing consistently but in smaller amounts to floristic variation were the relative importance of calcium in the soil exchange complex, topsoil salinity and subsoil phosphorus (Noy-Meir, 1974). In summary, Noy-Meir (1974) saw the major cause of vegetation variation in the area as an interaction between rainfall and texture, with increased sandiness operating in the same direction as increased rainfall (as expressed by the depth of wetting variables).

The fine degree of control operated by texture-related factors is illustrated by the very closely related species *Eucalyptus socialis* and *E. oleosa*. Sandy soils on crests and slopes of small dunes carry *E. socialis*, but this species gives way completely to *E. oleosa* with only slight increases in clay content on the intervening flats (Parsons & Rowan, 1968).

Some detailed autecological work is available on the mallees *E. incrassata* and *E. socialis* (Parsons, 1968a, 1969b). *Eucalyptus socialis* has drier upper and lower rainfall limits than *E. incrassata* (200–460 mm compared to 250–560 mm). Where the rainfall ranges overlap, *E. socialis* is absent from the sandiest soils lowest in nitrogen, phosphorus and calcium, which carry *E. incrassata*. Relatively fertile, less sandy soils carry *E. socialis*, whilst intermediate soils may carry both species. Thus the relative distribution of the two could be controlled by soil nutrients, soil water supply (inverse texture effect) or both.

Results of pot experiments (Parsons, 1968a, 1969) show that, when grown in monoculture, the species do not differ significantly in growth rate, or responses to varying levels of calcium, nitrogen and phosphorus, but that *E. socialis* has consistently higher root : shoot ratios. Results from competition experiments in sand culture showed no changes in relative competitive ability over a wide range of nitrogen and phosphorus levels, but *E. socialis* had a competitive advantage at high calcium levels. Similar competition experiments in soils carrying either or both species did not reproduce this high calcium effect, but it may operate on soils higher in calcium than those used. In the latter experiments, *E. incrassata* outcompeted *E. socialis* on all soils used at optimal water levels, but this advantage could be nullified on the fertile *E. socialis* soil by droughting. It was suggested that the higher percentage of plants wilting and the greater drought damage of *E. incrassata* than *E. socialis* on the *E. socialis* soil is caused, at least partly, by inferior drought avoidance of *E. incrassata* because of its faster growth in competition (faster water depletion rate) and its lower root : shoot ratio (Parsons, 1968a, 1969).

Thus, it may be suggested that *E. incrassata* can outcompete *E. socialis* on infertile soils with a high water-supplying capacity, whilst drier, more fertile soils carry *E. socialis* in part because of its superior drought avoidance (Parsons, 1969).

The finding that increased soil fertility has an important effect on drought susceptibility of some species by promoting faster growth can be compared with Noy-Meir's (1974) results. That is, not only does increased clay content reduce the amount of soil water available in the area (the inverse texture effect), but it is also correlated with higher fertility which may increase growth rate and water consumption and thus increase drought susceptibility in some species, thereby enhancing the inverse texture effect (Parsons, 1969; Noy-Meir, 1974).

Whilst the detailed information given in this section applies to mallee in southwestern New South Wales (Noy-Meir, 1971, 1974), much less is known about the large, disjunct area of mallee around Mt Hope (Carnahan, 1976) on the eastern edge of Noy-Meir's area. Here it appears that mallee dominated by *E. socialis*, *E. oleosa* and *E. dumosa* is widespread over acidic, non-calcareous loamy soils with a topsoil pH of about 5.6 (Beadle, 1948; Stannard, 1958; Holland, 1968, 1969*b*). This atypical relationship deserves further attention.

Wood (1929) has provided life form spectra (Raunkiaer, 1934) for the type of mallee dealt with in this section, stressing the high percentage of ephemeral species which occur only after rain. The percentage of ephemerals is higher still in the arid communities beyond the lower rainfall limit of mallee (Wood, 1929).

From detailed seasonal sampling of vascular plants, it is known that the percentage of species which are annuals is about 26% in a mature *E. socialis–E. dumosa* mallee community and can reach 44% in a mature *E. incrassata* community. Whilst some of these annuals can germinate in autumn (April), most germinate in winter (July–August). Maximum biomass for annuals occurs in spring or summer (October–early January). By March, towards the end of the dry season, biomass of annuals is insignificant. The normal life span of the annual species involved is from two to ten months (Holland, 1968, 1969*b*).

Perennial species in the field layer of these communities usually grow rapidly in the spring–early summer period (September–January), but produce no new shoots in the late autumn and winter period. By contrast, shoot production by mallee eucalypts and other tall shrubs is from mid-December to early May (Holland, 1968).

Western Australia. There are large areas of mallee mapped in Western Australia between the 250 and 380 mm annual isohyets (Carnahan, 1976). Further large areas were previously mapped as mallee (see e.g. Moore & Perry, 1970) but are now treated as woodland (Carnahan, 1976). This change will be treated in detail below.

The ecology of much Western Australian mallee is poorly understood, so that only a few general points can be made here. Mallee dominated by eucalypts including *Eucalyptus redunca* and *E. eremophila* is mapped as widespread (Beard, 1975) on areas of brown calcareous earths north of Esperance (Northcote *et al.*, 1967). Proceeding north on these soils into drier areas, where annual rainfall drops below about 300 mm, this vegetation is replaced by various communities dominated by appreciably taller eucalypts (more than 10 m tall) which are thus mapped as wood-lands (Beard, 1975). One common type is *E. oleosa–E. flocktoniae* woodland up to 18 m tall. There is continuous variation from this to typical mallee shrubland, *E. oleosa* varying greatly in size and occurring in both structural types (Beard, 1975).

Similar woodlands occur in the same climatic region on other loamy soils, for example the *E. transcontinentalis–E. flocktoniae* community commonly 12–18 m tall and the *E. salmonophloia* community commonly 18–27 m tall, in both cases the dominant eucalypts being single-stemmed. The lowest of these woodlands tend to have the highest eucalypt density and the tallest, the lowest density (Beard, 1969). These woodlands and similar ones make up the very large Goldfields area of Western Australia mapped as woodland at 1 : 6 000 000 by Carnahan (1976), which had been mapped as mallee on previous maps of Aus-tralian vegetation.

There is a similar transition from mallee shrubland to woodland east of the one just described. On the Nullarbor limestone southwest of Caiguna, there is a coastal strip of *E. socialis* mallee about 15 km wide. With decreasing rainfall inland on the same substrate, this changes to *E. oleosa–E. flocktoniae* woodland up to 18 m high (Beard, 1975).

These increases in eucalypt height with declining rainfall on compara-tively fertile soils are strongly reminiscent of the similar change from mallee to big mallee on such soils already mentioned in eastern Aus-tralia. It is again possible that one cause is less regeneration, lower eucalypt density and therefore greater height of individual eucalypts in the drier areas. Such an hypothesis has added credence in Western Australia, because in the examples described, soil uniformity gives greater assurance that fertility is similar throughout the sequence, and also, it is clear that eucalypt height and density are inversely corre-lated (Beard, 1969). In neither area is it known how community biomass changes with decreasing rainfall; clearly, more soil and vegetation data are desirable to clarify the position.

Some of the shorter vegetation mapped by Beard (1975) and sub-sequently by Carnahan (1976) as woodland in the 250 to 300 mm rainfall belt of Western Australia is likely to have a structure identical to 'big mallee' in eastern Australia; for example, the community named as woodland partly dominated by apparently multi-stemmed *Eucalyptus oleosa* in Plate 24, Beard (1975). However, the widespread dominance of eucalypt woodlands 15–27 m tall on non-floodplain sites in this rainfall belt is completely without parallel in eastern Australia, where similar soils in the same climate carry shrublands of the 'big mallee' type with a maximum height of about 9 m.

Whilst some of the smaller Western Australian woodland species, like *Eucalyptus flocktoniae* and *E. oleosa*, spread to eastern Australia, where their stature appears to be somewhat reduced, the tallest Western Australian species, like *E. dundasii*, *E. salmonophloia* and *E. transcontinentalis*, are endemic there (Beard, 1969; Chippendale, 1973). The reasons for the evolution of these species of strikingly larger stature in Western Australia are completely unknown.

In moving inland through the main eucalypt-dominated areas of eastern Australia, the driest of these areas carry mallee shrubland, whilst in Western Australia such areas usually carry eucalypt woodland (Carnahan, 1976).

In the Western Australian area considered in this section there is a range of mallee communities not discussed here (Beard, 1975 and references therein). Their detailed environmental relationships are not known.

It should be mentioned that in mallee 'core areas' in both western and eastern Australia, dunes predominantly made up of gypsum particles occur around margins of salt lakes and often carry mallee communities. Whilst the mallee eucalypts so far recorded are widespread species not confined to gypsum, it is not yet known whether any of the understorey species are obligate gypsophiles. Shrub and herb gypsophiles are already known from other communities on these deposits (Parsons, 1976).

At the drier rainfall limits of the core areas of mallee, species such as the widespread mallee *E. socialis* become confined to deep sandy soils along watercourses and occasionally skeletal soils on rocky slopes (Carrodus, Specht & Jackman, 1965), presumably because of the superior water-supplying characteristics of such soils in this climate (the inverse texture effect of Noy-Meir, 1974).

Other areas of mallee
Other types of mallee occur in wetter regions, often on acidic, non-calcareous soils and, especially in eastern Australia, often in regions where significant amounts of non-mallee vegetation occur, especially on the most productive sites.

Where average annual rainfall ranges from 380 mm to 430 mm, mallee usually predominates. In even wetter areas up to 660 mm, mallee becomes scarcer and is often restricted to a range of relatively infertile soils (Table 10.2).

The commonest of these types, and the one which usually adjoins the core areas of mallee on their wetter side, is mallee on either deep siliceous sands or on siliceous sands over sandy clays (Table 10.2). All the other communities listed in Table 10.2 can be found in comparatively wet areas as well, except for the final, arid category. Some Western Australian types not listed there include *Eucalyptus preissiana–E. lehmannii* mallee on the Barren Ranges quartzites and *E. gardneri–E. nutans* mallee on the basic, igneous 'greenstones' of the Ravensthorpe area (Beard, 1972*a*).

Large, continuous belts of mallee are usually absent once annual rainfall drops below about 200–230 mm and mallee occurs as discontinuous patches in a matrix of arid zone communities usually lacking eucalypts. These mallee communities are mapped as far north as 22°S (Carnahan, 1976), mainly in areas with annual rainfalls from 170 to 200 mm. They contain a distinctive suite of species mostly absent from the wetter mallee communities to the south (Table 10.2).

Production ecology and nutrient cycling
Biomass–time curves from *E. incrassata* mallee–broombush for 12 years after fire reflect rapid regeneration from lignotubers, and the eucalypts are still increasing in biomass after 12 years, with an average aerial biomass increment of 8–9% per year. The broombush species (*Baeckea behrii* and *Melaleuca uncinata*) contribute 20–30% of standing community biomass, reaching peak biomass eight years after fire and then declining (Specht, 1966). In other types of mallee, average aerial biomass increment for eucalypts is between 6 and 8% per year for stands up to about 35 years after burning (Holland, 1969*a*).

Productivity comparisons of 15-year-old mallee regeneration after clearing with mature mallee show above-ground net primary productivity of 5406 kg ha^{-1} yr^{-1} and 2379 kg ha^{-1} yr^{-1} respectively. The corresponding leaf area indices are 0.57 and 0.73. Understorey shrubs make up only 2.3% of the total standing biomass by about 55 years after fire. Peak productivity following destruction of all aerial plant parts by clearing or burning may be reached after about 15 years. Standing biomass may not reach a plateau for at least 30 years (Burrows, 1976). In the fire management context, it has been suggested that this time period to reach a plateau gives a rough guide to the length of time needed between fires to ensure maximal levels of species survival for species

Table 10.2. *Some mallee communities other than those listed in Table 10.1*

Category	Soil characteristics	Eucalypt species	Understorey type	Main references
Mallee on calcareous coastal soils				
(a) Eastern Australia	Calcareous beach sands; soils on aeolian calcarenite	*E. diversifolia* *E. rugosa*	Coastal shrubs	Parsons & Specht (1967); Parsons (1968a); Specht (1972).
(b) Western Australia	Calcareous beach sands; soils on aeolian calcarenite	*E. calcicola* *E. oraria*	Coastal shrubs	Brooker (1974); Beard (1976b).
Mallee on siliceous dunes and sandplains				
(a) Eastern Australia	Deep siliceous sand; siliceous sand over sandy clay	*E. incrassata* *E. foecunda*	Sclerophyllous shrubs, including 'broombush' type	Coaldrake (1951); French (1958); Specht (1966); Parsons & Specht (1967); Parsons (1968a); Specht (1972).
(b) Western Australia	Deep siliceous sand; siliceous sand over sandy clay	*E. incrassata* *E. tetragona* *E. redunca* *E. eremophila*	Sclerophyllous shrubs, including 'broombush' type	Beard (1972a, b, 1973a, b).
Mallee on ironstone soils				
(a) Eastern Australia	Various ironstone gravel soils; a gravelly sand over clay type is common	*E. cosmophylla* *E. diversifolia* *E. remota* *E. incrassata* *E. viridis*	Sclerophyllous shrubs, including "broombush" type	Cleland (1928); Baldwin & Crocker (1941); Northcote & Tucker (1948); Litchfield (1956); French (1958).

(b) Western Australia	Various ironstone gravel soils; a gravelly sand over clay type is common	*E. tetragona* *E. tetraptera* *E. incrassata* *E. eremophila*	Sclerophyllous shrubs, including 'broombush' type	Beard (1969, 1972a, 1973a).
Mallee on rocky, sandstone soils Eastern Australia	Skeletal or shallow loam-over-clay soils on Ordovician sandstone	*E. viridis* *E. polybractea* *E. behriana* *E. froggattii*	Sclerophyllous shrubs of 'broombush' type	Beadle (1948); Biddiscombe (1963); Rowan (1963); Hall (1970a, b).
Arid mallee communities[a]	Sandplains; interdune sands; skeletal, rocky soils	*E. gamophylla* *E. kingsmillii* *E. oleosa* complex *E. oxymitra* *E. pachyphylla* *E. youngiana* *E. gillii*	Shrubs and *Triodia* hummock grasses	Chippendale (1963); Boomsma (1972); Mabbutt *et al.* (1973); Beard (1975); Hall & Brooker (1976); Carnahan (1976).

[a]Mallee communities in areas where mean annual rainfall is less than 200–230 mm.

regenerating from underground lignotubers and rootstocks, as this time interval should be sufficient to prevent depletion of lignotuber and rootstock food reserves (Groves, 1977).

Regarding nutrient cycling, nutrient pool size for 15-year-old mallee regeneration after clearing has been compared with that for mature mallee. The results (Table 10.3) show that where the surface soil remains intact, as in this case, there is little difference in the total pool sizes, suggesting high resilience of the nutrient pool to massive disturbance. The ability to regenerate from lignotubers may exert a stabilising influence by allowing rapid initial uptake, after which community requirements (at least for mobile elements) may be largely met by recycling (Burrows, 1976).

Litter production is strongly seasonal with a pronounced mid-summer maximum, and in this regard is similar to other eucalypt communities occurring in southern Australia. Regarding nutrient withdrawal from leaves prior to abscission, the percentage weight loss per unit area is between 52 and 57% for phosphorus and between 37 and 48% for nitrogen. Whilst such figures indicate some conservation in the use of these nutrients, it is not yet clear whether they indicate any special adaptation to infertile soils or whether similar values also occur in vegetation on soils of higher fertility (Burrows, 1976).

Regeneration of mallee communities
The mallee eucalypts examined produce abundant germinable seed. Despite this, naturally occurring seedlings are only rarely observed and recovery from burning or felling is predominantly by the growth of new shoots from established lignotubers. Most observations of seedlings are in areas recently burned where disturbance has removed some of the established eucalypts. The scarcity of seedlings is attributed variously to rabbit grazing, competition from mature eucalypts or lignotuber regrowth, and climate (Sims, 1951; Parsons, 1968b and references therein; B. Wellington, personal communication). Given a stand that is not 'fully-stocked' and a scarcity or absence of rabbits, it is likely that a major determinant of seedling establishment will be spring and summer rainfall in the year following germination (Parsons, 1968b). Particularly in the drier mallee areas, it seems plausible that spring–summer rainfalls may only rarely be sufficient for this, leading to erratic pulses of seedling recruitment. In dry areas adjacent to mallee, rainfall has been sufficient to allow seedling regeneration of *Acacia sowdenii* only three times this century (Lange & Purdie, 1976).

Table 10.3. *Organic matter, nitrogen and phosphorus distribution in two stands of mallee dominated by Eucalyptus socialis, after Burrows (1976). 1, stand regenerated from clearing 15 years earlier; 2, stand undisturbed for about 55 years*

	Organic matter (kg ha^{-1})		Total nitrogen (kg ha^{-1})		Total phosphorus (kg ha^{-1})	
	1	2	1	2	1	2
Above-ground						
Total standing	19 553	40 164	75.4	88.1	5.8	5.0
Total litter	7 547	11 372	41.2	54.2	2.1	3.1
Total above-ground	27 100	51 536	116.6	142.3	7.9	8.1
Below-ground (to 1 m depth)						
Eucalypt lignotubers	15 288	13 860	25.4	23.1	2.5	2.2
Total lignotubers and roots	20 501	28 533	47.3	103.9	3.8	6.7
Soil	79 800	73 336	5074	5804	3423	4035
Total	127 401	153 405	5238	6050	3455	4050

Thus most observed regeneration is from lignotuber regrowth. In this regard it is worth noting that a 1 to 1.5-year-old mallee eucalypt plant can be completely defoliated 26 times before death ensues (Chattaway, 1958). Some regrowth shoots can produce seed about three years after fire (Gardner, 1957), although about eight years may be more usual (Parsons, 1968*b*). The lignotuber regeneration strategy is of course common in genera other than *Eucalyptus*, both in Australia and elsewhere (Gill, 1975).

Regarding the life span of mallee eucalypts, detailed radio-carbon dating of old plants shows that all plants examined are not demonstrably older than 200 yr B.P. If mallees perpetuate themselves through progressive tissue replacement, older tissues may be destroyed by decay, and true age may be greater than that shown by existing tissues (B. Wellington, personal communication).

In mallee eucalypts the importance of lignotuber regrowth, rather than regeneration from seed after burning, stands in marked contrast to the behaviour of *Callitris* spp., the native conifers which dominate large areas of woodlands within and adjacent to the main mallee areas. These species are killed by fire and regenerate solely from seed.

An equally interesting contrast can be seen in many mallee areas of Western Australia, where, interspersed among mallee communities, there are frequent areas dominated by single-stemmed low eucalypts which are killed by fire and which regenerate only from seed. The best known examples are *Eucalyptus annulata*, *E. platypus* and *E. spathulata* 5 to 7 m tall on low-lying clay soils (Beard, 1967, 1972*a*). These are referred to as 'thicket-formers' or 'marlocks' (Beard, 1967, 1972*a*), the latter name implying that lignotubers are absent or poorly developed (Burbidge, 1952). Further examples are *E. diptera*, *E. eremophila* and *E. forrestiana*, some of which occur in stands mixed with 'true' mallee species (Beard, 1975).

The same total dependence on seedling regeneration rather than lignotuber regeneration is seen in the tall, dry woodland species *E. salmonophloia* (Beard, 1972*b*), which occurs at low density over a sparse understorey, so that these woodlands 'burn only rarely and with difficulty if at all' (Beard, 1969). Beard (1969) indicated as a generalisation that the other eucalypt tree dominants of these arid woodlands in the Goldfields region of Western Australia also show this regeneration strategy, which is puzzling given that they include *E. oleosa*, which occurs widely as a mallee capable of lignotuber regrowth. Similarly, normally single-stemmed tree species, such as *E. falcata* and *E. gardneri*, are regarded as fire-tender seed-regenerators (Brockway & Hillis, 1955; Beard, 1972*b*), but are also said to occur in mallee forms capable of lignotuber regrowth after fire (Brockway & Hillis, 1955). More data on lignotuber incidence and behaviour are badly needed for these and similar species.

In summary, it is striking that Western Australia has a number of apparently strictly seed-regenerating non-mallee eucalypts in habitats which in eastern Australia would be entirely mallee-dominated. With the exception of flood-plain tree species like *Eucalyptus camaldulensis*, all the eucalypts in the mallee regions of eastern Australia have well-developed lignotubers which readily produce new shoots after burning. The apparent unimportance or absence of lignotubers in *E. salmonophloia* and similar woodlands at and around the main dry limits for eucalypts in Western Australia is an important exception to the generalisation that lignotubers are best developed in *Eucalyptus* where the genus is close to its physiological limits (Gill, 1975).

The only study of fire effects on community floristics is from small mallee with a *Triodia* understorey in a 260 mm annual rainfall area in Victoria. Here, unburnt areas carry 18 vascular plant species, but a year after fire a burnt area carried 63 species, including 26 annuals (Zimmer, 1940). This appears to be the largest fire-induced increase in species richness yet recorded in Australia (see Gill, 1977 and references therein).

From palynological data near the lower rainfall limit of mallee vegetation in the Nullarbor Plain, it is thought that firing and felling of mallee since 4000 to 6000 B.P. by Aborigines has led to some areas of mallee being replaced by *Acacia sowdenii–Myoporum platycarpum* tall open-shrubland because of restricted eucalypt regeneration in this dry area (Martin, 1973).

Regarding re-colonisation of mallee areas previously completely cleared for agriculture, the most important study is from the 300 to 330 mm rainfall zone in Victoria. Here, unstabilised deep sands were sown with *Secale cereale* (cereal rye). In one area, in the second year, dominance is shared by self-sown *S. cereale* and the alien annual *Brassica tournefortii*. In the third year, 95% of the plants are *B. tournefortii*, whilst in the fourth year this species declines markedly and 80% of the plants are the native annual composite *Myriocephalus stuartii* (Sims, 1949). The latter species was also one of the post-fire pioneers in Zimmer's (1940) study. These very marked and rapid changes in dominance by various annual species are strongly reminiscent of those recorded on abandoned farms in Idaho in a similarly semi-arid climate

(Piemeisel, 1951) and alien crucifers and *Salsola* are important in both areas. In Australia, thistles (*Onopordum, Carthamus*) and alien grasses (*Schismus, Hordeum*) are also important early colonisers, both in Victoria (Sims, 1949) and New South Wales (Beadle, 1948).

Little is known about colonisation by perennials. Mallee eucalypt seedlings can establish where mallee branches bearing fruit with seed are laid down on bare soil (Sims, 1951), presumably where rabbits are scarce. The wind-dispersed *Dodonaea angustissima* is an early colonist of abandoned mallee farmland in some areas (Beadle, 1948; Sims, 1951), as are *Acacia colletioides* (Ioannou, 1968), *A. brachybotrya, A. ligulata* (Sims, 1949) and *Cassia nemophila* (Sims, 1951), the last four presumably at least in part from soil-stored seed.

It is clear from work in adjacent communities (Hall, Specht & Eardley, 1964) and from general observations, that the most serious single threat to the future regeneration and conservation of mallee vegetation is the destruction of seedlings by grazing rabbits. Unless controlled, rabbits will certainly lead to the eventual disappearance of numerous species and communities. The evidence for this is clearest for the woody species extending to mallee areas dealt with by Hall, Specht & Eardley (1964). What few data are available suggest that the same is true of the mallee eucalypts themselves.

Concluding discussion

The overall relationship of understorey type to climate and soil can be summarised as follows (after Specht, 1972).

- (a). Infertile soils in the wettest areas have a dense, species-rich understorey of sclerophyllous shrubs usually less than 2 m high ('mallee–heath' type).
- (b). Infertile soils of areas of intermediate rainfall have a dense understorey of sclerophyllous shrubs with repeatedly-branching, erect stems terminating at about the same height, usually more than 2 m high and often species of *Melaleuca* ('mallee–broombush' type).
- (c). Infertile soils of the driest areas have a stratum of hummock grasses (*Triodia* spp.) with or without a range of shrubs.
- (d). Fertile soils have a range of understorey types including sparse grassy ones and various shrub mixtures. With increasing dryness, semi-succulent low shrubs (chenopods, *Cratystylis, Zygophyllum*) become progressively more important and finally predominant.

Regarding the general climatic and edaphic ranges of the eucalypts present within a given area, species are often very sensitive to environmental changes and show precise correspondence with particular micro-habitats, as was shown for *Eucalyptus incrassata* and *E. socialis* in Victoria. When total range is considered, however, a species can be found in a very wide range of habitats. For example, *E. incrassata* can occupy soils ranging from excessively drained to seasonally waterlogged and from extremely infertile siliceous sands and ironstone soils to relatively fertile calcareous loams. This, with the experimental data cited earlier, suggests that a number of species have wide physiological tolerances and that competition plays an important part in determining the precise correspondence with particular microhabitats observed within given areas.

Considering total species range, wide edaphic ranges are not uncommon, with species such as *Eucalyptus diversifolia*, *E. incrassata* and *E. eremophila* occurring over wide ranges of soil texture, pH and fertility. In these and other cases, climate (probably rainfall in particular) seems to be the most important factor determining the overall distribution limits. No mallee eucalypt has yet been shown to be physiologically specific to a narrow range of edaphic conditions. Narrow endemic species exist, however, but have not been investigated experimentally.

Within the climatically-determined overall range, in the only area studied in detail, soil texture is the most important determinant of species distribution. In most cases, this probably operates via water availability, fertility or an interaction between these two factors.

To give a rough idea of species richness in mallee eucalypts, north-western Victoria has 12 species of mallee eucalypts in an area of 22 362 km^2 which is mostly mallee-dominated, and any given community contains between one and four such species. Presumably, such species richness figures would be much higher in comparable areas of Western Australia.

Virtually all aspects of the ecology of mallee communities are in need of further work. For example, the very small amount of ecophysiological work available is virtually restricted to mineral nutrition, whilst so little is known of regeneration that grave concern should be felt concerning the management and conservation of these communities. Finally, it seems likely that work comparing mallee communities with the adjacent tall eucalypt woodlands of single-stemmed obligate seed-regenerators in very similar habitats in the Goldfields area of Western Australia would be capable of yielding a variety of valuable ecological insights.

I thank Dr M. Calder, Chairman, Botany School, University of Melbourne for generously supplying facilities during the study leave when this chapter was written. I have been greatly helped by stimulating discussions with Dr D.H. Ashton of the same department.

References

Atlas of Australian Resources, Second Series. (1973). Canberra: Australia, Department of Minerals & Energy.

Baldwin, J.G. & Crocker, R.L. (1941). The soils and vegetation of portion of Kangaroo Island, South Australia. *Transactions of the Royal Society of South Australia*, 65, 263–75.

Bamber, R.K. & Mullette, K.J. (1978). Studies of the lignotubers of *Eucalyptus gummifera* (Gaertn. & Hochr.). II. Anatomy. *Australian Journal of Botany*, 26, 15–22.

Beadle, N.C.W. (1948). *The Vegetation and Pastures of Western New South Wales*. Sydney: Government Printer.

Beard, J.S. (1967). A study of patterns in some West Australian heath and mallee communities. *Australian Journal of Botany*, 15, 131–9.

Beard, J.S. (1969). The vegetation of the Boorabbin and Lake Johnston areas, Western Australia. *Proceedings of the Linnean Society of New South Wales*, 93, 239–68.

Beard, J.S. (1972a). *The Vegetation of the Newdegate and Bremer Bay Areas, Western Australia*. Sydney: Vegmap Publications.

Beard, J.S. (1972b). *The Vegetation of the Southern Cross Area, Western Australia*. Sydney: Vegmap Publications.

Beard, J.S. (1973a). *The Vegetation of the Ravensthorpe Area, Western Australia*. Perth: Vegmap Publications.

Beard, J.S. (1973b). *The Vegetation of the Esperance and Malcolm Areas, Western Australia*. Perth: Vegmap Publications.

Beard, J.S. (1974). *Vegetation Survey of Western Australia – Great Sandy Desert*. Perth: University of Western Australia Press.

Beard, J.S. (1975). *Vegetation Survey of Western Australia–Nullarbor*. Perth: University of Western Australia Press.

Beard, J.S. (1976a). *The Vegetation of the Dongara Area, Western Australia*. Perth: Vegmap Publications.

Beard, J.S. (1976b). *The Vegetation of the Shark Bay and Edel Areas, Western Australia*. Perth: Vegmap Publications.

Biddiscombe, E.F. (1963). *A vegetation survey in the Macquarie region, New South Wales*. CSIRO, Australia, Division of Plant Industry, Technical Paper Number 18.

Blakely, W.F. (1955). *A Key to the Eucalypts*, 2nd edn. Canberra: Forestry & Timber Bureau.

Boomsma, C.D. (1972). *Native trees of South Australia*. South Australia, Woods and Forests Department Bulletin Number 19.

Brockway, G.E. & Hillis, W.E. (1955). Tan bark eucalypts of the semi-arid regions of south western Australia. *Empire Forestry Review*, 34, 31–41.

Brooker, M.I.H. (1974). Six new species of *Eucalyptus* from Western Australia. *Nuytsia*, 1, 297–314.

Brooker, M.I.H. & Blaxell, D.F. (1978). Five new species of *Eucalyptus* from Western Australia. *Nuytsia, 2*, 220–31.

Burbidge, N.T. (1952). The significance of the mallee habit in *Eucalyptus*. *Proceedings of the Royal Society of Queensland, 62*, 73–8.

Burbidge, N.T. (1960). The phytogeography of the Australian region. *Australian Journal of Botany, 8*, 75–212.

Burrows, W.H. (1976). Aspects of nutrient cycling in semi-arid Mallee and Mulga communities. Ph.D. thesis, Australian National University.

Carnahan, J.A. (1976). Natural vegetation. In *Atlas of Australian Resources*, Second Series. Canberra: Department of National Resources.

Carrodus, B.B. & Blake, T.J. (1970). Studies on the lignotubers of *Eucalyptus obliqua* L'Herit. I. The nature of the lignotuber. *New Phytologist, 69*, 1069–72.

Carrodus, B.B., Specht, R.L. & Jackman, M.L. (1965). The vegetation of Koonamore Station, South Australia. *Transactions of the Royal Society of South Australia, 89*, 41–57.

Chattaway, M.M. (1958). Bud development and lignotuber formation in eucalypts. *Australian Journal of Botany, 6*, 103–15.

Chippendale, G.M. (1963). Ecological notes on the 'Western Desert' area of the Northern Territory. *Proceedings of the Linnean Society of New South Wales, 88*, 54–66.

Chippendale, G.M. (1973). *Eucalypts of the Western Australian Goldfields*. Canberra: Australian Government Publishing Service.

Cleland, J.B. (1928). The plants of the Encounter Bay district, notes on the ecology. *South Australian Naturalist, 9*, 57–60.

Coaldrake, J.E. (1951). *The climate, geology, soils and plant ecology of a portion of the County of Buckingham (Ninety-Mile Plain), South Australia*. CSIRO, Australia, Bulletin Number 266.

Connor, D.J. (1966). Vegetation studies in north-west Victoria. I. The Beulah-Hopetoun area. *Proceedings of the Royal Society of Victoria, 79*, 579–95.

Costin, A.B. (1981). Alpine and sub-alpine vegetation. In *Australian Vegetation*, ed. R.H. Groves, pp. 361–376. Cambridge University Press.

Doing, H. (1981). Phytogeography of the Australian Floristic Kingdom. In *Australian Vegetation*, ed. R.H. Groves, pp. 3–25. Cambridge University Press.

French, R.J. (1958). *Soils of Eyre Peninsula*. South Australia Department of Agriculture Bulletin Number 457.

Gardner, C.A. (1957). The fire factor in relation to the vegetation of Western Australia. *Western Australia Naturalist, 5*, 166–73.

Gill, A.M. (1975). Fire and the Australian flora: a review. *Australian Forestry, 38*, 4–25.

Gill, A.M. (1977). Management of fire-prone vegetation for plant species conservation in Australia. *Search, 8*, 20–6.

Gillison, A.N. & Walker, J. (1981). Woodlands. In *Australian Vegetation*, ed. R.H. Groves, pp. 177–197. Cambridge University Press.

Groves, R.H. (1977). Fire and nutrients in the management of Australian vegetation. In *Proceedings of the Symposium on Environmental Consequences of Fire and Fuel Management in Mediterranean Ecosystems*, ed. H.A. Mooney & C.E. Conrad, pp. 220–9. Washington: US Department of Agriculture Forest Service General Technical Report WO–3.

Hall, E.A.A., Specht, R.L. & Eardley, C.M. (1964). Regeneration of the vegetation on Koonamore Vegetation Reserve, 1926–1962. *Australian Journal of Botany*, *12*, 205–64.

Hall, N. (1970*a*). *Blue-leaved mallee*, Eucalyptus polybractea *R.T. Bak.* Australia, Forestry & Timber Bureau Forest Tree Series Number 5.

Hall, N. (1970*b*). *Green mallee*, Eucalyptus viridis *R.T. Bak.* Australia, Forestry & Timber Bureau Forest Tree Series Number 6.

Hall, N. & Brooker, I. (1973). *Kybean mallee ash*, Eucalyptus kybeanensis *Maiden et Cambage.* Australia, Forestry & Timber Bureau Forest Tree Series Number 95.

Hall, N. & Brooker, I. (1974*a*). *Normanton box*, Eucalyptus normantonensis *Maiden et Cambage.* Australia, Forestry & Timber Bureau Forest Tree Series Number 160.

Hall, N. & Brooker, I. (1974*b*). *Sturt Creek mallee*, Eucalyptus odontocarpa *F. Muell.* Australia, Forestry & Timber Bureau Forest Tree Series Number 161.

Hall, N. & Brooker, I. (1975). *Blue mallee*, Eucalyptus gamophylla *F. Muell.* Australia, Forestry & Timber Bureau Forest Tree Series Number 179.

Holland, P.G. (1968). Seasonal growth of field layer plants in two stands of mallee vegetation. *Australian Journal of Botany*, *16*, 615–22.

Holland, P.G. (1969*a*). Weight dynamics of *Eucalyptus* in the mallee vegetation of southeast Australia. *Ecology*, *50*, 212–9.

Holland, P.G. (1969*b*). The plant patterns of different seasons in two stands of mallee vegetation. *Journal of Ecology*, *57*, 323–33.

Holland, P.G. (1969*c*). The maintenance of structure and shape in three mallee eucalypts. *New Phytologist*, *68*, 411–21.

Ioannou, N. (1968). A study of historical effects on the vegetation of the marginal lands. B.Sc. (Hons.) thesis, University of Adelaide.

Johnson, R.W. & Burrows, W.H. (1981). *Acacia* open-forests, woodlands and shrublands. In *Australian Vegetation*, ed. R.H. Groves, pp. 198–226. Cambridge University Press.

Jones, R. (1964). The mountain mallee heath of the McPherson Ranges. *University of Queensland, Department of Botany Papers*, *4*, 159–220.

Kelly, S. (1978). *Eucalypts*, Vol. 2. Melbourne: Nelson.

Kleinig, D. & Brooker, I. (1974). *Cliff mallee ash*, Eucalyptus rupicola *L. Johnson & D. Blaxell.* Australia, Forestry & Timber Bureau Forest Tree Series Number 169.

Lange, R. & Purdie, R. (1976). Western myall (*Acacia sowdenii*), its survival prospects and management needs. *Australian Rangelands Journal*, *1*, 64–9.

Lawrence, C.R. (1966). Cainozoic stratigraphy and structure of the Mallee region, Victoria. *Proceedings of the Royal Society of Victoria*, *79*, 517–53.

Leeper, G.W. (1970). Climates. In *The Australian Environment*, ed. G.W. Leeper, 4th edn (rev.), pp. 12–20. Melbourne: CSIRO, Australia & Melbourne University Press.

Litchfield, W.H. (1956). Species distribution over part of the Coonalpyn Downs, South Australia. *Australian Journal of Botany*, *4*, 68–115.

Mabbutt, J.A., Burrell, J.P., Corbett, J.R. & Sullivan, M.E. (1973). In *Lands of Fowler's Gap Station, New South Wales*, ed. J.A. Mabbutt, pp. 25–43. University of New South Wales Research Series Number 3.

Martin, H.A. (1973). Palynology and historical ecology of some cave excavations in the Australian Nullarbor. *Australian Journal of Botany*, *21*, 283–316.

Moore, R.M. & Perry, R.A. (1970). Map No. 3. Vegetation of Australia. In *Australian Grasslands*, ed. R.M. Moore. Canberra: Australian National University Press.

Mullette, K.J. (1976). Mallee and tree forms within *Eucalyptus* species. Ph.D. thesis, University of New South Wales.

Mullette, K.J. (1978). Studies of the lignotuber of *Eucalyptus gummifera* (Gaertn. & Hochr.) I. The nature of the lignotuber. *Australian Journal of Botany*, 26, 9–13.

Northcote, K.H. (1960). *Atlas of Australian Soils. Explanatory Data for Sheet 1 Port Augusta – Adelaide – Hamilton Area*. Melbourne: CSIRO, Australia.

Northcote, K.H., Bettenay, E., Churchward, H.M. & McArthur, W.M. (1967). *Atlas of Australian Soils. Explanatory Data for Sheet 5 Perth – Albany – Esperance Area*. Melbourne: CSIRO, Australia.

Northcote, K.H. & Tucker, B.M. (1948). *A Soil survey of the Hundred of Seddon and part of the Hundred of MacGillivray, Kangaroo Island, South Australia*. Council for Scientific and Industrial Research, Australia, Bulletin No. 233.

Noy-Meir, I. (1971). Multivariate analysis of the semi-arid vegetation in south-eastern Australia: nodal ordination by component analysis. *Proceedings of the Ecological Society of Australia, 6*, 159–93.

Noy-Meir, I. (1974). Multivariate analysis of the semi-arid vegetation in south-eastern Australia. *Australian Journal of Botany, 22*, 115–40.

Parsons, R.F. (1968*a*). Ecological aspects of the growth and mineral nutrition of three mallee species of *Eucalyptus. Oecologia Plantarum, 3*, 121–36.

Parsons, R.F. (1968*b*). An introduction to the regeneration of mallee eucalypts. *Proceedings of the Royal Society of Victoria, 81*, 59–68.

Parsons, R.F. (1969). Physiological and ecological tolerances of *Eucalyptus incrassata* and *E. socialis* to edaphic factors. *Ecology, 50*, 386–90.

Parsons, R.F. (1976). Gypsophily in plants – a review. *American Midland Naturalist, 96*, 1–20.

Parsons, R.F. & Rowan, J.N. (1968). Edaphic range and cohabitation of some mallee eucalypts in south-eastern Australia. *Australian Journal of Botany, 16*, 109–16.

Parsons, R.F. & Specht, R.L. (1967). Lime chlorosis and other factors affecting the distribution of *Eucalyptus* on coastal sands in southern Australia. *Australian Journal of Botany, 15*, 95–105.

Piemeisel, R.L. (1951). Causes affecting change and rate of change in a vegetation of annuals in Idaho. Ecology, *32*, 53–72.

Pryor, L.D. & Johnson, L.A.S. (1971). *A Classification of the Eucalypts*. Canberra: Australian National University.

Raunkiaer, C. (1934). *The Life Forms of Plants and Statistical Plant Geography*. Oxford: University Press.

Rowan, J.N. (1963). *A Study of the Mallee Lands around Wedderburn and Inglewood*. Melbourne: Soil Conservation Authority of Victoria (mimeographed report).

Rowan, J.N. (1971). *Salting on dryland farms in north-western Victoria*. Soil Conservation Authority of Victoria Technical Communication Number 7.

Rowan, J.N. & Downes, R.G. (1963). *A study of the land in north-western Victoria*. Soil Conservation Authority of Victoria Technical Communication Number 2.

Sims, H.J. (1949). Plant regeneration on stabilised sandhills in the Mallee. *Victorian Naturalist, 66*, 37–9.

Sims, H.J. (1951). The natural regeneration of some trees and shrubs at Walpeup, Victorian Mallee. *Victorian Naturalist*, *68*, 27–30.

Smith, D.F. (1963). The plant ecology of lower Eyre Peninsula, South Australia. *Transactions of the Royal Society of South Australia*, *87*, 93–118.

Specht, R.L. (1966). The growth and distribution of mallee-broombush (*Eucalyptus incrassata – Melaleuca uncinata* association) and heath vegetation near Dark Island Soak, Ninety-Mile Plain, South Australia. *Australian Journal of Botany*, *14*, 361–71.

Specht, R.L. (1969). A comparison of the sclerophyllous vegetation characteristic of mediterranean type climates in France, California, and southern Australia. I. Structure, morphology, and succession. *Australian Journal of Botany*, *17*, 277–92.

Specht, R.L. (1970). Vegetation. In *The Australian Environment*, 4th edn (rev.), ed. G.W. Leeper, pp. 44–67. Melbourne: CSIRO, Australia & Melbourne University Press.

Specht, R.L. (1972). *The Vegetation of South Australia*, 2nd edn. Adelaide: Government Printer.

Specht, R.L. (1973). Structure and functional response of ecosystems in the mediterranean climate of Australia. In *Ecological Studies. Analysis and Synthesis*, Vol. 7, ed. F. di Castri & H.A. Mooney, pp. 113–20. Berlin: Springer-Verlag.

Stannard, M.E. (1958). Erosion survey of the south-west Cobar peneplain. II. Soils and vegetation. *Journal of the Soil Conservation Service of New South Wales*, *14*, 30–45.

Wood, J.G. (1929). Floristics and ecology of the mallee. *Transactions of the Royal Society of South Australia*, *53*, 359–78.

Zimmer, W.J. (1937). *The flora of the far north-west of Victoria*. Forests Commission of Victoria Bulletin Number 2.

Zimmer, W.J. (1940). Plant invasions in the Mallee. *Victorian Naturalist*, *56*, 143–7.

11

Heathlands

R.L. SPECHT

The term 'heath' (or more appropriately 'heathland' when referring to the plant community) is a vernacular word describing wasteland in northwestern Europe (Rübel, 1914). By coincidence, much of the nutrient-poor wasteland in this region was covered with a degraded community dominated by *Calluna vulgaris* (heath or heather); but on apparently similar soils, this sclerophyllous, ericaceous species was absent and a stunted grassland dominated by *Agrostis* and *Festuca* (sometimes termed 'grass-heath') occurred. In European ecological literature use of the term heathland has been restricted to describe the low sclerophyllous communities of *Calluna*, *Erica*, *Rhododendron* and *Vaccinium* found on nutrient-deficient soils from lowland to alpine habitats (Gimingham, 1972).

Heathlands are well developed on nutrient-deficient sandstone soils in the southwestern Cape Province of South Africa, where they have been referred erroneously to the Mediterranean 'macchia', but are now termed more appropriately 'fynbos' (Kruger, 1979). Here the community contains a wealth of ericaceous species (including a great number of species of *Erica*) as an understorey to low sclerophyllous, often broader-leafed, shrubs belonging mainly to the families Proteaceae, Rhamnaceae and Rosaceae. Node-sedges (Restionaceae) and many members of the Liliales are common as a ground stratum.

The heathlands of Australia are closely allied to the fynbos of Cape Province, not to the macchia of Mediterranean Europe or the chaparral of California (Specht, 1969). The family Ericaceae is virtually absent, except in the sub-alpine heathlands where the genera *Agapetes*, *Gaultheria*, *Pernettya* and *Rhododendron* are present. The closely-allied heath-family Epacridaceae, with 25 genera, has taken the place of the family Ericaceae in lowland heath communities in southern Australia. As in South Africa, many broader-leafed, sclerophyllous shrubs (belonging to a wide range of families) overtop the epacridaceous understorey. Many genera of Cyperaceae, Liliales, Orchidaceae, Restionaceae, etc. are common in the ground stratum.

Attributes

Community structure
Over the last fifteen years, Australian ecologists have developed a two-way classification to describe the structure of Australian plant communities (Specht, 1970, 1981). This classification is based on two readily observable attributes of the plant community, life form and foliage projective cover of the uppermost stratum. The classification may be used to cover the same attributes of lower strata in the community.

By definition (Specht, 1981), Australian heathlands may be classified structurally as follows.

Tallest stratum – sclerophyllous shrubs 25 cm to 2 m tall.
   The height category may be subdivided if required into *tall* (1 to 2 m) and *low* (25 cm to 1 m).
   Foliage projective cover, 100 to 70% – *closed-heathland.*
   Foliage projective cover, 70 to 50% – *heathland.*
   Foliage projective cover, 50 to 30% – *open-heathland.*
Tallest stratum – low sclerophyllous shrubs, less than 25 cm tall.
   Foliage projective cover, 30 to 10% – *dwarf heathland* (fellfield).
   Foliage projective cover, < 10% – *dwarf open-heathland* (open-fellfield).
Pertinent aspects of the lower strata can be included as follows.
   Sclerophyllous shrubs and graminoids co-dominant – *graminoid-heathland,*
   Sclerophyllous shrubs and *Sphagnum* moss co-dominant – *bog-heathland.*

The above, purely structural terms, used to describe the range of heathland communities observed in Australia, avoid the inclusion of habitat terms such as 'wet' and 'dry' which previously have been applied to heathlands developed respectively on seasonally-waterlogged and seasonally-droughted soils deficient in most plant nutrients (Groves & Specht, 1965; Specht, 1972, 1979*b*). Also, the altitudinal terms, such as coastal, lowland, montane or sub-alpine, are not used.

True heathlands, as thus defined, are found in most parts of humid to sub-humid Australia, but usually only in small patches. These islands of heathland vegetation grade into structural formations in which the heathy strata are overtopped by a stratum of tall shrubs (a tall shrubland), low scattered trees (a woodland) or denser trees (an open-forest) usually of *Banksia*, *Casuarina* or *Eucalyptus*. Various terms, such as 'shrub-heath', 'mallee-heath', 'tree-heath', 'sclerophyll-woodland', 'sclerophyll-forest', have been used to describe these more complex communities, the terms 'heath' or 'sclerophyll' being used to designate the heathy understorey.

Observations made over several decades in southern Australia indicate that, apart from secondary successional changes following fire, the heathland and adjacent sclerophyll communities have not changed structurally unless the environment has been modified by man (Groves & Specht, 1981). In effect, most Australian heathland communities must be regarded as climax, not seral, vegetation.

Species-area

The number of species which are found in Australian heathlands varies from as few as 33 to as many as 131 species (Specht, 1979*b*; George, Hopkins & Marchant, 1979). The number of species found in the stand decreases continuously from a high value following fire; for example, 38 species recorded in a heathland in southeastern South Australia after a fire, decreased to 25 species at 15 years, 20 species at 25 years and 10 species at 50 years (Specht, Rayson & Jackman, 1958; Specht, 1980*b*).

The number of species, cited above, may be found in stands of over one hectare in area. On a smaller scale, species-area studies have shown that dry-heathlands on well-drained sites may support an average of 22 to 36 species on an area of 8 m². Fewer species (11 to 25) occupy the same area of wet-heathlands on seasonally-waterlogged soils in the same region (Specht, 1979*b*).

Life forms

Heathlands are dominated by low shrubs (up to 2 m tall) which possess small (< 225 mm²), evergreen, sclerophyllous leaves and extensive root systems, often arising from lignotubers (syn. 'burls'). Several shrubby species may be co-dominant, as shown in Figure 11.1, where *Banksia*, *Casuarina*, *Leptospermum* and *Xanthorrhoea* contribute the greatest biomass. In the open-heathland depicted in Figure 11.1, 33 of the 76 species recorded may be classed as shrubs (1 to 2 m) or subshrubs (25 cm to 1 m) at maturity. The genera *Casuarina* and *Phyllota*, which covered 28% of the ground six years after a fire, are relatively short-lived (15 to 25 years), whilst the co-dominants *Banksia* and *Xanthorrhoea* survive for well over 50 years (Specht *et al.*, 1958).

Many low evergreen plants (<25 cm tall) form a ground stratum together with evergreen hemicryptophytes (evergreen graminoid plants, a life form not recognised by Raunkiaer (1934), but introduced by Specht, 1979*a*). Seasonal grasses and geophytes are present, whilst annual herbs are rare, except where the community has been grazed and fired regularly.

Parasitic twining plants of the genus *Cassytha* may be common on the subshrubs.

Leaf characteristics
The leaves of most heathland species are sclerophyllous, with thick cuticles and sunken stomata. The cells of the leaf are usually thick-walled, often lignified (sometimes silicified) and contain tannins, resins

Figure 11.1. Profile sketch *a*, and coverage chart *b*, of a typical dry-heathland near Dark Island Soak, 16 km northeast of Keith, South Australia (see Specht & Rayson, 1957).

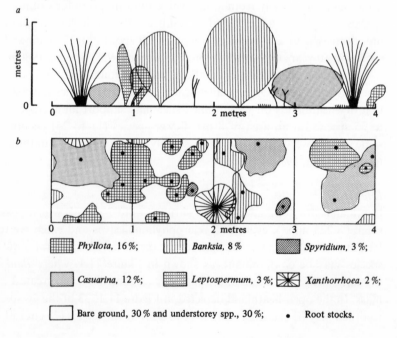

and essential oils thus increasing their flammability during periods of water stress.

Many genera characteristic of the upper stratum (1 to 2 m tall) possess broader leaves which may be classed as microphyll (over 225 mm^2 in area) on the Raunkiaer scale; low shrubs of *Acacia, Banksia, Eucalyptus, Xanthorrhoea*, etc. fall into this leaf-size class. Most subshrubs (25 cm to 1 m tall) possess small leaves in the nanophyll (25 to 225 mm^2) and leptophyll (<25 mm^2) size classes of Raunkiaer (1934); some of the smaller leaves are typically ericoid with the upper surface of the leaf hard and waxy and the lower surface deeply grooved.

The ground stratum (<25 cm tall) contains many evergreen hemicryptophytes with sclerophyllous, graminoid leaves, as well as dwarf shrubs usually with leptophyllous leaves. The leaves of the few seasonal plants are usually mesophyllous, rather than sclerophyllous, in texture.

Root systems

Underground organs such as lignotubers, rhizomes, bulbs, tubers are a distinctive attribute of heathlands. Although the above-ground part of the vegetation may be razed by fire, these underground organs survive and rapidly regenerate new aerial shoots. Heathlands in most humid parts of southern Australia possess so many root-stock regenerators that, within ten years from the fire, the above-ground vegetation contains about 95% (by dry weight) of these species, with only 5% of the biomass contributed by species regenerating by seed. As the environment becomes less humid, more gaps appear between the plants which form the upper stratum; this enables species which regenerate by seed to persist in the stand, as shown in the following equation (Specht, 1980*b*):

$$\text{R.S.R.} = 0.645 \text{ F.P.C.} + 41.4$$

where R.S.R. are root-stock regenerators (% total above-ground biomass), and F.P.C. the foliage projective cover (%) of the upper stratum.

In the nutrient-deficient conditions in which heathlands are found, peculiar rootlets develop in the interface between decomposing litter and the soil surface. Proteoid rootlets are characteristic of the family Proteaceae; restioid rootlets of the family Restionaceae; cyperoid or dauciform rootlets of the family Cyperaceae; mycorrhizal rootlets in the family Myrtaceae and many other families. A distinctive feature of all these different rootlets is their ability to synthesise polyphosphate from orthophosphate released in the decomposition of litter (Jeffrey, 1968).

Table 11.1. *Number of genera, listed under families, recorded in the six Australian heathland suites (see Figure 11.5). The sub-alpine suite has been subdivided further to show the differences between southeastern and northern Australia*

Family Nomenclature follows Willis & Airy Shaw (1973) except Liliiflorae, Huber (1969)	Total no. of genera	Sub-humid temperate (SW Australia)	Sub-humid temperate (S & SE Australia)	Humid SE Aust — Temperate	Humid SE Aust — Sub-tropical	Monsoonal N Australia	Sub-alpine SE Australia	Sub-alpine N Australia
Bryophyta								
Sphagnaceae	1	–	–	–	–	–	1	–
Pteridophyta								
Aspidiaceae	1	–	–	1	–	–	–	1
Aspleniaceae	1	–	–	–	–	–	1	–
Blechnaceae	2	–	–	1	–	–	2	1
Cyatheaceae	1	–	–	1	–	–	1	–
Dennstaedtiaceae	1	–	–	1	1	–	–	–
Gleicheniaceae	2	–	–	2	–	–	–	–
Grammitidaceae	2	–	–	–	–	–	–	2
Hemionitidaceae	1	–	–	–	–	–	–	–
Hymenophyllaceae	1	–	–	–	–	–	1	1
Lindsaeaceae	1	1	–	1	–	–	1	–
Lycopodiaceae	1	–	–	1	–	–	1	–
Marattiaceae	1	–	–	–	–	–	–	–
Platyzomataceae	1	–	–	–	–	1	–	–
Polypodiaceae	1	–	–	–	1	–	1	1
Schizaeaceae	1	–	–	–	1	–	1	–
Selaginellaceae	1	–	–	1	–	–	–	–
Total	19	1	–	9	3	1	9	6

Gymnospermae								
Cupressaceae	3	2	1	1	—	1	1	—
Phyllocladaceae	1	—	—	—	—	—	1	—
Podocarpaceae	3	—	1	—	1	—	3	—
Zamiaceae	1	1	—	—	—	—	—	—
Total	8	3	2	1	1	1	5	—
Monocotyledoneae								
Anthericaceae	15	15	5	4	5	1	2	3
Arecaceae	5	—	—	—	—	2	—	—
Asphodelaceae	1	1	1	1	—	—	—	—
Asteliaceae	2	—	—	—	—	—	2	—
Centrolepidaceae	5	5	—	—	—	1	2	—
Colchicaceae	3	3	1	1	1	—	—	—
Commelinaceae	1	—	—	—	—	—	—	—
Cyperaceae	23	6	4	7	6	8	8	2
Dasypogonaceae	8	8	1	1	1	1	—	—
Dianellaceae	2	2	2	2	2	—	2	—
Doryanthaceae	2	—	—	1	1	—	—	—
Ecdeiocoleaceae	1	1	—	—	—	—	—	—
Eriocaulaceae	—	—	—	—	—	1	—	—
Haemodoraceae	7	7	—	1	1	1	2	—
Hypoxidaceae	2	1	1	—	1	—	3	—
Iridaceae	4	2	1	2	1	1	2	—
Juncaceae	2	—	1	—	—	—	—	—
Orchidaceae	29	18	13	13	3	2	6	3
Pandanaceae	1	—	—	—	—	—	—	1
Philydraceae	—	—	—	—	—	11	7	—
Poaceae	27	6	5	7	8	2	2	—
Restionaceae	17	15	5	4	6	—	—	—
Smilacaceae	1	—	—	—	—	1	—	1
Taccaceae	1	—	—	—	—	1	—	—
Xanthorrhoeaceae	1	1	1	1	1	—	1	—
Xyridaceae	1	1	—	1	1	—	—	—
Zingiberaceae	1	—	—	—	—	1	—	1
Total	164	92	41	45	38	37	39	10

Family Nomenclature follows Willis & Airy Shaw (1973) except Liliiflorae, Huber (1969)	Total no. of genera	Sub-humid temperate (SW Australia)	Sub-humid temperate (S & SE Australia)	Humid SE Aust		Monsoonal N Australia	Sub-alpine	
				Temperate	Sub-tropical		SE Australia	N Australia
Dicotyledoneae								
Acanthaceae	3	—	—	—	—	3	—	—
Aizoaceae	3	1	1	—	—	2	—	—
Amaranthaceae	2	1	—	—	—	2	2	—
Apiaceae	4	—	1	—	—	1	4	—
Hydrocotylaceae	7	3	1	2	4	1	—	1
Apocynaceae	2	—	—	—	—	2	—	2
Araliaceae	4	—	—	—	—	2	1	—
Asclepiadaceae	2	—	—	—	—	2	—	—
Asteraceae	28	7	10	2	—	5	16	—
Balanopaceae	1	—	—	—	—	—	—	—
Baueraceae	1	—	—	1	1	—	1	1
Blepharocaryaceae	1	—	—	—	—	1	—	—
Boraginaceae	2	—	1	—	—	1	1	—
Brassicaceae	4	—	—	—	—	—	3	—
Brunoniaceae	1	—	—	1	1	—	—	—
Burseraceae	1	—	—	.	—	1	—	—
Byblidaceae	1	1	—	—	—	1	—	—
Caesalpiniaceae	2	—	—	—	—	2	—	—
Campanulaceae	1	—	1	1	1	—	1	—
Caryophyllaceae	2	—	1	—	—	1	1	—
Cassythaceae	1	1	1	1	—	1	—	—
Casuarinaceae	1	1	1	1	1	1	1	—
Celastraceae	2	—	—	—	—	1	1	1
Cephalotaceae	1	1	—	—	—	—	—	—

Family							
Chrysobalanaceae	—	—	1	—	—	—	1
Cleomaceae	—	—	1	—	—	—	1
Cochlospermaceae	—	—	1	—	—	—	1
Combretaceae	—	—	1	—	—	—	1
Convolvulaceae	—	—	1	—	1	—	1
Crassulaceae	—	—	—	—	—	—	1
Cucurbitaceae	—	—	—	—	—	8	9
Dicrastylidaceae	1	1	3	1	1	1	2
Dilleniaceae	—	1	2	—	—	—	1
Donatiaceae	—	—	—	—	—	—	—
Droseraceae	—	1	1	1	1	1	1
Ebenaceae	—	—	—	—	—	—	—
Ehretiaceae	—	1	1	1	—	—	1
Elaeocarpaceae	1	—	—	—	—	—	1
Epacridaceae	2	10	1	9	8	14	25
Ericaceae	2	2	—	—	—	—	4
Escalloniaceae	2	1	—	—	—	9	2
Euphorbiaceae	1	4	9	7	3	24	17
Fabaceae	—	1	12	14	13	—	27
Fagaceae	—	1	—	—	—	—	1
Flindersiaceae	—	—	1	—	1	—	1
Gentianaceae	—	1	—	1	1	—	2
Geraniaceae	—	2	—	2	2	—	2
Goodeniaceae	—	1	3	4	2	6	6
Gyrostemonaceae	—	—	—	2	—	2	2
Haloragaceae	—	1	—	2	1	3	3
Hypericaceae	—	2	—	2	—	—	1
Lamiaceae	—	1	3	2	3	6	7
Lauraceae	2	2	3	—	—	—	3
Lentibulariaceae	—	1	1	—	—	—	2
Linaceae	—	—	1	—	—	—	1
Lobeliaceae	—	1	—	1	1	1	3

Family — Nomenclature follows Willis & Airy Shaw (1973) except Liliflorae, Huber (1969)	Total no. of genera	Sub-humid temperate (SW Australia)	Sub-humid temperate (S & SE Australia)	Humid SE Aust — Temperate	Humid SE Aust — Sub-tropical	Monsoonal N Australia	Sub-alpine SE Australia	Sub-alpine N Australia
Loganiaceae	1	1	1	—	—	—	—	—
Loranthaceae	4	1	—	—	—	1	—	2
Lythraceae	1	—	—	—	—	2	—	—
Malvaceae	3	1	—	—	—	1	—	—
Melastomataceae	1	—	—	—	—	1	—	—
Meliaceae	1	—	—	—	—	1	—	—
Menispermaceae	1	—	—	—	—	1	—	—
Mimosaceae	1	1	1	1	—	1	1	2
Monimiaceae	2	—	—	—	1	—	—	—
Moraceae	1	—	—	—	—	1	—	—
Myoporaceae	2	2	2	—	—	1	—	1
Myrsinaceae	1	—	—	—	—	—	—	—
Myrtaceae	40	30	9	11	12	16	6	3
Nepenthaceae	1	—	—	—	—	1	—	—
Olacaceae	1	1	1	1	1	1	—	—
Oleaceae	2	—	1	—	—	1	—	—
Onagraceae	1	—	—	1	—	—	1	—
Opiliaceae	1	—	—	—	—	1	—	—
Oxalidaceae	1	—	1	1	—	—	1	—
Passifloraceae	1	—	—	—	—	1	—	—
Pittosporaceae	7	6	2	2	—	1	—	1
Plantaginaceae	1	—	—	—	—	—	1	—
Polygalaceae	1	1	1	1	1	1	1	—
Polygonaceae	1	—	—	—	—	1	—	—
Portulacaceae	3	1	—	—	—	2	1	—

Family								
Proteaceae	22	15	8	8	11	4	8	3
Rafflesiaceae	1	1	—	—	—	—	—	—
Ranunculaceae	4	1	—	2	—	—	—	—
Rhamnaceae	6	5	4	—	—	—	3	—
Rosaceae	3	1	1	—	1	6	—	—
Rubiaceae	12	9	8	4	6	3	4	—
Rutaceae	12	6	3	1	2	3	3	1
Santalaceae	6	1	1	—	1	5	1	—
Sapindaceae	5	—	—	—	—	3	—	—
Sapotaceae	3	—	—	—	—	—	—	1
Scrophulariaceae	5	1	3	1	—	1	3	—
Simaroubaceae	1	—	—	—	—	—	1	—
Solanaceae	3	3	1	—	—	1	1	—
Spigeliaceae	1	—	—	1	—	1	1	—
Stackhousiaceae	—	4	—	—	—	—	—	—
Sterculiaceae	9	2	1	1	2	5	—	—
Stylidiaceae	2	—	—	—	2	—	1	—
Tetracarpaeaceae	1	—	—	—	1	—	—	—
Thymelaeaceae	2	1	1	1	1	1	2	—
Tiliaceae	1	—	—	—	—	1	1	—
Tremandraceae	3	3	1	—	—	—	—	—
Ulmaceae	1	—	—	—	—	—	—	—
Verbenaceae	1	—	—	—	—	1	—	—
Violaceae	3	1	2	1	1	—	2	—
Winteraceae	1	—	—	—	—	1	1	—
Zygophyllaceae	1	—	—	—	—	—	—	—
Total	399	195	118	81	95	151	111	31
Grand Total	591	291	161	136	137	190	165	47

Nitrogen-fixing, nodulated rootlets are found in the families Casua-rinaceae, Fabaceae, Mimosaceae, and Zamiaceae.

Haustorial connections have been observed between the roots of.the semi-parasitic plants (families Olacaceae and Santalaceae, and the genus *Euphrasia* in the Scrophulariaceae) and the roots of other heathland species.

Bradysporous fruits

A number of characteristic species within heath vegetation possess hard woody fruit which retain their seed for many years, usually until released by heat during a bush fire (Gill & Groves, 1980; Specht, 1980*a*). Because of this bradysporous habit, regeneration of these species by seedlings is largely dependent on fire. The frequency of fire is important. A fire before the species sets viable seed may lead to extinction of the species; a fire too long delayed may lead to viable seed being destroyed in the woody fruits by boring insects and fungi.

Seeds

Bradyspory (discussed above) is a characteristic of heathland species of the families Casuarinaceae, Myrtaceae and Proteaceae. Seeds eventually released from the woody fruits of these species are relatively large and often winged in the families Casuarinaceae and Proteaceae, but are very small and wingless in the family Myrtaceae. These small seeds form an important food-source for harvester ants which abound in the heath-lands.

The other heathland species, belonging to a wide range of families (Table 11.1), release seed from their fruits as soon as it is mature. From 21 to 32% of the heath flora of southern Australia produce reasonably large seed with a thick hard testa, with a smooth and darkly coloured surface to which is attached a white or light-coloured appendage termed an 'elaiosome' (Berg, 1975, 1980). This relatively dry and hard appendage is attractive to ants which gather the seed into their nests, eat the elaiosome, but usually cannot pierce the hard testa to eat the embryo. The seed is then discarded either outside the nest or, if too large to move, stored in galleries. Myrmecochorous dispersal of seed is a unique characteristic of Australian heathland species. Approximately 1500 species of Australian vascular plants, representing 87 genera in 24 families are probably myrmecochorous, compared with less than 300 myrmecochorous species known for the rest of the world (Berg, 1980).

Distribution

Heathland vegetation, with or without an overstorey of sclerophyllous trees or tall shrubs, is widespread across Australia in all humid and sub-humid areas (Figure 11.2). Remnants of the flora are found even in the arid centre of Australia.

By far the greatest concentration of heathland species is found on the sandplain and lateritic soils of southwest Western Australia where 50% of 3700 typical Australian heathland species are located (Figure 11.3). The Hawkesbury sandstone soils of the Sydney area of New South Wales support 20% of the Australian heath flora. Both these areas have a warm temperate climate. The proportion of heathland species falls to 3 to 6% in the tropics, and 9 to 14% in the cool temperate climate of southeastern Australia.

The heathland flora forms a distinct suite of plant communities (Figure 11.4), clearly separated from closed-forest (rainforest) vegetation and open-communities containing a wealth of grasses, herbs, ground-

Figure 11.2. Distribution of heathland vegetation ($\pm$ trees or tall shrubs) in Australia (after Carnahan, 1976).

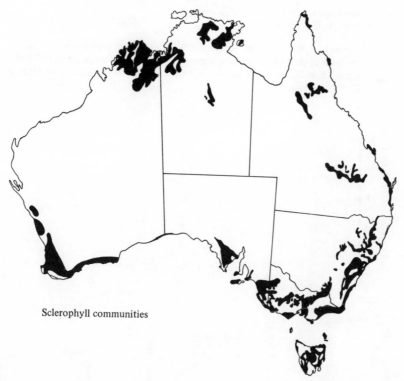

Sclerophyll communities

Figure 11.3. Regional distribution of 3700 typical Australian heathland species. The circled numbers give the percentages of heathland species in each area, and the dashed lines indicate the mean annual temperature.

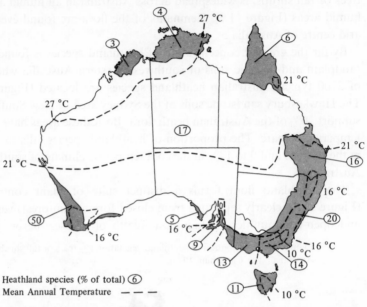

Heathland species (% of total) ⑥
Mean Annual Temperature — — —

Figure 11.4. Relationships of 32 major Australian plant formations classified on the basis of the 1398 component genera by the classificatory program DIVINF (Lance & Williams, 1968), run on the CSIRO computer in Canberra by Dr M.B. Dale (September, 1978).

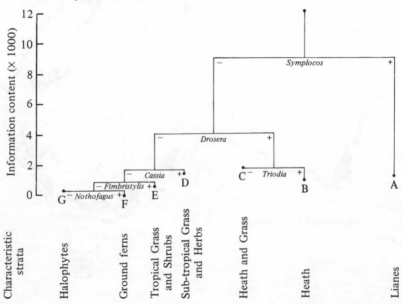

Figure 11.5. Relationship of 32 Australian heathland communities classified on the basis of the 591 component genera by the classificatory program DIVINF (Lance & Williams, 1968), run on the CSIRO computer in Canberra by Dr M.B. Dale (September, 1978). Group A, Southwest (Western Australia) dry-heathlands; Group B, Dark Island (South Australia) dry-heathland, Dark Island (South Australia) mallee-broombush, Billiatt (South Australia) mallee-broombush, Mt Compass (South Australia) mallee-heathland, Hincks – Murlong (South Australia) heathland/mallee-broombush; Group C, Shoalwater Bay (Qld) dry-heathland, Toolara (Qld) wet-heathland, Beerwah (Qld) wet-heathland, Sydney (NSW) dry-heathland, Sydney (NSW) wet-heathland, Mt Hardgrave, North Stradbroke Island (Qld) dry-heathland; Group D, Dave's Creek, Lamington N.P. (Qld) wet-heathland, Bonny Hills (NSW) dry heathland, Tidal River, Wilson's Promontory (Victoria) dry-heathland, Darby River, Wilson's Promontory (Victoria) dry-heathland, Barry's Creek, Wilson's Promontory (Victoria) wet-heathland, Frankston (Victoria) dry-heathland, Lower South East (South Australia) wet-heathland; Group E, Oenpelli, Arnhem Land (NT) dry-heathland, Jardine River (north Qld) wet-heathland, Cape Flattery – Lizard Island (north Qld) dry-heathland, Burra Ranges (north Qld) dry-heathland, Forty-Mile Scrub (north Qld) dry-heathland; Group F, Mt Bellenden Ker (north Qld) alpine heathland, Mt Kosciusko (NSW) alpine heathland, Mt Kosciusko (NSW) alpine bog-heathland, Lake Mountain (Victoria) alpine heathland, Lake Mountain (Victoria) alpine bog-heathland, Cradle Mountain (Tasmania) alpine heathland, Mt Wellington (Tasmania) alpine heathland (for further site details see Specht, 1979*b*).

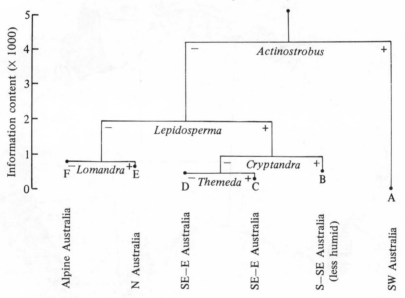

ferns or halophytes. The two heathland subgroups may be distinguished: Group B includes communities with a preponderance of sclero-phyllous plants (true heathlands, mallee-heath, sclerophyll-woodland and dry sclerophyll-forest); Group C includes communities in tem-perate Australia where more herbaceous elements (grass, herbs, ferns) are intermingled with the heathland genera (alpine heathlands, wet sclerophyll-forest on oligotrophic soils, savannah/sclerophyll-woodland).

If the 591 genera recorded in Australian heathlands *sensu stricto* are analysed by the same classificatory program (DIVINF) which was used to show the relationships of the heathland flora to other Australian plant formations (Figure 11.4), the relationships shown in Figure 11.5 result. The Australian heathlands fall into six natural groups: Group A in southwest Western Australia; Group B in South Australia extending into the drier side of the Great Dividing Range in southeastern Australia;

Figure 11.6. Distribution of the major Köppen climatic types (after Dick, 1975) in which the heathland vegetation groups (defined in the classificatory program, Figure 11.5) are found in Australia.

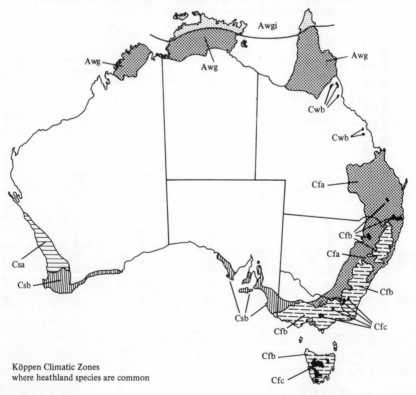

Köppen Climatic Zones
where heathland species are common

Group C in warm temperate to sub-tropical eastern Australia; Group D in temperate areas of southeastern Australia extending into the highlands of eastern Australia; Group E in northern Australia; and Group F in alpine/sub-alpine areas of southeastern Australia with a small outlier on the mountain summits in northern Queensland (Figure 11.6).

Many of the heathland genera have a considerable distributional range throughout these six major suites; some show disjunct distribution patterns occurring in isolated pockets found either in eastern and southwestern heathlands or in northern and southwestern, sometimes also eastern heathlands (Specht, 1980*c*); other genera may be confined to one heathland suite; some are invaders from adjacent communities. The number of genera recorded in each of the heathland suites is listed under families in Table 11.1. As the heathland genera are not listed in Table 11.1, because of space, an idea of the genera in five major heathland families is given below.

Epacridaceae (28 genera): *Acrotriche, Andersonia, Archeria, Astroloma, Brachyloma, Choristemon, Coleanthera, Conostephium, Cosmelia, Cyathodes, Dracophyllum, Epacris, Leucopogon, Lissanthe, Lysinema, Melichrus, Monotoca, Needhamiella, Oligarrhena, Pentachondra, Prionotes, Richea, Rupicola, Sphenotoma, Sprengelia, Styphelia, Trochocarpa, Woollsia.*

Fabaceae (24 genera): *Aotus, Bossiaea, Brachysema, Burtonia, Chorizema, Daviesia, Dillwynia, Eutaxia, Gastrolobium, Gompholobium, Goodia, Hardenbergia, Hovea, Isotropis, Jacksonia, Kennedia, Mirbelia, Oxylobium, Phyllota, Platylobium, Pultenaea, Sphaerolobium, Templetonia, Viminaria.*

Myrtaceae (38 genera): *Actinodium, Agonis, Angophora, Astartea, Austromyrtus, Baeckea, Balaustion, Beaufortia, Callistemon, Calothamnus, Calytrix, Calythropsis, Chamaelaucium, Conothamnus, Darwinia, Eremaea, Eucalyptus, Eugenia, Fenzlia, Homalocalyx, Homoranthus, Hypocalymma, Kunzea, Lamarchea, Leptospermum, Lhotskya, Melaleuca, Micromyrtus, Phymatocarpus, Pileanthus, Regelia, Scholtzia, Sinoga, Thryptomene, Tristania, Verticordia, Wehlia, Xanthostemon.*

Proteaceae (22 genera): *Adenanthos, Banksia, Bellendena, Cenarrhenes, Conospermum, Dryandra, Franklandia, Grevillea, Hakea, Isopogon, Lambertia, Lomatia, Musgravea, Orites, Persoonia, Petrophile, Stirlingia, Strangea, Symphionema, Synaphea, Telopea, Xylomelum.*

Restionaceae (17 genera): *Alexgeorgia, Anarthria, Chaetanthus,*
    Coleocarya, Dielsia, Empodisma, Harperia, Hopkinsia,
    Hypolaena, Lepidobolus, Leptocarpus, Lepyrodia, Loxocarya,
    Lyginia, Meeboldina, Onychosepalum, Restio.

Ecological Relationships

Climate

Six major suites of heathland communities were distinguished by the
classificatory program depicted in Figure 11.5. The distribution of these
groups coincides with major Köppen climatic types defined for humid
and sub-humid areas of Australia by Dick (1975).

The Köppen climatic types, in which each heathland group is developed, are shown on a map of Australia (Figure 11.6) and may be
described as follows.

Group A: Csa/Csb (warm climate with a long dry summer, ranging from
    hot (over 22°C) to mild (below 22°C)).
Group B: Csb (warm climate with a long mild (under 22°C) dry
    summer).
Group C: Cfa (warm climate, with uniform rainfall, long hot summer
    (over 22°C)).
Group D: Cfb (warm climate, with uniform rainfall, long mild summer
    (under 22°C)).
Group E: Aw (hot climate with a long dry winter).
Group F: Cfc in southeastern Australia (warm climate, with uniform
    rainfall, short mild summer (over 10°C, under 22°C)).
    Cwb in northern Queensland (warm climate, with a dry winter
    and a long mild summer (over 10°C, under 22°C)).

Soils

Heathland vegetation is found from temperate to tropical areas of
Australia, from lowland to alpine sites. Wherever the vegetation occurs,
the major ecological factor controlling its distribution appears to be the
nutrient status of the soil.

As shown in Figure 11.7, heathland soils are invariably low in essential
plant nutrients such as phosphorus and nitrogen. Other plant nutrients
(potassium, sulphur, copper, zinc, manganese, molybdenum) may be of
limited availability in some heathland soils. These soils may be developed
either on parent material (sandstone, quartzite, acid granites, wind-

distributed sand) low in plant nutrients, or on highly leached soils
(podzols, lateritic podzols and solonised sandy soils). Poorly drained,
peaty soils, also low in available nutrients, may support a bog-heathland.

The soil analyses presented in Figure 11.7 illustrate the tenuous
interface between heath and savannah floras. Intense leaching of plant

Figure 11.7. Edaphic relationships (based on the levels of total nitrogen and
total phosphorus in the surface soil, Stace *et al.*, 1968) of heathy and *Triodia*
communities (open circles; zone A) compared with grassy communities
(triangles; zone C), chenopod (open squares) and *Stipa* (filled squares) communities (zone B) and grassy (filled circles) and ferny (asterisks) communities
(zone D).

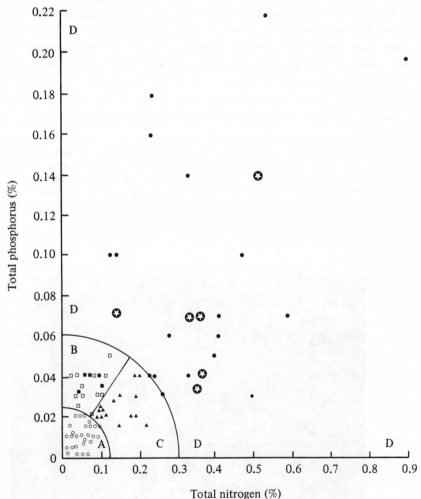

nutrients will lead to the development of heath vegetation. Slightly more fertile soils or the deficient soils contaminated with nutrients will lead to the extinction of the heath flora and the development of a savannah flora, as shown in the manipulation experiments reported by Specht (1963), Heddle & Specht (1975), and Specht, Connor & Clifford (1977).

Fire

Fire is an integral factor in Australian heathlands. The dense structure of the mature community and the presence of flammable volatile oils and resins in the sclerophyllous leaves of the component species make the heathlands a potential fire hazard during periods of drought.

The vegetation is well-adapted to the ravages of fire (Gill & Groves, 1980). The aerial stems of many species may be destroyed by fire, but many plants sprout again from lignotubers, rhizomes, tubers and bulbs protected in the soil (Specht, 1980*b*). Other species release large numbers of seed held above-ground in hard woody fruits which spring open during a fire. A wealth of regenerating plants and seedlings soon appears (Figure 11.8), thus repairing the catastrophic effect of the fire.

Figure 11.8. Dry-heathland, 12 years since the last fire, near Keith, South Australia. The taller (2 m), large-leafed shrubs are *Banksia ornata*; the specimens with the grass-like leaves are *Xanthorrhoea australis*; the low, small-leafed shrubs are *Casuarina pusilla*, *Phyllota pleurandroides* and *Leptospermum myrsinoides*.

Animals

Native animals such as the kangaroo (*Macropus* spp.), wallaby (*Macropus* spp.) and emu (*Dromaius novaehollandiae*), and a few introduced rabbits (*Oryctolagus cuniculus*), all essentially visitors from adjacent communities, exert a minor grazing pressure on heathlands in southern Australia. Black cockatoos (*Calyptorhynchus funereus*) may visit the ecosystem to extract grubs from the inflorescences of *Banksia* spp. or *Xanthorrhoea* spp., the latter species tending to flower only after a bushfire. The introduced house mouse (*Mus musculus*) invades the heathlands after a fire, but tends to disappear as the community ages.

In mature stands of heath, honeyeaters (Meliphagidae) and wrens (Maluridae) are resident birds, feeding on nectar and insects and probably aiding cross-fertilization of many of the heathland species.

Insects inhabiting the heathland fall into classes such as foliage-feeders, nectar-feeders, seed-gatherers, wood-suckers and wood-borers; other insects and spiders are predators and parasites on these first-order consumers. The populations of each of these classes of insects fluctuate seasonally in response to the seasonal growth and flowering rhythms of the heathland species (Edmonds & Specht, 1980). Ants are particularly active in seed removal and dispersal, especially attracted by the large number of seeds which possess fleshy elaiosomes (Berg, 1975).

Collembola, mites, worms, nematodes and protozoa act with decomposing fungi and bacteria in the decomposition of litter, returning mineral nutrients to the soil where they tend to be trapped by a network of fine rootlets at the interface of litter and soil.

References

Berg, R.Y. (1975). Myrmecochorous plants in Australia and their dispersal by ants. *Australian Journal of Botany*, *23*, 475–508.

Berg, R.Y. (1980). The role of ants in seed dispersal in Australian lowland heathland. In *Ecosystems of the World*, vol. 9B, ed. R.L. Specht, pp. 51–60. Amsterdam: Elsevier Scientific Publishing Company.

Carnahan, J.A. (1976). Natural vegetation. In *Atlas of Australian Resources*, Second Series. Canberra: Department of National Resources.

Dick, R.S. (1975). A map of the climates of Australia: According to Köppen's principles of definition. *Queensland Geographical Journal*, *3*, 33–69.

Edmonds, S.J. & Specht, M.M. (1980). Dark Island heath, South Australia: Faunal rhythms. In *Ecosystems of the World*, vol. 9B, ed. R.L. Specht, pp. 15–28. Amsterdam: Elsevier Scientific Publishing Company.

George, A.S., Hopkins, A.J.M. & Marchant, N.G. (1979). The heathlands of Western Australia. In *Ecosystems of the World*, vol. 9A, ed. R.L. Specht, pp. 211–30. Amsterdam: Elsevier Scientific Publishing Company.

Gill, A.M. & Groves, R.H. (1980). Fire regimes in heathlands and their plant ecological effects. In *Ecosystems of the World*, vol. 9B, ed. R.L. Specht, pp. 61–84. Amsterdam: Elsevier Scientific Publishing Company.

Gimingham, C.H. (1972). *Ecology of Heathlands*. London: Chapman & Hall.

Groves, R.H. & Specht, R.L. (1965). Growth of heath vegetation. I. Annual growth curves of two heath ecosystems in Australia. *Australian Journal of Botany*, *13*, 261–80

Groves, R.H. & Specht, R.L. (1981). Seral considerations in heathland. In *Vegetation Classification in the Australian Region*, ed. A.N. Gillison & D.J. Anderson, (in press). Canberra: CSIRO, Australia & Australian National University Press.

Heddle, E.M. & Specht, R.L. (1975). Dark Island heath (Ninety-Mile Plain, South Australia). VIII. The effect of fertilizers on composition and growth, 1950–1972. *Australian Journal of Botany*, *23*, 151–64.

Huber, H. (1969). Die Samenmerkmale und Verwandtschaftsverhältnisse der Liliifloren. *Mitteilungen Botanische Staatssammlung München*, *8*, 219–538.

Jeffrey, D.W. (1968). Phosphate nutrition of Australian heath plants. II. The formation of polyphosphate by five heath species. *Australian Journal of Botany*, *16*, 603–13.

Kruger, F.J. (1979). South African heathlands. In *Ecosystems of the World*, vol. 9A, ed. R.L. Specht, pp. 19–80. Amsterdam: Elsevier Scientific Publishing Company.

Lance, G.N. & Williams, W.T. (1968). A note on a new information statistic classificatory program. *Computing Journal*, *11*, 195.

Raunkiaer, C. (1934). *The Life Form of Plants and Statistical Plant Geography*. Oxford University Press.

Rübel, E.A. (1914). Heath and steppe, macchia and garigue. *Journal of Ecology*, *2*, 232–7.

Specht, R.L. (1963). Dark Island heath (Ninety-Mile Plain, South Australia). VII. The effect of fertilizers on composition and growth, 1950–1960. *Australian Journal of Botany*, *11*, 67–94.

Specht, R.L. (1969). A comparison of the sclerophyllous vegetation characteristic of mediterranean type climates in France, California and southern Australia. I. Structure, morphology and succession. II. Dry matter, energy and nutrient accumulation. *Australian Journal of Botany*, *17*, 277–92, 293–308.

Specht, R.L. (1970). Vegetation. In *The Australian Environment*, 4th edn (rev.), ed. G.W. Leeper, pp. 44–67. Melbourne: CSIRO, Australia & Melbourne University Press.

Specht, R.L. (1972). *The Vegetation of South Australia*. 2nd edn. Adelaide: Government Printer.

Specht, R.L. (ed.) (1979a). *Ecosystems of the World*, vol. 9A. *Heathlands and Related Shrublands*. Amsterdam: Elsevier Scientific Publishing Company.

Specht, R.L. (1979b). The sclerophyllous (heath) vegetation of Australia: The eastern and central States. In *Ecosystems of the World*, vol. 9A, ed. R.L. Specht, pp. 125–210. Amsterdam: Elsevier Scientific Publishing Company.

Specht, R.L. (ed.) (1980a). *Ecosystems of the World*, vol. 9B. *Heathlands and Related Shrublands*. Amsterdam: Elsevier Scientific Publishing Company.

Specht, R.L. (1980b). Responses of selected ecosystems: heathlands and related shrublands. In *Fire and the Australian Biota*, ed. A.M. Gill, R.H. Groves & I.R. Noble, pp. 395–415. Canberra: Australian Academy of Science.

Specht, R.L. (1980c). Major vegetation formations in Australia. In *Ecological Biogeography in Australia*, ed. A. Keast, pp. 163–298. The Hague: Junk.

Specht, R.L. (1981). Structural attributes – foliage projective cover and standing biomass. In *Vegetation Classification in the Australian Region*, ed. A.N. Gillison & D.J. Anderson, (in press). Canberra: CSIRO, Australia & Australian National University Press.

Specht, R.L., Connor, D.J. & Clifford, H.T. (1977). The heath-savannah problem: the effect of fertilizer on sand-heath vegetation of North Stradbroke Island, Queensland. *Australian Journal of Ecology*, 2, 179–86.

Specht, R.L. & Rayson, P. (1957). Dark Island heath (Ninety-Mile Plain, South Australia). I. Definition of the ecosystem. *Australian Journal of Botany*, 5, 52–85.

Specht, R.L., Rayson, P. & Jackman, M.E. (1958). Dark Island heath (Ninety-Mile Plain, South Australia). VI. Pyric succession: changes in composition, coverage, dry weight, and mineral nutrient status. *Australian Journal of Botany*, 6, 59–88.

Stace, H.C.T., Hubble, G.D., Brewer, R., Northcote, K.H., Sleeman, J.R., Mulcahy, M.J. & Hallsworth, E.G. (1968). *A Handbook of Australian Soils*. Glenside, S. Aust.: Rellim Technical Publications.

Willis, J.C. & Airy Shaw, H.K. (1973). *A Dictionary of the Flowering Plants and Ferns*, 8th edn. Cambridge University Press.

12

Chenopod shrublands

J.H. LEIGH

Distribution

Approximately six per cent (433 800 km²) of the total land area of
mainland Australia is occupied by Chenopod shrublands dominated
by various annual and perennial species of the family Chenopodiaceae.
The terminology of these communities has varied over the years, having
been variously referred to as 'shrub steppe' (Williams, 1955; Wood &
Williams, 1960), 'saltbush – xerophytic midgrass grazing lands' (Moore,
1969), and 'low shrubland' (Specht, 1970), but in this chapter they are
referred to as Chenopod shrublands.

These communities lie mostly south of the Tropic of Capricorn in an
environment receiving rainfall ranging from 125 to 266 mm per annum,
30–50% of which falls in winter. The most extensive areas occur in South
Australia (247 500 km²), western New South Wales (77 000 km²) and
southeast Western Australia (75 800 km²). Smaller areas (approximately
16 000 km²) are found in Queensland and Northern Territory. Victoria
contains less than 1500 km² of this vegetation type. These areas were
mapped by Dr R.M. Moore and modified by F.T. Bullen and C.V. Soles
as part of a revised map of Australian Grazing Lands (*Atlas of
Australian Resources*, 1980). The distribution of the Chenopod shrubland
together with isohyets is reproduced in Figure 12.1.

Floristics

The shrubland is composed of xeromorphic halophytes which are
drought- and salt-tolerant. A total of 280 species in the family
Chenopodiaceae is endemic to Australia and a further 17 species have
become naturalised (P.G. Wilson, personal communication). Of this
total the most widespread and economically important genera, with the

number of endemic species in parentheses, are *Sclerolaena* (80), *Maireana* (60), *Atriplex* (40), *Chenopodium* (20), *Rhagodia* (12), *Salsola* (3) and *Enchylaena* (2). Of less importance and with more restricted occurrence to areas such as salt marshes are *Halosarcia* (24), *Sclerostegia* (5) and *Sarcocornia* (3). Most naturalised species are forbs and belong in the genus *Chenopodium* (10); they occur as weeds of habitation, waste places and crop lands.

Occurrence and present condition

The great majority of the endemic species enumerated above occur as shrubs or subshrubs. These shrubs may occur as extensive monospecific stands presenting a simple and uniform appearance, or they may occur

Figure 12.1. Distribution of Chenopod shrublands (hatched areas) in Australia in relation to the isohyets representing median annual rainfalls of 100, 200 and 400 mm (dashed lines). Figure modified from *Atlas of Australian Resources*, 1980.

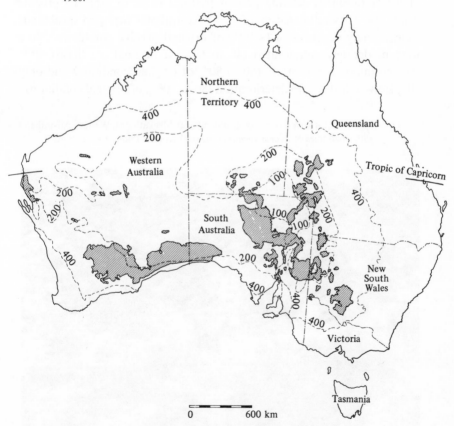

(cf. *Atriplex vesicaria*) as vegetated bands aligned along topographical contours, or as mixed stands of two or more species. Generally, the plants are less than 1.5 m high, although *Atriplex nummularia* may attain a height of 3 m. Bushes may be congregated in groups or occur as individuals separated by areas of bare soil (Figure 12.2). Most species may also be present as an understorey to low woodlands with an overstorey of species of *Eucalyptus*, *Casuarina*, *Acacia*, *Heterodendrum* and *Myoporum* (see Chapters 8, 9 & 10; Gillison & Walker, 1981; Johnson & Burrows, 1981; Parsons, 1981).

Newman & Condon (1969) assessed the extent to which the original rangeland types may have degenerated over the last 100 to 150 years, since settlement by European man and the introduction of domestic livestock, particularly sheep. There is much evidence to suggest that many Australian rangelands have altered and degenerated. In particular, the shrublands have fared badly. Newman & Condon assessed the extent of deterioration by relating present pastoral value to expected pristine condition. They estimated that 25% of Australian Chenopod shrublands were severely degenerated (< 40% of original pristine conditions), 40% were moderately deteriorated (40 to 60% of original condition), 25% showed minor degeneration (60 to 80% of original condition) and only 10% showed little or no deterioration (80 to 100% of original condition).

Figure 12.2 *Atriplex vesicaria* shrubland near Meekatharra, Western Australia, with small trees of *Acacia aneura* in the background. Photo: C.J. Totterdell

In considering Chenopod species individually, data collated by Hartley & Leigh (1979) indicate that some 13% (28) of all endemic species (280) can be considered at risk, and of these 3% (7 species, 3 *Atriplex*, 2 *Sclerolaena*, 1 *Maireana* and 1 *Malacocera*) are endangered (in serious risk of disappearing from the wild state within a decade if present land use and other causal factors continue to operate) and 6% (15 species, 4 *Atriplex*, 7 *Sclerolaena*, 2 *Maireana*, 2 *Roycea*) are vulnerable and at risk over a longer period, i.e. 20 years, if present land use patterns were to be altered. Of all endemic Chenopods, 6.6% (14 species) are restricted endemics known to occur only in areas with a maximum range of less than 100 km; because of their narrow range this group could be at great risk from factors such as cropping, fire, grazing, or mining. Most of the species at risk occur within South Australia (57%) and Western Australia (21%), with only 13% recorded for Victoria, 9% for New South Wales and none for the other States.

Description and distribution of economically important species

The original distribution of individual species was determined largely by factors such as physical and chemical properties of the soil, and soil moisture status, particularly as influenced by topography. A secondary but nevertheless significant factor affecting the present-day occurrence of the shrubs has been the introduction by man of domestic and feral animals. The saltbush country was most easily settled and was one of the first regions to be occupied in the early days of pastoral expansion. At the present time Chenopod shrublands support approximately 2.75 $\times$ 10^6 sheep at an average stocking rate of 6 sheep km^{-2} and 84 000 cattle at an average stocking rate of 2 head km^{-2} (R.D. Graetz, personal communication).

From the viewpoint of resource stability and the pastoral industry, the six most valued shrubs are *Atriplex nummularia*, *A. vesicaria*, *Maireana aphylla*, *M. astrotricha*, *M. pyramidata* and *M. sedifolia*. A brief description of these six species follows.

Atriplex nummularia (*Oldman saltbush*)

A blue-grey shrub, *A. nummularia* grows to a height and diameter of about 3 m, and has large, oval or almost circular leaves 1 to 2 cm long. The leaves are covered with a scaly layer; their margins may be uninterrupted or slightly toothed. The plants are usually of one sex, the female flowers occurring in dense clusters in the leaf axils at the end of the branchlets. The fruiting body consists of two flat paper-like bracteoles, hemispherical in outline, toothed or smooth-edged, pressed closely together but united only at the base, where the seed is enclosed, the whole being 5 to 8 mm across.

Atriplex nummularia is the largest endemic species of *Atriplex*, and is more common in southeastern Australia, becoming restricted to local drainage areas in more inland regions. This shrub occurs either in association with *Acacia pendula* (boree or myall) (an association considered to be once far more widespread than it is today), or as an understorey to *Eucalyptus largiflorens* (black box) or in association with *A. vesicaria*. Although it shows a preference for clay and clay–loam soils, *A. nummularia* can be found on most soil types.

Atriplex vesicaria (*Bladder saltbush*)
This perennial shrub, with woody brittle stems, reaches a height and diameter of 60 to 70 cm. Its leaves are roughly oval in outline, up to 25 mm in length, and have a silvery-grey appearance because of vesicular hairs on the surface. The seed is enclosed within two bracteoles, on each of which generally there occurs a large membranous bladder-like appendage, the whole being up to 12 mm in diameter.

The most extensive tracts of *A. vesicaria* occur in parts of western New South Wales and the northern and eastern regions of South Australia. Other areas are found on the Nullarbor Plain, extending into Western Australia (Figure 12.2), in northwestern Victoria, southwestern Queensland and southern-central Australia.

This shrub grows on a wide range of soil types and in a variety of habitats. On the riverine plain of central southwestern New South Wales it is found on flat terrain of brown and grey soils of heavy texture. Here it may occur in pure stands or in association with *Maireana aphylla* or *A. nummularia*. In South Australia and southwestern New South Wales, *A. vesicaria* occurs on red-brown soils, some with a clay subsoil, others with a subsoil containing limestone nodules. In the latter case, the associated shrubs are usually *Maireana sedifolia* or *M. astrotricha*. In northwestern New South Wales and northern South Australia *A. vesicaria* is found on the loams of the gibber plains and on the stony desert soils of the undulating and hilly country surrounding the Barrier and Grey Ranges (29°03'S; 141°57'E). The communities in most of these areas are in a severely degraded state. Possibly the most vigorous stands of *A. vesicaria* occur in the vicinity of Hay, in central New South Wales.

Maireana aphylla (*Cotton bush*)
Maireana aphylla is a much-branched perennial shrub growing to a height and diameter of about 1 m, with slender striate branches which become rigid and spiny with age. The leaves, short (1–2 mm long), thick, and crowded along the branchlets, fall off at an early stage. A membranous wing, pink when young, later maturing to brown, surmounts the short tube of the fruiting body, the whole being up to 12 mm across. Clusters of white cotton-like galls, caused by small grubs, often occur along the branches, which give the shrub its common name of 'cotton bush'.

Maireana aphylla is distributed widely throughout much of arid and semi-arid Australia. It prefers clay and clay–loam soils, and reaches its maximum density on the southern riverine plain of New South Wales where it may be the dominant species or it may be associated with other Chenopod shrubs, particularly *A. vesicaria*. Elsewhere, associated plants may be *M. pyramidata* or *Eremophila maculata*.

Maireana astrotricha (*Southern bluebush*)
Similar in habit and appearance to *M. sedifolia*, although greyish rather than blue, *M. astrotricha* can be distinguished by its flatter leaves, which taper to a conspicuous petiole. The lower leaves may be up to 15 mm long and 3 to 4 mm broad. The winged fruits are 6 to 15 mm across, with the prominently lobed wing surmounting a large hard obconical tube 3 to 5 mm long.

The major areas of *M. astrotricha* occur on sandy soils in the far west of New South Wales and in South Australia (where the similar *M. planifolia* also occurs). Minor areas of *M. astrotricha* are also found in central Australia, Queensland and Western Australia. This shrub may be associated with *A. vesicaria* and various other *Maireana* species.

The relative abundance of each of the *Maireana* shrubs so far discussed is largely determined by the depth at which the limestone layer occurs beneath the soil surface. With the layer close to the surface *M. sedifolia* is the predominant species, and *M. pyramidata* and then *M. astrotricha* take precedence as the depth to the layer increases.

Maireana pyramidata (*Black bluebush*)
Maireana pyramidata is the largest of the Australian *Maireana* species with a height of almost 2 m. Its greenish-grey, fleshy leaves are more or less obovoid, 2 to 4 mm long, and shorter than the other species. The

characteristic fruiting body has a pyramidal spongy extension above the narrow horizontal wing which encircles the short basal tube. Before maturity the fruits are bright green; when mature and dry they turn black.

The distribution of *M. pyramidata* is similar to that of *M. sedifolia*, being confined to the calcareous red and red-brown soils. *Maireana pyramidata* is perhaps the more prevalent species of the two in the eastern areas and is a more common constituent of the shrub component of communities dominated by *Eucalyptus* (mallee), *Casuarina cristata–Heterodendrum oleifolium* (inland rosewood), and *Acacia aneura* (mulga).

Maireana sedifolia (*Pearl bluebush*)

A much-branched perennial with thick woody stems, *M. sedifolia* grows to a height and diameter of about 1 m. The young branches and foliage have a dense covering of white hairs which imparts a pale-blue, almost white appearance to the bush. The leaves are soft, thick, almost cylindrical in outline and 3 to 10 mm in length. The short tube of the fruiting body is encircled by a reddish or light-brown horizontal wing, 8–10 mm in diameter, finely veined, unbroken or with one radial slit.

Maireana sedifolia extends over much of western New South Wales and South Australia and is one of the major shrubs on the Nullarbor Plain. It occurs either in pure stands or in association with *A. vesicaria*, *M. pyramidata* or *M. astrotricha*, or as an understorey to *Myoporum platycarpum* (sugarwood) or *Casuarina cristata* (belah [New South Wales], black oak [South Australia]); it occupies possibly the largest area of any of the Chenopod browse shrubs in Australia.

Maireana sedifolia prefers calcareous soils in which the limestone lies close to the surface. These soils are particularly susceptible to erosion if the surface cover is removed, either by wind in the case of the level alluvial desert loams or by both wind and water in the case of the undulating stony desert-soils.

Additional Chenopods which may be widespread but of little economic value, or may be of pastoral value but of minor occurrence in relation to the associated vegetation, or may be only locally common but pastorally important, include the following species: *Atriplex semibaccata*, *Chenopodium auricomum*, *C. nitrariaceum*, *Enchylaena tomentosa*, *Maireana brevifolia*, *M. georgei*, *M. tomentosa*, *M. triptera*, *M. villosa*, and *Rhagodia spinescens*. Many species of *Sclerolaena* (copperburrs) have also been shown to be of pastoral importance.

Community relationships

As mentioned previously it is generally accepted, mostly in the absence of sound scientific evidence, that the distribution of most Chenopod shrubs is determined by factors such as the physical and chemical properties of the soil and the soil moisture status.

The most thorough studies on factors affecting inter-species relationships, namely those which relate to the relative distribution of *Atriplex vesicaria* and *Maireana sedifolia*, were investigated by Carrodus & Specht (1965). They showed that the distribution of the two species was correlated with the depth to which the soil was wetted by normal rainfall, *M. sedifolia*, the deeper-rooted species, occurring on soils which could be wetted to a depth of 50 cm or more, and *A. vesicaria*, the shallow-rooted species, on soils in which a heavy clay subsoil or hardpan impeded root penetration beyond 30 cm. Both species could grow successfully in pots of the surface soil associated with the other species. Plants of *A. vesicaria* were also shown to be capable of reducing the percentage moisture in the soil to a significantly lower level than could *M. sedifolia* when subjected to drought. The practical consequence of these attributes therefore is that *M. sedifolia* occupies deep soils and, being long-lived, once established is capable of exploiting all the soil moisture available, whereas *A. vesicaria* with its shallower root system can exploit a smaller soil volume. With the onset of drought *A. vesicaria* could be expected to succumb more rapidly and so *M. sedifolia* is capable of maintaining the community in a closed state. In shallower soil, however, *A. vesicaria* has the advantage of more rapid re-establishment when drought breaks. If a community of *M. sedifolia* is destroyed, it takes a very long time to re-establish itself (Hall, Specht & Eardley, 1964) and in the meantime the community may be occupied by *A. vesicaria*.

On superficial examination of most Chenopod-dominated communities, particularly in the summer, the overriding impression is of Chenopod shrubs. With the possible exception of some of the salt marsh communities, all Chenopod shrubland communities characteristically support a ground cover, the main grass genera of which include *Danthonia*, *Stipa*, *Eragrostis* and *Aristida*. In the autumn, winter and spring months a number of annual grasses, mostly naturalised, including *Lolium*, *Hordeum*, *Vulpia* and *Bromus*, are common, as are a number of forbs, both native (e.g. *Brachycome*, *Calotis*, *Helichrysum*, *Helipterum*) and naturalised (e.g. *Medicago* and *Trifolium*). These naturalised grasses and forbs have changed significantly the productivity of such communities, and the legumes probably the soil fertility. The importance to the pastoral industry of these associated groups of cover grasses and forbs will be discussed more fully later in this chapter.

A high intensity of sheep grazing has led to a reduction, in some instances, in the numbers of desirable Chenopod shrubs and an increase in the numbers of unpalatable spiny shrubs, chiefly species of *Nitraria* and *Sclerolaena*. Under extreme conditions, complete baring or 'scalding' of the surface may occur, so that only bare soil remains.

A characteristic pattern of vegetated bands develops on areas of gently sloping pediments aligned along topographical contours and separated by bare interbands (Valentine & Nagorcka, 1979). This creates a number of interesting microtopographical features which enable some rain to run off the bare areas and infiltrate the vegetated areas. Should the bare areas be enlarged and the bands be breached by overgrazing then the fetch of the run-off water may be sufficient to encourage water erosion (Valentine & Nagorcka, 1979). Charley & Cowling (1968) described a process for the lateral distribution of salt in a similar shrub community, where chlorides were leached out of the bands and accumulated in the interbands, thereby allowing the growth of salt-intolerant species and stabilisation along the bands.

The results of early studies of Osborn, Wood & Paltridge (1932), and more recently, of Lange (1969), Barker & Lange (1969, 1970) and Barker (1979) and Lay (1979) have shown clearly the influence that location of watering points, and associated pattern of grazing by domestic stock, have on the biomass, density and species composition of Chenopod shrubs and associated species located in the Chenopod shrubland communities of South Australia. For example, Osborn *et al.* (1932) showed that *A. vesicaria* populations were denuded in the first 400 m from the watering point but beyond this, growth was stimulated. Similarly, Barker (1979) showed that within a wide range of Chenopod shrub–steppe communities only one species (*A. vesicaria*) disappeared close to watering points as a result of sheep grazing, whereas two species (*M. pyramidata* and *Sclerolaena patenticuspis*) were stimulated and spread into the area of the watering point from other parts of the paddock, one species (*Maireana excavata*) was stimulated to grow at an intermediate distance from the watering point and a further seven species (*Atriplex eardleyae, A. spongiosa, Carthamus lanatus, Inula graveolens, Maireana brevifolia, Marrubium vulgare* and *Sclerolaena paradoxa*) invaded from outside the paddock into the region of the watering point.

Trumble & Woodroffe (1954) described the effects of eleven years of sheep grazing in a study of the *Maireana sedifolia* association in South

Australia. Food intake was measured in a number of grazing treatments by assessing the amounts of green edible forage present in spring. Moderate to heavy grazing, particularly in favourable seasons, was more beneficial to growth of *M. sedifolia* and animal production from it was greatest under this regime.

Palatability, pastoral value and response to grazing
Palatability
No single index of acceptability to livestock in general can be given for a particular species because acceptability is dependent on a number of interacting factors. The more important of these are class of livestock, and level of availability and growth stage, both of the species in question and that of other associated species.

In general it has been observed that most Chenopod shrubs are of moderate to low palatability and that stock prefer associated grasses and forbs when these are available. On the riverine plain of southwestern New South Wales it was found that *Atriplex vesicaria* and *Maireana aphylla* shrubs were ungrazed or only lightly grazed (10% of the diet or less) by sheep when grasses and forbs were present, even though these grasses and forbs were often only minor constituents of the pastures (Leigh, Wilson & Mulham, 1968; Wilson, Leigh & Mulham, 1969). In times of feed scarcity and an absence of associated species, *A. vesicaria* comprised 70–90% of the diet eaten, whereas *M. aphylla* seldom exceeded 10% of the diet eaten.

More recently, a comparison has been made into the diets selected by sheep and goats grazing identical low woodland communities (*Casuarina cristata–Heterodendrum oleifolium*) near Ivanhoe, New South Wales, at light stocking rates (Wilson, Leigh & Mulham, 1975). This community characteristically has a tree overstorey, many species of large shrub (chiefly species of *Acacia*, *Apophyllum*, *Cassia* and *Eremophila*) and a ground stratum of grasses (chiefly *Stipa variabilis*), forbs and Chenopod subshrubs (chiefly *Sclerolaena diacantha*). Results of diet analyses over four seasons between May 1971 and February 1972 showed that trees and large shrubs (excluding Chenopods) made up between 58 and 80% of the goat diets, compared to 3–24% of the sheep diets, grasses made up less than 3% of goat diets and up to 49% of sheep diets, and *Sclerolaena* species (chiefly *S. diacantha*) less than 8% in goat diets and up to 48% in sheep diets.

Whilst palatability comparisons as presented above may be generally valid, it is also true that in certain circumstances considerable differences in palatability within individual species may exist, even when the plants are growing on the same soil type. In *A. vesicaria* shrublands growing on the riverine plain of New South Wales, Williams, Anderson & Slater (1978) demonstrated changes in the sex ratio between shrub populations subject to different grazing intensities. Sheep were observed to graze selectively on the green bladder fruits of the female plants, and thereby reduce their reproductive vigour with consequential losses to seed production, and they tended to avoid the granular catkins of the male plants.

Reaction to grazing
The effects of grazing on various plants differ from species to species but it is generally agreed that some degree of grazing is beneficial. The point at which defoliation by grazing has a detrimental effect on the plant varies with species. The degree to which *A. vesicaria* is grazed has a direct bearing on its chances of survival. Moderate grazing leads to a more compact, leafier bush (Osborn, Wood & Paltridge, 1932); complete defoliation will almost certainly result in its death (Leigh & Mulham, 1971). On the other hand, *A. nummularia* and *Maireana* species appear able to withstand periods of total defoliation. None of the plants will survive extended periods of continual defoliation, however.

The stems of the two major *Atriplex* species, *A. vesicaria* and *A. nummularia*, are particularly brittle, and they may be severely damaged by livestock trampling. Large stands of *A. nummularia* have been destroyed by the combined effects of trampling, especially by cattle, and grazing.

Selectivity differences between classes of livestock and subsequent regrowth pattern may be of some significance for pasture management and weed control. Wilson (1976) observed, in a comparison between the effects of cattle and sheep grazing, that immature plants of *Salsola kali* (soft roly-poly) were grazed closely by sheep and thereby prevented from forming their normal upright habit and prickly stems characteristic of mature bushes, whereas cattle ignored the young plants so that many mature plants of *S. kali* occurred.

Chenopods appear to vary considerably in their response to fire, according to fire intensity, both between species and also within species. One of the earliest records concerning the effects of fire on Chenopod shrublands is that by Murry (1931) when she reported on the 1922 wildfire in the Lake Torrens district of South Australia. She found large areas of what was originally good quality shrubland to be covered with dead bushes of myall, bluebush, mulga and saltbush, although some scattered plants of black bluebush (*Maireana pyramidata*) had survived the fire.

More recently, Lay (1976) found most species of bluebush including *Maireana pyramidata*, *M. sedifolia* and *M. astrotricha*, to be susceptible to wildfires in 1975 and 1976 although in one case he found 50% had survived a low-intensity fire. He also recorded saltbush as being extremely fire-sensitive, particularly when there was complete scorching of the canopy, even though no part of the plant was actually burnt. Similarly, in Western Australia, Mitchell (1976) also found pearl bluebush (*M. sedifolia*) to vary considerably in its regeneration after wildfires and no plants of saltbush were observed to recover. Some bladder saltbush (*A. vesicaria*) stands on the riverine plain of New South Wales were burnt by wildfires in 1956/57 and have since recovered, presumably through seedling recruitment (J.C. Noble, personal communication).

Mineral content and forage value

In comparison with non-Chenopods, these shrubs are high in nitrogen, sodium, potassium and chloride salts (Wilson, 1966a). Typical results of analyses showing the range in these constituents, determined over a number of localities, are presented in Table 12.1. Species of *Atriplex* accumulate ions to approximately the same concentration despite wide variations in the mineral composition of the soils on which they grow. High sodium levels are usually accompanied by high nitrogen levels. From an animal husbandry viewpoint these high mineral contents may

Table 12.1. *Published values for sodium, potassium, chloride and crude protein content and dry matter digestibility of several chenopod shrubs, expressed as a percentage of dry weight. After Leigh, 1972*

	Sodium	Potassium	Chloride	Crude protein	Digestibility
Atriplex angulata	3.3	1.8	—	17.5	58
A. nummularia	3.8–8.2	1.14–3.9	7.8–13.6	17.0–21.9	68–74
A. semibaccata	3.5	4.2	—	10.0–19.9	—
A. vesicaria	3.2–6.5	1.2–7.2	3.9–14.3	11.1–18.4	52–54
Maireana brevifolia	—	—	—	24.8	—
M. pyramidata	6.6	1.5	3.4	15.1–22.0	58
M. sedifolia	—	—	—	17.2	—
Sclerolaena diacantha	—	—	—	12.5	41

have a number of deleterious effects. High sodium levels reduce the water use of Chenopods per unit of dry weight and increase the water needs of grazing animals. Salt intakes approaching 200 g NaCl sheep^{-1} day^{-1} have been measured for sheep grazing *A. vesicaria* in summer (Wilson, 1967). Penned sheep fed on a diet of *A. vesicaria* leaf were shown to consume 11.3 l water day^{-1} (Wilson, 1966*b*), sheep grazing on a natural *A. vescaria* community drank 7.5 l day^{-1}, whereas similar sheep grazing on grassland consumed less than 3.2 l day^{-1} (Wilson, 1967). The increased water intake is necessary for urinary excretion of the high intakes of salt. Substantial reductions in feed intake also resulted when sheep consuming Chenopod leaf had only saline water to drink. This situation, where sheep grazing saltbush pastures in summer have no feed other than saltbush leaf and have access only to highly saline bore water, is by no means uncommon in Australia. It has been suggested for such situations, because of the need to make frequent visits to water, that the grazing range of sheep is likely to be severely decreased and that this may result in over-exploitation of the vegetation growing nearest to the water source (Squires & Wilson, 1971).

Chenopod shrubs generally have been regarded over the years as fodder of high quality on the basis of analyses of the crude protein content and dry matter digestibility of the type given in Table 12.1. It is true that the crude protein values are high relative to those of pasture grasses and legumes and are maintained throughout the year. The ability of these species to remain green under moisture stress, together with their high crude protein contents, has led numerous workers to suggest that Chenopod shrubs could serve as valuable protein supplements to sheep and cattle when the associated grasses were dry and possibly deficient in protein. Weight losses of pen-fed sheep have been reduced when *A. nummularia* and *A. vesicaria* were given as a supplement to a protein-deficient diet (Wilson, 1966*a*), but similar supplements have not been shown to improve weight gain or wool growth of animals grazing natural grasslands (Wilson, 1966*c*). It may be concluded that protein levels of the species available within the pasture have been adequate.

Field experiments designed to assess the pastoral value of *Atriplex nummularia*, *A. vesicaria* and *Maireana aphylla* have been conducted near Deniliquin in southwestern New South Wales (Leigh, Wilson & Williams, 1970). Comparisons were made of growth rates of wool and changes in body weight of flocks of sheep grazing established native perennial grass communities in which Chenopod shrubs were either dominant constituents or absent. It was found that the presence of the three Chenopod shrub species was of little or no value in reducing seasonal fluctuations in animal production and it was concluded that in terms of wool production the introduction of these shrubs into a perennial grassland was of doubtful economic value. The reason for the lack of a positive response would in all cases be because of a relatively low intake of the Chenopod shrubs themselves in relation to the total nutritional needs of the animals. In areas like the riverine plain of New South Wales, where elimination of Chenopods leads to the establishment of a stable subclimax grassland with a high component of perennial grasses (Chapter 14, Love, 1981), the disclimax grasslands have been shown to be capable of sustaining higher levels of animal production than adjacent climax communities dominated by Chenopod shrubs (Wilson & Leigh, 1970).

In other areas of Australia where the bushes are not replaced by stable disclimax communities, the Chenopod shrubs are regarded as valuable to ensure an amount of feed for use during dry periods, as an aid to increased rainfall infiltration, for nutrient cycling and also for the possible reduction in wind erosion.

Conclusions

Chenopod shrublands have significantly captured the curiosity, interest and imagination of a great number of research scientists over the past fifty years. In comparison to other Australian plant communities they are relatively well understood and have received a disproportionate amount of study in relation to their land area or value. Proceedings of two symposia held in Australia over the last decade, the first in Deniliquin in 1969 (Jones, 1970) and the second held more recently in 1975 (Graetz & Howes, 1979) summarised much of our knowledge. From those symposia and from a digest of the many other published papers it could be concluded that relatively little remains to be learnt of their physiology, nutrition, general ecology, pastoral value or reaction to fire and grazing.

As a broad generalisation it can be assumed that pastoralists with significant areas of Chenopod shrublands know how to manage and utilise them, that adequate extension guidance is available from State personnel, and that the future of Australia's Chenopod shrublands very largely depends on economic pressures, rather than any lack of vital biological knowledge. It is imperative that properties remain large enough to be economically viable. Owners or lessees must be able to manage their Chenopod shrublands in a conservative way, free from economic pressures which might otherwise force them to overstock for financial reasons. Perhaps the most important lesson which has been learnt is that such lands once they have been depleted of vegetation and once erosion has occurred are often both costly and difficult to re-habilitate. No one wishes to see a return to the eroded state of these shrublands in the 1930s so eloquently described by Ratcliffe (1936).

References

Atlas of Australian Resources (1980). 3rd Series. Pastures. Canberra: Department of National Development.

Barker, S. (1979). Shrub population dynamics under grazing – within paddock studies. In *Chenopod Shrublands, Studies in the Australian Arid Zone* IV, ed. R.D. Graetz & M.W. Howes, pp. 83–106. Perth: CSIRO, Australia, Division of Land Resources Management.

Barker, S. & Lange, R.T. (1969). Effects of moderate sheep grazing on plant populations of a black oak–bluebush association. *Australian Journal of Botany*, *17*, 527–37.

Barker, S. & Lange, R.T. (1970). Population ecology of *Atriplex* under sheep grazing. In *The Biology of Atriplex. Studies in the Australian Arid Zone I*, ed. R. Jones, pp. 105–20. Canberra: CSIRO, Australia, Division of Plant Industry.

Carrodus, B.B. & Specht, R.L. (1965). Factors affecting the relative distribution of *Atriplex vesicaria* and *Kochia sedifolia* (Chenopodiaceae) in the arid zone of South Australia. *Australian Journal of Botany*, *13*, 419–33.

Charley, J.L. & Cowling, S.W. (1968). Changes in soil nutrient status resulting from overgrazing and their consequence in plant communities of semi-arid areas. *Proceedings of the Ecological Society of Australia*, *3*, 28–38.

Gillison, A.N. & Walker, J. (1981). Woodlands. In *Australian Vegetation*, ed. R.H. Groves, pp. 177–197. Cambridge University Press.

Graetz, R.D. & Howes, K.M.W. (ed.) (1979). *Chenopod Shrublands. Studies in the Australian Arid Zone IV*. Perth: CSIRO, Australia, Division of Land Resources Management.

Hall, E.A.A., Specht, R.L. & Eardley, C.M. (1964). Regeneration of the vegetation on Koonamore Vegetation Reserve, 1926–1962. *Australian Journal of Botany*, *12*, 205–64.

Hartley, W. & Leigh, J.H. (1979). *Plants at Risk in Australia*. Australian National Parks and Wildlife Service Occasional Paper No. 3.

Johnson, R.W. & Burrows, W.H. (1981). *Acacia* open-forests, woodlands and shrublands. In *Australian Vegetation*, ed. R.H. Groves, pp. 198–226. Cambridge University Press.

Jones, R. (ed.) (1970). *The Biology of Atriplex. Studies in the Australian Arid Zone I*. Canberra: CSIRO, Australia, Division of Plant Industry.

Lange, R.T. (1969). The piosphere: sheep track and dung patterns. *Journal of Range Management*, *22*, 396–400.

Lay, B.G. (1976). Fire in the pastoral country. *Journal of Agriculture of South Australia*, *79*, 9–14.

Lay, B.G. (1979). Shrub population dynamics under grazing – a longterm study. In *Chenopod Shrublands. Studies of the Australian Arid Zone IV*, ed. R.D. Graetz & K.M.W. Howes, pp. 107–24. Perth: CSIRO, Australia, Division of Land Resources Management.

Leigh, J.H. (1972). Saltbush and other browse shrubs. In *The Use of Trees and Shrubs in the Dry Country of Australia*, ed. N. Hall, pp. 284–98. Canberra: Australian Government Publishing Service.

Leigh, J.H. & Mulham, W.E. (1971). The effect of defoliation on the persistence of *Atriplex vesicaria*. *Australian Journal of Agricultural Research*, 22, 239–44.

Leigh, J.H., Wilson, A.D. & Mulham, W.E. (1968). A study of merino sheep grazing a cotton-bush (*Kochia aphylla*)–grassland (*Stipa variabilis–Danthonia caespitosa*) community of the Riverine Plain. *Australian Journal of Agricultural Research*, 19, 947–61.

Leigh, J.H. Wilson, A.D. & Mulham, W.E. (1979). A study of sheep grazing a Belah (*Casuarina cristata*) – Rosewood (*Heterodendrum oleifolium*) shrub woodland in western New South Wales. *Australian Journal of Agricultural Research*, 30, 1223–36.

Leigh, J.H., Wilson, A.D. & Williams, O.B. (1970). An assessment of the value of three perennial chenopodiaceous shrubs for wool production of sheep grazing semi-arid pastures. *Proceedings of the XIth International Grassland Congress*, pp. 55–9.

Love, L.D. (1981). Mangrove swamps and salt marshes. In *Australian Vegetation*, ed. R.H. Groves, pp. 319–334. Cambridge University Press.

Mitchell, A.A. (1976). *Regeneration of shrubs after fire in the Goldfields district of Western Australia*. Australian Arid Zone Research Conference, Kalgoorlie Working Papers 3 c, 21–3.

Moore, R.M. (1969). Grazing lands of Australia. In *Australian Grasslands*, ed. R.M. Moore, p. 86. Canberra: Australian National University Press.

Murry, B.J. (1931). A study of the vegetation of the Lake Torrens plateau, South Australia. *Transactions of the Royal Society of South Australia*, 55, 91–112.

Newman, J.C. & Condon, R.W. (1969). Land use and present condition. In *Arid Lands of Australia*, ed. R.O. Slatyer & R.A. Perry, pp. 105–32. Canberra: Australian National University Press.

Osborn, T.G.B., Wood, J.G. & Paltridge, T.B. (1932). On the growth and reaction to grazing of the perennial saltbush, *Atriplex vesicaria*: an ecological study of the biotic factor. *Proceedings of the Linnean Society of New South Wales*, 57, 377–402.

Parsons, R.F. (1981). Eucalyptus scrubs and shrublands. In *Australian Vegetation*, ed. R.H. Groves, pp. 227–252. Cambridge University Press.

Ratcliffe, F.N. (1936). *Soil drift in and pastoral areas of South Australia*. Council for Scientific and Industrial Research, Australia, Pamphlet No. 64.

Specht, R.L. (1970). Vegetation. In *The Australian Environment*, 4th edn (rev.), ed. G.W. Leeper, pp. 44–67. Melbourne: CSIRO, Australia & Melbourne University Press.

Squires, V.R. & Wilson, A.D. (1971). Distance between food and water supply and its effect on drinking frequency, and food and water intake of Merino and Border Leicester sheep. *Australian Journal of Agricultural Research*, 22, 283–90.

Trumble, H.C. & Woodroffe, R. (1954). The influence of climatic factors in the reaction of desert shrubs to grazing by sheep. In *Biology of Deserts*, ed. J.L. Cloudsley–Thompson, pp. 129–47. London: Institute of Biology.

Valentine, I. & Nagorcka, B.N. (1979). Contour patterning in *Atriplex vesicaria* communities. In *Studies in the Australian Arid Zone IV*, ed. R.D. Graetz & R.M.W. Howes, pp. 61–74. Perth: CSIRO, Australia, Division of Land Resources Management.

Williams, D.G., Anderson, D.J. & Slater, K.R. (1978). The influence of sheep on pattern and process in *Atriplex vesicaria* populations from the Riverine Plain of New South Wales. *Australian Journal of Botany, 26*, 381–92.

Williams, R.J. (1955). Vegetation regions of Australia. In *Atlas of Australian Resources*, 2nd series. Canberra: Department of National Development.

Wilson, A.D. (1966a). The value of *Atriplex* (saltbush) and *Kochia* (bluebush) species as food for sheep. *Australian Journal of Agricultural Research, 17*, 147–53.

Wilson, A.D. (1966b). The intake and excretion of sodium by sheep fed on species of *Atriplex* (saltbush) and *Kochia* (bluebush). *Australian Journal of Agricultural Research, 17*, 155–63.

Wilson, A.D. (1966c). Saltbush and irrigated pasture as supplements to a native pasture. *CSIRO, Australia, Division of Plant Industry, Field Station Record, 5*, 71–6.

Wilson, A.D. (1967). Observations on the adaption of sheep to saline drinking water. *Australian Journal of Experimental Agriculture and Animal Husbandry, 7*, 321–4.

Wilson, A.D. (1976). Comparison of sheep and cattle grazing on a semi-arid grassland. *Australian Journal of Agricultural Research, 27*, 155–62.

Wilson, A.D. & Leigh, J.H. (1970). Comparisons of the productivity of sheep grazing native pastures of the Riverine Plain. *Australian Journal of Experimental Agriculture and Animal Husbandry, 10*, 549–54.

Wilson, A.D., Leigh, J.H. & Mulham, W.E. (1969). A study of Merino sheep grazing a saltbush (*Atriplex vesicaria*)–cotton-bush (*Kochia aphylla*) community on the Riverine Plain. *Australian Journal of Agricultural Research, 20*, 1123–36.

Wilson, A.D., Leigh, J.H. & Mulham, W.E. (1975). Comparison of the diets of goats and sheep on a *Casuarina cristata–Heterodendrum oleifolium* woodland community in western New South Wales. *Australian Journal of Experimental Agriculture and Animal Husbandry, 15*, 45–53.

Wood, J.G. & Williams, R.J. (1960). Vegetation. In *The Australian Environment*, 3rd edn (rev.), pp. 67–84. Melbourne: CSIRO, Australia & Melbourne University Press.

13

Natural grasslands

R.H. GROVES & O.B. WILLIAMS

Terminology

The term grassland can have various meanings. In a strict sense 'a natural grassland is a plant community in which the dominant species are perennial (native) grasses, there are few or no shrubs, and trees are absent' (Moore, 1964). In a wider sense, any plant community, whether natural or developed by man, in which grasses provide a substantial proportion of the feed for domestic stock can be called a grassland and Moore (1970a) adopted this wider interpretation in his book on Australian grasslands.

For the purposes of the present chapter we define natural grasslands as those communities in Australia dominated by indigenous perennial grasses, but which nearly always include some introduced and alien species. Only some arid grasslands dominated by plants of the genus *Triodia* seem to be native (cf. natural) in the sense of having few or no aliens. Once the proportion of the latter increases, and especially the proportion of the introduced leguminous element, the vegetation becomes, in Australian terminology, an introduced and 'improved' pasture, the 'improvement' arising from the addition of superphosphate and the deliberate promotion of the introduced grass and legume component.

Some other terms need defining before we can describe the different types of natural grasslands occurring in Australia. A tussock grass is a dense, erect clump of tillers, usually tufty in appearance, the term being synonymous with that used by North Americans, bunch grass. Individual tussocks of the grasses *Triodia* and *Plectrachne* are usually so large and almost hemispherical in form (Figure 13.1) that they have been given a special term 'hummock' by Australian ecologists. Thus a hummock grassland is one dominated by the hummock grasses *Triodia* and *Plectrachne* in arid and semi-arid Australia, to which they are confined.

Grassland types and their distribution

Moore & Perry (1970) recognised four basic types of natural grasslands in Australia.

1. Arid tussock grassland. The areas of Mitchell grasses (*Astrebla* spp.) extensively distributed in inland Queensland, Northern Territory and northern Western Australia in a zone receiving between 200 and 500 mm average annual rainfall and characterised by predominantly summer rain.
2. Arid hummock grasslands. Dominated by *Triodia* and *Plectrachne* in areas with less than 200 mm average annual summer and/or winter rainfall. They have a very extensive distribution throughout arid Australia.
3. Coastal grasslands. Dominated by *Sporobolus* and *Xerochloa* and confined to the tropical summer rainfall region.

Figure 13.1. Hummock grassland of *Trioda* in central Australia. Photo: C.J. Totterdell.

4. Sub-humid grasslands. These were subdivided further.

   a. Tropical. Grasslands dominated by *Dichanthium* and *Eulalia*, and sometimes by *Bothriochloa* and *Heteropogon*, in eastern and northern Queensland in regions with predominantly summer rain.

   b. Temperate. Grasslands with an irregular distribution from north of Adelaide around the zone of 500 to 1000 mm average annual winter and/or summer rainfall of southeastern Australia to northern New South Wales. Dominant genera are *Themeda*, *Poa* and *Stipa*.

   c. Sub-alpine. Confined to the cold and wet mountain regions of the Monaro region of southern New South Wales, northeastern Victoria and the central plateau of Tasmania. Dominant genera are *Poa* and *Danthonia*.

As in any such classification there is inevitable overlap in terminology and in the distribution patterns of dominant genera. Moore (1970*b*) and Coaldrake (1979) both discussed the latter aspect for sub-humid grasslands in eastern Australia. Blake (1938) even went so far as to refer to two of the above basic types, arid tussock grassland dominated by *Astrebla* and tropical grassland dominated by *Dichanthium*, as a 'fluctuating climax', the direction of the fluctuation being influenced by rainfall and presumably also by the previous grazing regime. We shall refer to this theme subsequently.

The four major grassland types of Moore & Perry (1970), although sometimes with different terminologies, have been described in a number of different papers, some details of which are summarised in Table 13.1. Their distribution has been mapped by Wood & Williams (1960), whose terminology was revised by Specht (1970), and also by Moore & Perry (1970) and Carnahan (1976, 1977).

Grasslands other than those listed in Table 13.1 occur in both northern and southern Australia. They occur as the perennial grass understorey to open *Eucalyptus* woodland in northern Australia (Chapter 8, Gillison & Walker, 1981); in southern Australia they have formed as a result of removing or thinning the original tree cover and repeated burning and grazing of the understorey. In the latter case, such grasslands are not always natural, although they now consist of native perennial grasses with a tussock habit. Dominant genera in the understorey of the northern communities are *Themeda* and *Sorghum* near Katherine, Northern Territory (Arndt & Norman, 1959; Norman, 1962,

1963*a*, *b*, 1969) and *Heteropogon* in eastern Queensland (Shaw & Bisset, 1955; Shaw, 1957; Tothill, 1971*a*). A northern outlier of *Themeda*-dominant understorey to eucalypt woodland was described by Heyligers (1966) for lowland Papua. In temperate Australia *Danthonia* is the dominant genus in similar communities on the Northern Tablelands of New South Wales (Robinson, 1976) and *Poa* and *Themeda* predominate on the Southern Tablelands of New South Wales (Story, 1969). Other 'disclimax' grasslands occur in the Riverina of New South Wales (Moore, 1953; Williams, 1956, 1961), which after removal of the original Chenopod shrub overstorey (Chapter 12, Leigh, 1981) are now dominated by *Danthonia* and *Stipa*.

It is certainly not our intention to describe all these grassland types again in detail; rather, we wish in this chapter to draw attention to some of the inter-relationships within these different grasslands at the three levels of the community, the species population, and that of individual plants.

Community and species dynamics
The two main ecological factors affecting community and species dynamics of Australian grasslands seem to us to be interactions between previous grazing and fire regimes, and soil type as it influences the soil moisture regime. Other factors such as fire alone, grazing alone, soil nutrient status, and temperature as related to seasonal changes have also been shown to alter the proportions of dominant grassland species in the short or long term. In this section we shall discuss briefly one or two selected examples of each of these changes but shall refer to others in more detail as they relate to survival of individual plants in the next section.

The interaction between fire and grazing is especially important in maintaining the herbaceous character of those grasslands which may have once been woodlands (see Chapter 8, Gillison & Walker, 1981; also Tothill, 1971*b*) or shrublands. In general, grasslands, because of the protected apices of the major perennial species, are able to withstand fairly frequent fires in the non-growing season and light, regular grazing during the growing season. One major reason why the composition of some grasslands in Australia has changed so dramatically since European settlement is not because of previous fire or grazing regimes *per se* but rather because of the interaction between these two factors. If grazing occurs soon after a fire and continues then the regenerative capacities of some grasses are so severely depleted that the botanical

Table 13.1. *Previous studies on natural grasslands in Australia*

Grassland type	Main genera	Locality	Reference
Arid tussock	*Astrebla*	Inland southern Queensland	Everist (1935); Blake (1938); Roe (1941); Roe & Allen (1945); Davidson (1954); Holland & Moore (1962); Orr (1975); Williams & Roe (1975).
		Central Australia	Perry & Lazarides (1962).
		Northern Territory	Perry (1960, 1970).
Arid hummock	*Triodia, Plectrachne*	Northwest Western Australia	Burbidge (1943, 1945, 1959); Suijdendorp (1955, 1969, 1980).
		Northern Territory	Perry (1960, 1970).
		Central Australia	Perry & Lazarides (1962); Winkworth (1967).
		South Australia	Specht (1972).
		New South Wales	Beadle (1948).
Coastal	*Sporobolus*	Northern Territory	Perry (1960).
	Xerochloa	Western Cape York, Queensland	Pedley & Isbell (1971).
Sub-humid tropical	*Dichanthium, Eulalia, Bothriochloa*	Northwestern Queensland	Blake (1938); Bishop (1977).
		Central Queensland	Story (1967).
		Southeastern Queensland	Tothill & Hacker (1973).
temperate	*Themeda, Poa, Stipa*	Northeastern NSW	Roe (1947); Biddiscombe (1953); Begg (1959); Robinson & Lazenby (1976).
		Southeastern NSW	Pryor (1939); Story (1969); Doing (1972).
		Western Victoria	Patton (1936); Willis (1964); Groves (1965); Connor (1966); Stuwe & Parsons (1977).
		South Australia	Specht (1972).
sub-alpine	*Lomandra*[a]	South Australia	Specht (1972).
	Poa, Danthonia	Southeastern NSW	Costin (1954).

[a] Family Liliaceae.

composition may change. This seems to us a major reason for the difference in status of grasslands dominated, for example, by *Themeda* in southern Australia, almost a 'relict' grassland (Stuwe & Parsons, 1977), and the extensive distribution of fairly similar grasslands in southern Africa. The latter seem to be maintained by resting them periodically during the flowering–seeding–postseeding season, although like their Australian counterparts they have been burnt and grazed for many years.

An example of the effects of the interaction between fire and grazing is provided by the grassy understorey to eucalypt woodland in eastern Queensland presently dominated by *Heteropogon contortus*. These grasslands have been burnt annually in spring for about 100 years and this regime has induced and maintained the dominance of *Heteropogon*, a relatively undesirable species for cattle grazing (Shaw, 1957). In areas burnt annually but protected from grazing (as in railway enclosures) *Themeda australis*, probably the dominant grass in pre-settlement times, still remains dominant. Thus an annual burning regime together with continual grazing has led to a *Heteropogon*-dominant understorey. In parts of South Africa a regime of less frequent burning (every two years?) and the inclusion of some 'rest' periods in the grazing regime has retained a balance between *Themeda* and *Heteropogon* more favourable for animal production. The change from *Themeda* to *Heteropogon* dominance in the Australian example is no doubt exacerbated by the selective effects of grazing soon after a fire which may effectively increase the grazing intensity; in the same way Costin (1970) argued that some highly palatable herbs were at risk from grazing in sub-alpine and alpine vegetation.

Another example of the ecological significance of the fire and grazing interaction on grassland composition was described by Suijdendorp (1955, 1969, 1980) for arid tussock grassland in Western Australia. Summer (November) burning and deferred grazing controlled the growth of soft spinifex, *Triodia pungens*, a relatively undesirable species for animal production, and promoted the growth of other, more desirable species.

The effects of the interaction between fire and grazing by native and feral (cf. domestic) animals remains unexplored for Australian grasslands, though it has been shown recently to be very significant for understorey composition of some eucalypt forests and woodlands (Leigh & Holgate, 1979). In view of the ubiquity of rabbit grazing in southern

Australian grasslands for almost as long as sheep and cattle grazing, this is a surprising omission.

Lazarides, Norman & Perry (1965) reported the effects of fire alone on the short-term developmental pattern of a range of native grasses comprising the understorey to eucalypt woodland at Katherine, Northern Territory. Reproductive development in plants of *Themeda australis* and *Chrysopogon* spp. was depressed by burning, *Sehima nervosum* was unaffected in this regard, whilst burning in September, towards the end of the dry season, accelerated the development of reproductive tillers in *Heteropogon contortus*. Although these results were obtained for only one season, and that season was drier than the long-term average, they emphasise the responses between different species of the ultimate units of a grassland, namely, individual tillers. A similar approach was followed by Tainton, Groves & Nash (1977) who described longer term survival of different-aged tillers of *Themeda triandra* in relation to different times of burning in Natal, South Africa. Other examples of the effects of fire on Australian grasslands (Smith, 1960; Norman, 1963*a*, 1969; Groves, 1974) use grosser measures of plant performance, such as shoot biomass, and show that burning usually decreases shoot biomass in the growing season after treatment. Fire even in the absence of grazing by domestic stock may influence, therefore, plant performance and, in the long term, botanical composition of some grasslands.

Results of a pasture survey in the vicinity of Trangie, New South Wales, showed that grazing at a level of 2.5 sheep ha^{-1} on either of two soil types of different textures caused a change from a mixed grass understorey composed mainly of *Stipa falcata*, *Enneapogon flavescens* and *Paspalidium gracile* to one dominated by *Chloris truncata* and an increased population of winter and summer annuals (Biddiscombe, 1953). For example, on severely grazed areas on a coarse-textured soil the percentage basal area of *C. truncata* increased tenfold, and the increase in relation to the total basal area of perennial grasses was from 4.7% *C. truncata* in the ungrazed pasture to 88.0% *C. truncata* in the pasture grazed severely. Robinson & Dowling (1976) suggested the botanical changes which may occur in natural pastures in the area of Glen Innes, New South Wales, in relation to sheep grazing, although in this example the effects of grazing are confounded by regular additions of superphosphate. Some of their suggested relationships are reproduced in Figure 13.2 to illustrate the complex interactions between grasses within a grazed community.

Grazing, unaccompanied by fire, can lead to large changes in grass dominance as the above two examples show, the direction of the changes being influenced partly by soil type and seasonal conditions. Williams (1956) showed that soil type, as expressed by soil–plant moisture relations, may have a major effect on grassland composition. On the Riverine Plain of New South Wales (34°30′; 145°00′E), sandy loam soil supports grassland dominated by *Stipa falcata*, clay loam soil is characterised by co-dominance of *Stipa* and *Danthonia caespitosa* and a clay soil supports *Danthonia*-dominant grassland (Moore, 1953). Williams (1956) quantified this relationship and related it to soil moisture characteristics, some data for which are shown in Table 13.2. More water was available under *Stipa* grassland than under the other grasslands but this level of moisture, and especially that from light falls of rain, was used more rapidly because of a presumed greater pattern of exploration by *Stipa* roots of the more permeable profile. On the heavier soils with less moisture available in the surface soil over the year, ephemeral herbs were

Figure 13.2. Relationships suggested by Robinson & Dowling (1976) between frequency of natural grassland components (*Themeda*, thick line; *Sporobolus*, dotted line; *Danthonia*, open circles; *Bothriochloa*, open triangles) and stocking rate in the Glen Innes area, New South Wales.

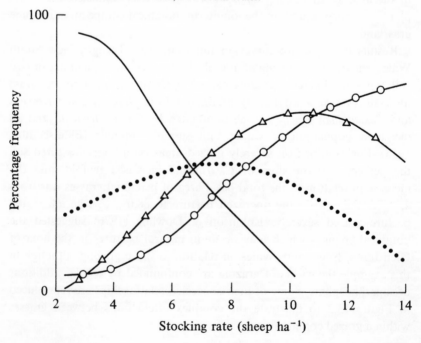

common. Moisture in the subsoil was recharged through deep cracks in the heavier soil profile, and this could then be tapped in summer by xeromorphic plants such as Chenopod shrubs. Thus two different grasslands and a third, intermediate between them, may occur within a short distance of one another because of the over-riding effect of soil type as it affects soil moisture availability.

Soil nutrient status is a factor which greatly influences the distribution of some Australian vegetation types (see, for example, Chapter 11, Specht, 1981), but it has not been studied as extensively in grasslands. We wish to discuss one example in which soil nutrient status is a major factor, though not the only one, influencing botanical change. In some areas of the Southern Tablelands of New South Wales, tussock grasslands are dominated by *Themeda australis*. Some of these have probably formed as a result of clearing the original eucalypt forest or woodland, repeatedly burning the understorey and grazing at a low intensity (Story, 1969). Another indigenous perennial grass, *Poa labillardieri*, becomes dominant when the *Themeda* grasslands are fertilised and grazed more intensively by sheep. In an economic sense, *Poa* is a weed, in that its dominance soon reduces the stocking capacity of the land for sheep. By growing *Themeda* and *Poa* plants alone and in mixture in pots to which different levels of phosphorus and nitrogen were added, Groves, Keraitis & Moore (1973) showed that *Poa* was better able to absorb nutrients and grow in response to a low level of added phosphorus than *Themeda*, when the two grasses were mixed. When the two were grown alone, both

Table 13.2. *Botanical composition of grassland and some soil properties at three sites on the Riverine Plain, New South Wales (adapted from Williams, 1956)*

	Perennial species Composition (%)		Surface soil texture	Soil air space 0–7.5 cm (% vol)	Soil permeability
Grassland	*Stipa*	*Danthonia*			
A	82.3	2.2	sandy loam	23.5	good
B	36.4	31.7	clay loam	14.9	poor
C	3.4	69.6	light clay (depression[a])	5.5	very poor

[a]Refers to 'gilgai' pattern of microrelief, of which the depression has the most clay at the surface (Williams, 1955).

responded similarly to increasing nutrients. Although obtained with glasshouse-grown plants, this result is similar to that observed in the field, with other factors, such as selective grazing, interacting with the changed soil nutrient status to induce changes in grass composition.

Just as there may be changes in grassland composition because of different patterns of seasonal rainfall and hence different soil moisture regimes there may also be botanical changes because of seasonal temperature conditions, especially if combined with wet growing seasons. These seasonal effects will average out over a period of years but in the short term they may be significant. Because different grasses possess different pathways for photosynthesis (the so-called C_3 and C_4 pathways (Black, 1971) some grasses (C_3) found in southern Australia grow mainly in the spring and autumn months, and others (C_4 grasses) grow mainly in summer when temperatures are higher. This fundamental difference in growth, especially in grasses, has been recognised for some time and is implicit in the terminology 'cool season' and 'warm season' grasses (see, for example, Moore, 1970b).

In southern Australian grasslands usually both groups of grasses are present as dominants, at least in the pre-settlement condition. Generally, wet summers favour the growth of the C_4 grasses, usually of tropical origin, and wet spring and autumn periods favour the C_3 element, originating from temperate regions. The seasonal dominance of one group of grasses over the other depends largely on moisture regimes and temperature, C_3 grasses (such as *Danthonia*) usually having an optimum temperature for growth of 20° to 25° (Hodgkinson & Quinn, 1976) and C_4 grasses (such as *Themeda*) having an optimum of about 25° to 35° (Groves, 1975). In mixtures, C_4 grasses usually dominate over a range of temperatures higher than that for C_3 grasses and *vice versa*, as Groves, Williams & Hagon (unpublished observations) recently showed for *Danthonia–Themeda* mixtures under both phytotron and field conditions.

Throughout this section we have discussed results of the effects of single factors on grass growth and grassland composition, although we have stressed the importance of interactions, for example, that between fire and grazing, in altering grassland dynamics. Over the past 100 to 150 years all the factors we have mentioned have been interacting to produce present day grasslands. Hopefully, the approach and methodology of systems ecology will, in the near future, enable us to understand more completely these complex interactions.

Plant dynamics

Descriptions of observed population changes in natural grasslands are replete with phrases such as 'controlled the growth of species A', 'encouraged the development of a stand of species B', 'becomes dominant for a short period only to be replaced by species C' or 'disappears'. Supporting evidence for these statements is seldom given. In this section we provide some examples of the way that the various populations of a species in natural grasslands respond to climate and to management.

In the course of a long-term demographic study (Williams & Roe, 1975) an *Astrebla*-dominant grassland became *Dichanthium*-dominant for some years in the early 1950s before returning to *Astrebla*- dominance, or so it appeared. This was a classical example of the 'fluctuating climax' (Blake, 1938) we referred to earlier. Holland & Moore (1962) provided historical and contemporary descriptions of this phenomenon. The data presented in Table 13.3 show, however, that there is a population explosion of *Dichanthium sericeum* against a background of recruitment and normal survivorship of *Astrebla*. The data do not represent the performance of a replaced dominant so much as a prolonged 'aspect', with *Dichanthium*, a less-adapted species in this particular environment, profiting from a sequence of mid- to late-summer rainfall seasons.

In terms of population performances it is difficult to differentiate between long-lived grasses such as *Astrebla* and short-lived grasses such as *Dichanthium*. In the *Astrebla–Dichanthium* example the population of *Astrebla* exploded in 1941 to reach a peak of 210 plants per 5 m^2; but in 1958 the age composition was: 1 plant > 18 years, 4 plants = 12 years, 1 plant = 8 years, 1 plant = 5 years; at the 30-year-stage in 1972 the

Table 13.3. *Demography of* Dichanthium sericeum *and* Astrebla *spp. in arid tussock grassland at 'Gilruth Plains', Cunnamulla, Queensland. Density in plants per 40m²*

Sampling time	1947 September		1953 July		1958 July	
	No.	Age (years)	No.	Age (years)	No.	Age (years)
Dichanthium sericeum	53	< 1 (1 cohort)	943	0.5–6.5 (5 cohorts)	3	< 1–4.5 (3 cohorts)
Astrebla spp.	81	0.8–> 5 (2 cohorts)	76	0.4–> 7.5 (6 cohorts)	75	0.4–> 12.5 (9 cohorts)

Astrebla population was only three and the ground was largely bare. A second example of population explosions in a much shorter-lived perennial grass, viz. *Stipa variabilis* at Ivanhoe, New South Wales (Wilson, Leigh, Hindley & Mulham, 1975), showed only four of 255 plants per 10 m² in a cohort attaining two years of age and none surviving to three years. At a less xeric site, distant 300 km southeast at Deniliquin, New South Wales, in a *Danthonia caespitosa* grassland (Williams, 1956), only one of 57 plants of *S. variabilis* per 10 m² in a cohort reached five years of age.

We conclude from these three examples for which there is population information that the recruitment of many Australian grass species is generally by population explosion and is markedly episodic. Sometimes, the time interval between substantial cohorts (age groups) may exceed the life span of all but a few of the older plants.

Grazing by domestic, indigenous or feral herbivores is the major 'treatment' applied to Australia's natural grasslands. Under grazing, compared with non-grazing, species populations could be expected to decrease, increase, or be unaffected. Such a comparison in the Australian context requires, however, an appreciation of what is meant by the terms 'non-grazing', 'ungrazed' and 'protected from grazing'. These three terms refer to 'release from grazing' because most species in grasslands have a history of only 100 to 150 years of heavy to excessive utilisation, often accompanied by soil erosion and severe drought. This direct and indirect European onslaught on plants also meant less fuel and fewer fires. This combination of factors was unique compared with the preceding millennia (at least 50 000 years B.P) of light grazing by small, non-domesticated herbivores and the fire regimes of the Aboriginal population.

The differing survivorship of *Danthonia caespitosa* and *Enteropogon acicularis* cohorts when grazed and not grazed (Williams, 1970) illustrate the favoured 'slightly-to-unaffected-by-grazing' type of performance represented in Figure 13.3a, compared with the disadvantaged model (Figure 13.3b).

The use of the *Danthonia* and the *Enteropogon* demographic performance as models in relation to grazing must be tempered by our present understanding (or lack of it) of the location-specific nature of such performances. Both perennial grasses occur along a north–south cline with the experimental site at Deniliquin located in the middle of the *Danthonia* range and towards the southern extremity of the *Enteropogon* range. We infer from the results of studies by Hodgkinson & Quinn (1976) and by Michalk & Herbert (1978) that *Enteropogon* conforms more to the *Danthonia* model as one proceeds northwards; the *Enteropogon* model is appropriate for *Danthonia* at *Danthonia*'s northern and southern extremity. Presumably the associated perennial grass and non-grass species in the *Danthonia* grassland at Deniliquin, with their known reaction to grazing and release from grazing (Williams, 1970, 1974) but with an unknown demography, are performing in a similar manner along unknown clines.

Astrebla populations 1300 km apart at the extremities of the north–south cline, viz. Julia Creek and Cunnamulla in Queensland, appear to behave similarly in terms of episodic recruitment and the survivorship of cohorts under grazing and release from grazing, but there appear to be marked differences in tiller longevity and flux; tillers in the north are predominantly annual unlike those of southern plants. Northern plants are markedly rhizomatous and, with age, lose the compact tussock or bunch grass form so characteristic of the southern plants. This rhizomatous characteristic is more pronounced on friable, self-mulching clays, than on non-cracking soils.

Initially the difference between north and south in the proportion of summer rainfall in the annual total was expected to be expressed in annual or near-annual northern recruitment. The early summer rainfall now understood to be influential in the production of large cohorts is scarce overall at both sites, however.

Drought takes two forms in Australia. There is the normal seasonal drought which occurs in the summer of southern Australia and in the winter of northern Australia. The non-seasonal or severe drought occurs

Figure 13.3 Graphical representations of different responses of grass density to grazing: *a*, species only slightly affected or unaffected; *b*, species markedly disadvantaged by grazing. Full lines, grazed grasslands; dashed line, ungrazed grasslands.

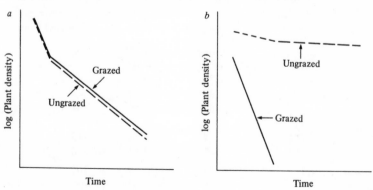

spasmodically when the rain-bearing weather systems fail. Occasionally rain-bearing systems penetrate from the north or the south of the continent in the normal seasonal drought period and deposit substantial precipitation on a dry landscape. All three of these rainfall events are part of the Australian environment and elicit a series of plant responses which we shall now consider in terms of survivorship.

The survivorship curves of the annual cohorts of a species are of the same general form, with an initial substantial fall and a low half-life followed by a slow decline and a much longer half-life (see Williams, 1970). A corollary of this performance is that small cohorts are short-lived and large cohorts are long-lived (see Fig. 1a of Williams, 1970). However, there are some cohorts in which the initial decline is less severe or in which the decline is arrested temporarily. Upon analysis we note that the first example represents a normal sequence of rainfall season, seasonal drought and rainfall season; the second example represents the sequence of rainfall season, seasonal drought interrupted by unseasonal rainfall, then seasonal rainfall.

Severe drought means no recruitment, but unexpectedly does not appear to have as marked an effect on survivorship of established plants as we had anticipated (see Wright & Van Dyne, 1976, for another example). The example of severe mortality in *Danthonia caespitosa* described by Williams (1968) is puzzling not because it occurred in both grazed and ungrazed treatments or because three such events have occurred over a 30-year-period, but because these deaths appear to be caused by the late arrival of the seasonal cool-season rains after an almost rainless summer; this could not be termed a drought. Further,

Table 13.4. *Age groups of* Danthonia caespitosa *plants in grazed and ungrazed quadrats before and after drought in 1966 at Deniliquin, New South Wales (plant density per 10 m²)*

Age (years)	Grazed		Released from grazing	
	1965	1968	1965	1968
0–0.5	272	75	41	46
0.5–2.5	133	962	8	410
2.5–4.0	69	44	31	5
4.0–8.0	7	91	10	25
8.0–11.0	4	4	13	8
>11.0	2	11	6	18
Total	487	1181	109	512

this mortality is in marked contrast to the improvement of *D. caespitosa* populations in these same quadrats over the severe (1st decile) non-seasonal drought of 1967 (see Table 13.4). Before this drought there were 21 plants m^{-2} in grazed quadrats and 7 plants m^{-2} in ungrazed quadrats that were capable of flowering; this category increased after the drought to 149 and 34 respectively.

We conclude from our studies of the post-drought populations that pre-drought cohorts *can* contribute substantially, in which case the immediate post-drought cohorts are small. When pre-drought plants are few, leaving the ground bare, the immediate post-drought cohorts may be larger (and so live longer) than cohorts which start at other times.

The bare ground resulting from the absence of *Astrebla* recruitment has been ascribed erroneously to grazing or to a combination of drought and grazing. Ungrazed treatments at both extremities of the *Astrebla* formation can proceed to bare ground (see Table 13.5 for the northern site). The apparent necessity for a certain amount of unoccupied ground and co-incident rainfall before there can be recruitment is illustrated in Table 13.5. In this extreme case the ground was cultivated and cropped between 1958 and 1966 at which stage the paddock reverted to sheep grazing. Recruitment of *Astrebla* occurred in 1968 and in July 1974 the number of plants in this cropped portion, 4.38 ± 2.88 (s.d.) m^{-2}, exceeded those in the adjacent unploughed portion, $1.67 \pm 1.86 \, m^{-2}$, and the basal areas were 0.26 ± 0.23 and $0.10 \pm 0.20\%$ respectively; this basal area had been reduced from values of 1.89 and 0.15% respectively, by denudation following excessive rainfall of 760 mm in January 1974. This monthly rainfall value is almost twice the long-term median *annual* rainfall and is a further example of massive climatic events of low probability that directly affect plant populations.

Table 13.5. *Density and basal area of plants of* Astrebla *spp. in a paddock excluded from grazing since 1962 at Julia Creek, Queensland*

Year	Plant density (no. $100m^{-2}$)	Basal area (%)
1962	400	3.2
1966	414	—
1967	350	—
1975	56	—
1978	52	0.3

Finally, in Figure 13.4 we set out half-life estimates for the two groups of grasses (*Astrebla, Danthonia* and *Enteropogon*) and contrast them with half-life estimates for a number of Australian shrub species. It is noteworthy that many long-lived shrubs in southern Australia are probably of pre-settlement age and that recruitment has been prevented over the past 100 years by the effects of sheep and rabbit grazing. Grasses such as *Cymbopogon bombycinus* and *Enteropogon* could also be represented by pre-settlement individuals.

Burning is the second major treatment applied to Australia's natural grasslands (see Tothill, 1971*a* and our previous discussion) but its effect on plant dynamics as distinct from changes in botanical composition (Norman, 1969) is largely unknown. From preliminary results of a burning experiment on natural grassland dominated by *Heteropogon contortus* but with scattered remnant *Themeda australis*, A.O. Nicholls & G.J. Burch (personal communication) concluded that *Heteropogon* plants are fire-sensitive but the profuse seeding, frequent and substantial recruitment, and rapid growth offset this mortality. The high turnover rate for *Heteropogon* populations is in marked contrast to that of *Themeda* populations which produce little viable seed, have few seedlings and rely on a tiller surplus with expansion from the few initial plants to effect its dominance.

Figure 13.4. Survivorship of key perennial species in the absence of grazing.

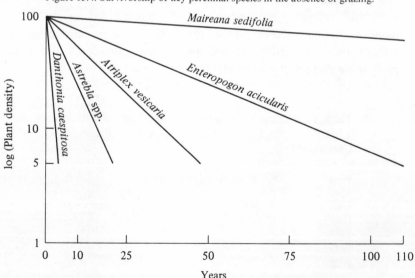

Grazing ecology and management

Australian experience in the pastoral industry and that gained from results of grazing experiments on natural grasslands has shown little benefit accruing to the vegetation and little economic benefit when seasonal deferments have been applied. Substantial differences in stocking rates (sheep or cattle per unit area per year) cause some differences in botanical composition but usage has to be severe and prolonged for permanent changes in vegetation to occur.

One reason for the lack of clear differences attributable to grazing, at least in southern Australia, is the heavy use that was given the grasslands by sheep and by rabbit plagues during the period from about 1870 to 1950; the period from 1895 to 1946 was also notable for the frequency of severe droughts. Treatments applied in experiments of a few years' duration are unlikely to be as influential now as they were in the first

Figure 13.5. Hypothetical representation of the importance of different systems of early management on survivorship of a cohort of a population.

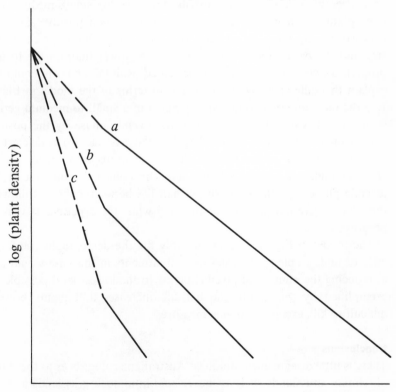

phase of exploitation. Further, long-term exclosure (release from grazing) for even 20 years could hardly be expected to aid species which had been eliminated and it is now not surprising that sheep die without causing further deterioration of droughted grasslands.

Grasslands in northern Australia do show differences in vegetative composition and plant performance which can be ascribed to management, although the management required to kill plants and create bare ground is brutal rather than severe.

Although we can explain some of the observed botanical changes in both southern and northern grasslands in terms of survivorship models, such as are represented in Figures 13.3*a*, *b* and 13.4, it is with the model in Figure 13.5 that possible success or failure in grassland management may lie. Figure 13.5 is derived from Figure 1 of Williams (1970) and incorporates an observation from a set of estimates made in the course of a survey of *Astrebla* spp. size-classes and densities on sheep properties in southwestern Queensland (R. Roe, personal communication). A few properties showed a size-class distribution covering small, medium and large plants; management of these grasslands was a positive aspect of property management. Most properties had the medium size-class missing and the basal area of *Astrebla* spp. was lower than on the former properties where all classes were represented. Rainfall and soils could not explain this difference. We suggest that in terms of the model in Figure 13.5 the medium-sized cohort(s) originated in a small to medium germination event which was favoured in its juvenile phase by the positive management exemplified by (*a*) as against the 'average' (*b*) or deleterious (*c*). This emphasis on the juvenile phase, our appreciation of the lasting effect of high and low population levels in cohorts at the end of the juvenile phase, and the similarity in half-life between cohorts following this phase, were not part of our thinking while the experiments were in progress.

The model in Figure 13.5 presumably fits the demographic performance of undesirable species; desirable treatments in this case would aim at reducing the number of juvenile plants to the lowest level possible, as exemplified by (*c*), before the half-life increases and plants become difficult to kill, except by unique measures.

Conclusions
There is little disagreement amongst Australian ecologists as to the status of 'climax' grasslands such as the *Triodia* hummock grasslands and the *Astrebla* tussock grasslands described earlier in this chapter. The 'dis-

climax' or degraded grasslands that have developed from the climax grasslands and from the shrubland and woodland understoreys have been much studied in terms of their development since European settlement and there is substantial agreement as to the biophysical causal factors involved. 'Succession', both as a concept and as a tool for management, has been difficult to apply, however, to a number of Australian vegetation types, e.g. the arid shrublands (Beadle, 1948, p. 229) and the coastal sclerophyll communities (Beadle, 1951; Groves & Specht, 1981). A similar situation applies to grasslands.

In this chapter we have stressed that each of a number of grass and non-grass communities may exhibit several 'aspects', with occupancy for three to five years by nearly pure stands of one of several short-lived perennials on a seemingly random basis. There seems to us to be no generalised and ordered sequence (i.e. succession) of these so-called 'aspects' moving towards a community dominated by an extremely long-lived species which can then restrain the subordinate species. Several schemes have been postulated (for example, those of Moore, 1970*b*) to account for the changes in botanical composition of natural grasslands in southern Australia in response to a complex of factors including compaction of soil by sheep, soil erosion and the redistribution of all or part of the surface horizon, fertilizer application and the consequent upsurge in alien, herbaceous species.

We conclude that this absence of a formal successional framework has not adversely influenced grassland research or management. On the contrary, it has freed the Australian grassland ecologist to concentrate on the all-important relationships existing both between species and between individual plants within a species, as this review has attempted to show.

The information on *Astrebla* populations given by G.R. Lee, E.J. Weston and J.C. Scanlon of the Queensland Department of Primary Industries is acknowledged.

References

Arndt, W. & Norman, M.J.T. (1959). *Characteristics of native pasture on Tippera clay loam at Katherine, NT.* CSIRO, Australia, Division of Land Research and Regional Survey Technical Paper No. 3.

Beadle, N.C.W. (1948). *The Vegetation and Pastures of Western New South Wales, with Special Reference to Soil Erosion.* Sydney: Government Printer.

Beadle, N.C.W. (1951). The misuse of climate as an indicator of vegetation and soils. *Ecology,* 32, 343–5.

Begg, J.E. (1959). Annual pattern of soil moisture stress under sown and native pastures. *Australian Journal of Agricultural Research, 10*, 518–29.

Biddiscombe, E.F. (1953). A survey of the natural pastures of the Trangie district, New South Wales, with particular reference to the grazing factor. *Australian Journal of Agricultural Research, 4*, 1–28.

Bishop, H.G. (1977). The response to nitrogen and phosphorus fertilizer of native pasture on the Balbirini land system in north-west Queensland. *Tropical Grasslands, 11*, 257–62.

Black, C.C. (1971). Ecological implications of dividing plants into groups with distinct photosynthetic production capacities. *Advances in Ecological Research, 1*, 87–114.

Blake, S.T. (1938). The plant communities of western Queensland and their relationships, with special reference to the grazing industry. *Proceedings of the Royal Society of Queensland, 49*, 156–204.

Burbidge, N.T. (1943). Ecological succession observed during regeneration of *Triodia pungens* R.Br. after burning. *Journal and Proceedings of the Royal Soceity of Western Australia, 28*, 149–56.

Burbidge, N.T. (1945). Ecological notes on the De Grey–Coogan area, with special reference to physiography. *Journal and Proceedings of the Royal Society of Western Australia, 29*, 151–61.

Burbidge, N.T. (1959). *Notes on plants and plant habitats observed in the Abydos-Woodstock area, Pilbara district, Western Australia*. CSIRO, Australia, Division of Plant Industry Technical Paper No. 12.

Carnahan, J.A. (1976). Natural vegetation. In *Atlas of Australian Resources*, Second Series. Canberra: Department of National Resources.

Carnahan, J.A. (1977). Natural vegetation. In *Australia: A Geography*, ed. D.N. Jeans, pp. 175–95. Sydney: University Press.

Coaldrake, J.E. (1979). The natural grasslands of Australasia. In *Ecology of Grasslands and Bamboolands in the World*, ed. M. Numata, pp. 133–40. The Hague: Junk.

Connor, D.J. (1966). Vegetation studies in north-west Victoria. II. The Horsham area. *Proceedings of the Royal Society of Victoria, 79*, 637–53.

Costin, A.B. (1954). *A Study of the Ecosystems of the Monaro Region of New South Wales*. Sydney: Government Printer.

Costin, A.B. (1970). Ecological hazards of the Snowy Mountains scheme. *Proceedings of the Ecological Society of Australia, 5*, 87–98.

Davidson, D. (1954). The Mitchell grass association of the Longreach district, with notes on gidgee and coolibah communities. *University of Queensland, Department of Botany Papers, 3*, 45–59.

Doing, H. (1972). *Botanical composition of pasture and weed communities in the Southern Tablelands region, south-eastern Australia*. CSIRO, Australia, Division of Plant Industry Technical Paper No. 30.

Everist, S.L. (1935). Inland pastures. II. Response during 1934 season of Mitchell and other grasses in western and central Queensland. *Queensland Agricultural Journal, 46*, 374–87.

Gillison, A.N. & Walker, J. (1981). Woodlands. In *Australian Vegetation*, ed. R.H. Groves, pp. 177–197. Cambridge University Press.

Groves, R.H. (1965). Growth of *Themeda australis* tussock grassland at St. Albans, Victoria. *Australian Journal of Botany, 13*, 291–302.

Groves, R.H. (1974). Growth of *Themeda australis* grassland in response to firing and mowing. *CSIRO, Australia, Division of Plant Industry Field Station Record, 13*, 1–7.

Groves, R.H. (1975). Growth and development of five populations of *Themeda australis* in response to temperature. *Australian Journal of Botany, 23*, 951–63.

Groves, R.H., Keraitis, K. & Moore, C.W.E. (1973). Relative growth of *Themeda australis* and *Poa labillardieri* in pots in response to phosphorus and nitrogen. *Australian Journal of Botany, 21*, 1–11.

Groves, R.H. & Specht, R.L. (1981). Seral considerations in heathland. In *Vegetation Classification in the Australian Region*, ed. A.N. Gillison & D.J. Anderson, (in press). Canberra: CSIRO, Australia & Australian National University Press.

Heyligers, P.C. (1966). Observations of *Themeda australis–Eucalyptus* savannah in Papua. *Pacific Science, 20*, 477–89.

Hodgkinson, K.C. & Quinn, J.A. (1976). Adaptive variability in the growth of *Danthonia caespitosa* Gaud. populations at different temperatures. *Australian Journal of Botany, 24*, 391–6.

Holland, A.A. & Moore, C.W.E. (1962). *The vegetation and soils of the Bollon district in south-western Queensland.* CSIRO, Australia, Division of Plant Industry Technical Paper No. 17.

Lazarides, M., Norman, M.J.T. & Perry, R.A. (1965). *Wet-season development pattern of some native grasses at Katherine, N.T.* CSIRO, Australia, Division of Land Research & Regional Survey Technical Paper No. 26.

Leigh, J.H. & Holgate, M.D. (1979). Responses of the understorey of forests and woodlands of the Southern Tablelands to grazing and burning. *Australian Journal of Ecology, 4*, 25–45.

Michalk, D.L. & Herbert, P.K. (1978). The effects of grazing and season on the stability of *Chloris* spp. (Windmill grasses) in natural pasture at Trangie, New South Wales. *Australian Rangelands Journal, 1*, 106–11.

Moore, C.W.E. (1953). The vegetation of the south-eastern Riverina, New South Wales. II. The disclimax communities. *Australian Journal of Botany, 1*, 548–67.

Moore, C.W.E. (1964). Distribution of grasslands. In *Grasses and Grasslands*, ed. C. Barnard, pp. 182–205. London: Macmillan.

Moore, R.M. (ed) (1970a). *Australian Grasslands*. Canberra: Australian National University Press.

Moore, R.M. (1970b). Australian grasslands. In *Australian Grasslands*, ed. R.M. Moore, pp. 87–100. Canberra: Australian National University Press.

Moore, R.M. & Perry, R.A. (1970). Vegetation. In *Australian Grasslands*, ed. R.M. Moore, pp. 59–73. Canberra: Australian National University Press.

Norman, M.J.T. (1962). Response of native pasture to nitrogen and phosphate fertilizer at Katherine, N.T. *Australian Journal of Experimental Agriculture and Animal Husbandry, 2*, 27–34.

Norman, M.J.T. (1963a). The short-term effects of time and frequency of burning on native pastures at Katherine, N.T. *Australian Journal of Experimental Agriculture and Animal Husbandry, 3*, 26–9.

Norman, M.J.T. (1963b). The pattern of dry matter and nutrient content changes in native pasture at Katherine, N.T. *Australian Journal of Experimental Agriculture and Animal Husbandry, 3*, 119–24.

Norman, M.J.T. (1969). The effect of burning and seasonal rainfall on native pasture at Katherine, N.T. *Australian Journal of Experimental Agriculture and Animal Husbandry, 9*, 295–8.

Orr, D.M. (1975). A review of *Astrebla* (Mitchell grass) pastures in Australia. *Tropical Grasslands, 9*, 21–36.

Patton, R.T. (1936). Ecological studies in Victoria. IV. Basalt plains association. *Proceedings of the Royal Society of Victoria, 48*, 172–90.

Pedley, L. & Isbell, R.F. (1971). Plant communities of Cape York Peninsula. *Proceedings of the Royal Society of Queensland, 82*, 51–74.

Perry, R.A. (1960). *Pasture lands of the Northern Territory, Australia.* CSIRO, Australia, Land Research Series No. 5.

Perry, R.A. (1970). Arid shrublands and grasslands. In *Australian Grasslands*, ed. R.M. Moore, pp. 246–59. Canberra: Australian National University Press.

Perry, R.A. & Lazarides, M. (1962). *Vegetation of the Alice Springs area.* CSIRO, Australia, Land Research Series No. 6, pp. 208–36.

Pryor, L.D. (1939). The vegetation of the Australian Capital Territory: a study of synecology. M.Sc. thesis, University of Adelaide.

Robinson, G.G. (1976). Productivity and response to nitrogen fertilizer of the native grass *Danthonia racemosa* (Wallaby grass). *Australian Rangelands Journal, 1*, 49–52.

Robinson, G.G. & Dowling, P.M. (1976). Management of natural pastures on the Northern Tablelands of New South Wales – a survey. *Australian Rangelands Journal, 1*, 70–4.

Robinson, G.G. & Lazenby, A. (1976). Effect of superphosphate, white clover and stocking rate on the productivity of natural pastures, Northern Tablelands, N.S.W. *Australian Journal of Experimental Agriculture and Animal Husbandry, 16*, 209–17.

Roe, R. (1941). Studies on the Mitchell grass association in south-western Queensland. I. Some observations on the response of Mitchell grass pastures to good summer rains following the 1940 drought. *Journal of the Council for Scientific and Industrial Research, Australia, 14*, 253–9.

Roe, R. (1947). *Preliminary survey of the natural pastures of the New England district of New South Wales and a general discussion of their problems.* Council for Scientific and Industrial Research, Australia, Bulletin No. 210.

Roe, R. & Allen, G.H. (1945). *Studies on the Mitchell grass association in south-western Queensland. II. The effect of grazing on the Mitchell grass pasture.* Council for Scientific and Industrial Research, Australia, Bulletin No. 185.

Shaw, N.H. (1957). Bunch spear grass dominance in burnt pastures in south-eastern Queensland. *Australian Journal of Agricultural Research, 8*, 325–34.

Shaw, N.H. & Bisset, W.J. (1955). Characteristics of a bunch spear grass (*Heteropogon contortus* (L.) Beauv.) pasture grazed by cattle in sub-tropical Queensland. *Australian Journal of Agricultural Research, 6*, 539–52.

Smith, E.L. (1960). Effects of burning and clipping at various times during the wet season on tropical tall grass range in northern Australia. *Journal of Range Management, 13*, 197–203.

Specht, R.L. (1970). Vegetation. In *The Australian Environment*, 4th edn (rev.), ed. G.W. Leeper, pp. 44–67. Melbourne: CSIRO, Australia & Melbourne University Press.

Specht, R.L. (1972). *The Vegetation of South Australia*, 2nd edn. Adelaide: Government Printer.

Specht, R.L. (1981). Heathlands. In *Australian Vegetation*, ed. R.H. Groves, pp. 253–275. Cambridge University Press.

Story, R. (1967). *Vegetation of the Isaac-Comet area.* CSIRO, Australia, Land Research Series No. 19, pp. 108–28.

Story, R. (1969). *Vegetation of the Queanbeyan-Shoalhaven area.* CSIRO, Australia, Land Research Series No. 24, pp. 113–33.

Stuwe, J. & Parsons, R.F. (1977). *Themeda australis* grasslands on the Basalt Plains, Victoria: floristics and management effects. *Australian Journal of Ecology, 2*, 467–76.

Suijdendorp, H. (1955). Changes in pastoral vegetation can provide a guide to management. *Journal of Agriculture, Western Australia, 3rd Series, 4*, 683–7.

Suijdendorp, H. (1969). Deferred grazing improves soft spinifex association. *Journal of Agriculture, Western Australia, 4th Series, 10*, 487–8.

Suijdendorp, H. (1980). Responses of hummock grasslands to fire in north-western Australia. In *Fire and the Australian Biota*, ed. A.M. Gill, R.H. Groves & I.R. Noble, pp. 417–424. Canberra: Australian Academy of Science.

Tainton, N.M., Groves, R.H. & Nash, R. (1977). Time of mowing and burning veld: short term effects on production and tiller development. *Proceedings of the Grassland Society of Southern Africa, 12*, 59–65.

Tothill, J.C. (1971a). A review of fire in the management of native pasture with particular reference to north-eastern Australia. *Tropical Grasslands, 5*, 1–10.

Tothill, J.C. (1971b). Grazing, burning and fertilizing effects on the regrowth of some woody species in cleared open-forest in south-east Queensland. *Tropical Grasslands, 5*, 31–4.

Tothill, J.C. & Hacker, J.B. (1973). *The Grasses of Southeast Queensland.* Brisbane: University of Queensland Press.

Williams, O.B. (1955). Studies in the ecology of the Riverine Plain. I. The gilgai microrelief and associated flora. *Australian Journal of Botany, 3*, 99–112.

Williams, O.B. (1956). Studies in the ecology of the Riverine Plain. II. Plant-soil relationships in three semi-arid grasslands. *Australian Journal of Agricultural Research, 7*, 127–39.

Williams, O.B. (1961). Studies in the ecology of the Riverine Plain. III. Phenology of a *Danthonia caespitosa* Gaudich. grassland. *Australian Journal of Agricultural Research, 12*, 247–59.

Williams, O.B. (1964). Energy flow and nutrient cycling in ecosystems. *Proceedings of the Australian Society of Animal Production, 5*, 291–300.

Williams, O.B. (1968). Studies in the ecology of the Riverine Plain. IV. Basal area and density changes of *Danthonia caespitosa* Gaudich. in a natural pasture grazed by sheep. *Australian Journal of Botany, 16*, 565–78.

Williams, O.B. (1970). Population dynamics of two perennial grasses in Australian semi-arid grassland. *Journal of Ecology, 58*, 869–75.

Williams, O.B. (1974). Vegetation improvement and grazing management. In *Studies in the Australian Arid Zone, II. Animal Production*, ed. A.D. Wilson, pp. 127–43. Melbourne: CSIRO, Australia.

Williams, O.B. & Roe, R. (1975). Management of arid grasslands for sheep: plant demography of six grasses in relation to climate and grazing. *Proceedings of the Ecological Society of Australia, 9*, 142–56.

Willis, J.H. (1964). Vegetation of the Basalt Plains in western Victoria. *Proceedings of the Royal Society of Victoria, 77*, 397–418.

Wilson, A.D., Leigh, J.H., Hindley, N.L. & Mulham, W.E. (1975). Comparison of the diets of goats and sheep grazing on a *Casuarina cristata–Heterodendrum oleifolium* woodland community in western New South Wales. *Australian Journal of Experimental Agriculture and Animal Husbandry, 15*, 45–53.

Winkworth, R.E. (1967). The composition of several arid spinifex grasslands in central Australia in relation to rainfall, soil water relations, and nutrients. *Australian Journal of Botany, 15*, 107–30.

Wood, J.G. & Williams, R.J. (1960). Vegetation. In *The Australian Environment*, 3rd edn (rev.), pp. 67–84. Melbourne: CSIRO, Australia & Melbourne University Press.

Wright, R.G. & Dyne, G.M. van (1976). Environmental factors influencing semidesert grassland perennial grass demography. *The Southwestern Naturalist*, *21*, 259–74.

VEGETATION OF EXTREME HABITATS

14

Mangrove swamps and salt marshes

LESLEY D. LOVE

Mangrove swamps and salt marshes develop on sediments deposited in sheltered estuaries, inlets and bays, and in the lee of islands and off-shore reefs. They are subject to periodic inundation and are usually intersected by a dendritic creek system. Mangrove swamps are woody communities which occur on tropical shores; salt marshes, dominated by samphires and grasses, occur on temperate shores. Both communities are found in many areas of the Australian coast.

Plants in these communities are subject to high salt levels in the soil solution, to waterlogging of the soil and to tidal scour, with the intensity of these influences varying seasonally over a wide range. Most of the plants have unusual morphological and physiological adaptations to their extreme habitat, such as the existence of pneumatophores, viviparity, salt excretion, succulence, high internal osmotic pressures in their cells and the presence of aerenchymatous tissue. The majority of the species are classified as halophytes, although some species can grow in the absence of salt.

A striking feature of many mangrove swamps and salt marshes is the patterning of the vegetation in zones approximately parallel to the shore, corresponding to the rise of the land with increasing distance from the sea and the consequent gradient in environmental conditions. This is generally described as succession, each distinct zone of the vegetation being equivalent to a successional stage or sere.

Factors affecting colonisation

There are a number of primary factors which may determine the suitability of an area for colonisation by species characteristic of mangrove swamps and salt marshes. These include the rate of sediment accretion, the stability of the sediment layer, the composition of the sediments, physiography, the velocity of the prevailing winds, wave energy, tidal range, the strength and temperature of ocean currents, and the size and number of rivers and their consequent capacity for transport, which in turn is related to the aridity of the climate. Most of these factors influence the ratio of accretion to erosion, which is the most important single factor in the suitability of an area for colonisation.

Much of the Australian coastline lacks the extensive swamps and marshes found in other countries at similar latitudes. Jennings & Bird (1967) gave some reasons for this. Australian estuaries are generally arid, with low run-off coefficients and few perennial rivers, which flood only spasmodically. This results in less fluvial sediments available to form deltas and barriers. More than half the coast of Australia (from Shark Bay, Western Australia, southwards and across to Fraser Island, Queensland, and much of Tasmania) is subject to high wave energy and low to moderate tidal ranges. Eastern Queensland, protected by the Great Barrier Reef, has the highest rainfall, a coast typified by low wave energy, a fairly high tidal range and, because of the concurrence of these factors, has the most extensive areas of mangrove swamps in Australia. The high tidal range and low wave energy of the northwestern coast make it suitable for mangroves, although a lower rainfall and the sandy or rocky nature of the coastline are limiting factors for mangrove colonisation. The coasts in tropical Australia tend to have wide, open estuaries with fewer dunes, and barriers may form deltas. A further factor influencing the distribution of mangroves in Australia is that the main drift is from south to north along both east and west coastlines, which is unfavourable for the supply of mangrove seedlings to the south coast (Bird, 1972).

Distribution
Overall distribution patterns of mangrove swamps and salt marsh species are dependent on the species available for colonisation, regional climates and past history. The tallest and most extensive swamps are in eastern Queensland, although salt marshes are very limited in extent in this region (Jones, 1971). All twenty-nine mangrove species recorded for Australia occur north of Daintree River, Queensland (latitude 16°17'S), whereas Macnae (1966) recorded only seven species in the salt marshes in that region. Further south, the size of the trees and the number of species are both reduced. Only nine mangrove species occur south of Sarina, Queensland (latitude 21°30'S), and only three occur south of Ballina,

New South Wales (latitude 28°53′S). Of these, *Excoecaria agallocha* is not found south of Coff's Harbour (latitude 30°18′S) and *Aegiceras corniculatum* extends only to Merimbula, New South Wales (latitude 36°50′S). Mangroves are rarely extensive along the New South Wales coast, but usually border salt marshes dominated by *Salicornia quinqueflora* and *Sporobolus virginicus* at suitable localities.

Beyond Corner Inlet, Victoria (latitude 38°45′S), the southernmost location of the mangrove *Avicennia marina*, its distribution continues sporadically westward along the southern coastline, being found at Shallow Inlet, Anderson's Inlet, Port Phillip Bay and Barwon Heads in Victoria (Bird, 1971) and at twenty-two locations in South Australia between Adelaide and Ceduna, with the most extensive stands being at

Fig. 14.1. *Arthrocnemum australasicum* and *Sporobolus virginicus* salt marsh in north coastal New South Wales. Shrubs are *Avicennia marina* (Photo: E. Slater).

Port Adelaide and Port Pirie, the heads of Gulf St Vincent and Spencer's Gulf, Franklin Harbour and Ceduna (Butler, Depers, McKillup & Thomas, 1977). Gulf St Vincent and Spencer's Gulf are less affected by refracted ocean swell, have a higher tidal range toward their heads, less wind and fewer barriers than other southern estuaries and, hence, favour the development of more extensive swamps and marshes (Jennings & Bird, 1967).

The southern mangroves were considered by Macnae (1966) to be relicts of a period when seas were warmer and patterns of drift more favourable, although Butler *et al.* (1977) found no evidence of a north-wards retreat of mangroves. Bird (1972), on the other hand, preferred to explain *Avicennia* distribution in terms of coastal evolution.

In these southern localities, extensive salt marshes dominated by samphires and grasses usually occur behind a fringe of mangroves. Up to thirty-four salt marsh species have been recorded in temperate localities (Saenger, Specht, Specht & Chapman, 1977). In the Victorian marshes the samphires *Arthrocnemum arbusculum*, *A. halocnemoides* and *Salicornia quinqueflora* are dominant (Patton, 1942; Bridgewater, 1975), and in the South Australian marshes *Arthrocnemum arbusculum*, *A. halocnemoides* and the salt bush *Atriplex paludosa* are dominant (Osborn & Wood, 1923; Wood, 1937).

On the coast of Western Australia there are isolated occurrences of *Avicennia marina* at Bunbury (latitude 33°30′S) and the Abrolhos Islands (latitude 28°40′S) (Sauer, 1965), but north of Carnarvon (latitude 24°53′S) mangroves are more abundant, and the number of species increases to fifteen along the northwestern coast, where extensive stands occur (Gardner, 1923; Beard, 1967b; Saenger *et al.*, 1977). The samphires *Arthrocnemum* and *Salicornia* also dominate the salt marshes of Western Australia, with extensive marshes containing up to twenty-nine species at higher latitudes, whilst the more northerly marshes are mostly small fringes of a few species behind the mangroves.

Along the northern coast of Australia salt marshes dominated by the samphires *Tecticornia australasica* and *Arthrocnemum* spp. are poorly developed behind the more extensive mangroves. Mangroves do best on the coast of Arnhem Land and in the southern part of the Gulf of Carpentaria (Lear & Turner, 1977). Bare salt flats are common in the more arid areas. Saenger *et al.* (1977) reported up to twenty mangrove species in Arnhem Land and up to twelve species in the lower part of the Gulf. Wells (1979), who examined seventy-seven river systems from the Kimberleys to the western shoreline of the Gulf of Carpentaria, recorded a total of twenty-five mangrove species for the Northern Territory, one of which (*Avicennia officinalis*) is previously unrecorded for Australia.

In Tasmania, no mangroves occur, probably because of frost. Only the salt marshes in the vicinity of Launceston and Hobart have been described (Curtis & Somerville, 1947; Kratochvil, Hannon & Clarke, 1973). The dominant species are *Salicornia quinqueflora*, *Distichlis distichophylla* and *Arthrocnemum arbusculum*. Saenger *et al.* (1977) recorded up to twenty-six species for Tasmania.

One inland occurrence of the mangrove *Avicennia marina* was noted by Beard (1967*a*) 40 km from the Western Australian coast behind Eighty Mile Beach. The mangroves line a salt creek which has no present tidal connection with the ocean and they appear to be a relict stand. In

Table 14.1. *Species list of Australian mangroves and their occurrence. Those species marked with an asterisk extend south of the Tropic of Capricorn*

Acanthaceae	Plumbaginaceae
Acanthus ilicifolius (shrub)*	*Aegialitis annulata**
Arecaceae	Rhizophoraceae
Nypa fruticans	*Bruguiera cylindrica*
Avicenniaceae	*B. exaristata*
Avicennia eucalyptifolia	*B. gymnorrhiza**
A. marina var. *resinifera**	*B. parviflora*
	B. sexangula
Bombacaceae	*Ceriops decandra*
Camptostemon schultzii	*C. tagal* var. *australis**
Caesalpiniaceae	*C. tagal* var. *tagal*
Cynometra iripa	*Rhizophora apiculata*
Combretaceae	*R. mucronata*
Lumnitzera littorea	*R. stylosa**
*L. racemosa**	Rubiaceae
Euphorbiaceae	*Scyphiphora hydrophyllacea*
*Excoecaria agallocha**	Sonneratiaceae
Meliaceae	*Sonneratia alba*
Xylocarpus australasicus	*S. caseolaris*
X. granatum	Sterculiaceae
Myrsinaceae	*Heritiera littoralis*
*Aegiceras corniculatum**	
Myrtaceae	
Osbornea octodonta	

South Australia and Western Australia salt marshes commonly grade into salt deserts in the more arid areas. Relict communities of salt marsh species also occur up to 500 km from the coast. Saenger *et al.* (1977) listed seven such species – *Arthrocnemum halocnemoides*, *A. leiostachyum*, *Frankenia pauciflora*, *Maireana brevifolia*, *Nitraria schoberi*, *Salicornia quinqueflora* and *Salsola kali*.

Flora

The exact number of mangrove species in Australia is uncertain, owing to problems in classification of the families Avicenniaceae, Sonneratiaceae and Rhizophoraceae. A list of the most-recognised species, representing fourteen families, is given in Table 14.1, based on the records of Macnae (1966), Jones (1971), Gill (1975), Lear & Turner (1977) and Saenger *et al.* (1977). The majority of these species have a much wider distribution in mangrove communities to the north of Australia, and only *Osbornea* is endemic. A list of understorey plants, lianes, epiphytes and algae associated with mangroves was given by Saenger *et al.* (1977).

The structure of the vegetation varies with the favourability of the environment. Closed-forest, 10–30 m tall, occurs in tropical areas with a high rainfall. Low closed-forest, 5–10 m tall, is the usual form in sub-tropical areas, with low open-forest or low woodland growing on some temperate shores. In less favourable areas of the temperate region, *Avicennia* is reduced to open-scrub (as little as 1 m high) or tall shrubland.

The salt marsh flora is much richer in species. Saenger *et al.* (1977) recorded fifty-five salt marsh species, excluding a further forty-nine temperate and sub-tropical species bordering marshes, at least some of which are inundated by spring tides. The more widely distributed, non-introduced species from this list are presented in Table 14.2. The two families most widely represented in salt marshes are the Chenopodiaceae (21 species) and the Poaceae (7 species). Australian marshes lack the extensive development of grasses so common in marshes in the northern hemisphere (Chapman, 1960). *Spartina anglica* has been introduced to several estuaries and Boston (1973) reported that it is spreading at Anderson's Inlet in Victoria, where it has colonised the seaward edge of the *Avicennia* fringe. Endemic salt marsh species include *Selliera radicans* and *Wilsonia* spp. The majority of salt marsh species have a wide distribution over temperate and sub-tropical areas, and a few extend into the tropics as well. Only two species are confined to the tropics and thirteen are primarily temperate. Around Sydney and Adelaide several species recorded by earlier workers are now rare or absent, because of the encroachment on their habitat by housing or industrial developments.

Salt marshes dominated by samphires 0.5–2 m tall are low shrublands to open-shrublands (Saenger *et al.*, 1977), the former being when the dominants are grasses, herbs and *Salicornia quinqueflora* (cf. a closed-herbland or herbland formation, which is dominated by grasses and herbs only).

Table 14.2. *Species list of Australian salt marsh plants and their occurrence, the latter recorded as tropical (Tr), sub-tropical (St) and temperate (Te)*

Aizoaceae
Disphyma blackii (Te)
Sesuvium portulacastrum (Tr, St)

Apiaceae
Apium prostratum (St, Te)

Asteraceae
Angianthus preissianus (Te)
Cotula coronopifolia (St, Te)

Batidaceae
Batis argillicola (Tr)

Caryophyllaceae
Spergularia media (Te)

Chenopodiaceae
Arthrocnemum arbusculum (St, Te)
A. halocnemoides var. *pergranulatum* (Tr, St, Te)
A. leiostachyum (Tr, St, Te)
Atriplex paludosa (Te)
Enchylaena tomentosa (Tr, St, Te)
Hemichroa pentandra (Te)
Rhagodia baccata (St, Te)
Salicornia blackiana (Te)
S. quinqueflora (Tr, St, Te)
Salsola kali (Tr, St, Te)
Suaeda australis (Tr, St, Te)
Tecticornia australasica (Tr)
Theleophyton billardieri (Te)
Threlkeldia diffusa (Te)

Convolvulaceae
Wilsonia backhousei (St, Te)
W. humilis (Te)
W. rotundifolia (Te)

Cyperaceae
Baumea juncea (St, Te)
Scirpus maritimus (St, Te)

Frankeniaceae
Frankenia pauciflora (St, Te)

Goodeniaceae
Selliera radicans (St, Te)

Juncaginaceae
Triglochin striata (St, Te)

Malvaceae
Lawrencia spicata (Te)

Plumbaginaceae
Limonium australe (St, Te)

Poaceae
Distichlis distichophylla (Te)
Puccinellia stricta (St, Te)
Sporobolus virginicus (Tr, St, Te)
Xerochloa barbata (Tr)

Primulaceae
Samolus repens (Tr, St)

Scropholariaceae
Mimulus repens (Tr, St)

Morphological and physiological adaptations of species

Mangrove and salt marsh species have a number of adaptations to high salinity and to waterlogged and unstable substrates. Most mangrove and salt marsh plants can adjust their osmotic pressure to that of the external solution in order to obtain water. Scholander (1968) found that all ten Australian mangrove species he tested had high negative sap pressures of 30 to 60 atmospheres in the xylem (cf. 25 atmospheres for seawater). The samphires *Salicornia* and *Tecticornia* gave even higher values. Various mechanisms for coping with these high salt concentrations are known. Scholander concluded that these plants can separate fresh water from seawater by a simple non-metabolic ultrafiltration process combined with ion transport. He showed that nearly fresh water could be obtained from the roots. Another mechanism is that most halophytes accumulate ions, which may be translocated to different parts of the plant. Salt often accumulates in the shoot, whence it may be secreted by salt glands, through the cuticle or in guttation fluid, retransported through the phloem to the roots and soil, or else concentrated in special leaf hairs.

Salt secretion is known in the mangrove genera *Acanthus, Aegialitis, Aegiceras, Avicennia, Bruguiera, Ceriops* and *Sonneratia* and the salt marsh genera *Frankenia* and *Limonium*. Field (1976) found that *Aegiceras corniculatum* takes in 10–20% of the seawater salt and secretes from the salt glands on both surfaces of the leaf a solution ten to twenty times as concentrated as the sap. He postulated an active transportation process which is sensitive to illumination and probably to other variables. Van Steveninck, Armstrong, Peters & Hall (1976) suggested that organic solutes in one type of mesophyll cells may provide intercellular osmotic adjustments while other cells, which store chloride ions, regulate the flow or provide the pathway for transport.

Atkinson *et al.* (1967) compared salt regulation in *Aegialitis annulata*, which secretes salt, with *Rhizophora mucronata*, which does not, and found that the latter excludes salt from its xylem at 17 m-equiv. $Cl\, l^{-1}$ of sap, but *Aegialitis* only excludes it at 85 to 122 m-equiv. and balances the extra output by secretion. Scholander, Hammel, Hemmingsen & Carey (1962) found that the xylem sap of salt-secreting mangroves carries 0.2–0.5% NaCl, about ten times higher than that of mangroves which do not secrete salt and about one hundred times higher than that of ordinary land plants.

Other methods of salt excretion have not been studied in Australian maritime halophytes, but European *Salicornia* species retransport salt through the phloem to the roots, and species of *Atriplex* can concentrate salt in special leaf hairs (Waisel, 1972).

Juncus sheds its leaves when the salt concentration builds up to too high a level, and *Salicornia* sheds its stems; they die back and produce new shoots. Probably many Australian halophytes fit into this category.

Halophytes have a general problem of balancing their nutrient uptake. Evidently some ions can be absorbed selectively, but it is not clear how. Rains & Epstein (1967) showed that *Avicennia marina* can absorb potassium preferentially at high sodium chloride concentrations. It is possible that some mangrove and salt marsh species have an actual requirement for sodium or chloride ions.

The majority of halophytes are facultative halophytes, but some species (*Rhizophora mucronata, Avicennia marina* and *A. alba*) are probably obligate halophytes (Chapman, 1976). Connor (1969) found that up to 1.5% NaCl increased growth of *Avicennia marina*, and Clarke & Hannon (1970) found that *Avicennia marina, Aegiceras corniculatum, Salicornia quinqueflora* and *Suaeda australis* all grew better in the presence of seawater up to concentrations of, respectively, 100, 60, 20 and 60%.

Many salt marsh species are leafless, articulated succulents. This life form is a response to salinity, as are the thick, succulent leaves of many mangrove and salt marsh plants.

The most striking adaptation to waterlogging is the presence of pneumatophores in mangroves. These aerial breathing roots have a wide variety of unusal forms, which are listed below.

> From simple knee roots which emerge above the soil surface
> and curve below again (*Lumnitzera littorea, Aegiceras,
> Excoecaria*) to curving, thickened knee roots which may be very
> large (*Bruguiera, Ceriops*).
> Erect, slender, conical pneumatophores (*Avicennia, Sonneratia,
> Xylocarpus australasicum, Lumnitzera racemosa*).
> Long, looping, bow-shaped pneumatophores (*Rhizophora*).

Other interesting roots which probably have the primary function of anchorage but also assist in aeration are given below.

> Aerial prop or stilt roots descending from the lower branches
> (*Rhizophora*) or stem base (*Acanthus,* and sometimes *Bruguiera*
> and *Avicennia marina*).
> Buttress roots (*Ceriops, Xylocarpus, Heritiera*) which meander
> from the trunk in conspicuous ridges (in *Xylocarpus granatum*
> and *Heritiera*).
> Thickened, erect portions of the horizontal roots emerge above
> the soil surface (*Camptostemon*).

Aerial anchoring and breathing roots tend to be best developed in the species most frequently inundated. Troll & Dragendorff (1931), as cited by Gill (1975), showed that *Sonneratia* grew new pneumatophores as the older ones became covered by sediment. I have observed that the number and size of pneumatophores in *Avicennia* varies considerably with location and this may well be true for other mangroves. In the rapidly accreting sediments at Salt Pan Creek, near Sydney, trunks have apparently developed from buried branches of mature trees.

Scholander, van Dam & Scholander (1955) showed that *Avicennia* pneumatophores and *Rhizophora* stilt roots can supply oxygen stored in their aerenchyma to the radial roots, even in completely anaerobic conditions. Most mangroves have shallow, but extensive, root systems to avoid the very poorly aerated soils encountered at depth.

Aerenchyma tissue is also well developed in other organs, for instance *Nypa* petioles (Gill, 1975), stems and leaves of many salt marsh species and in viviparous mangrove seedlings.

The reason for viviparity in several mangrove species (the family Rhizophoraceae and the genera *Aegiceras*, *Avicennia*, and *Nypa*) is unknown, although the most popular theory is that the more mature seedlings can grow rapidly after settling and so have a better chance of survival in extreme conditions. The observations of G. Carey (personal communication) suggest that a need for oxygen is the most essential feature of *Avicennia* and *Aegiceras* embryos and there is some support for this idea. For example, Brown, Outred & Hill (1969) found that the respiratory gaseous exchange of detached seedlings of *Avicennia*, *Bruguiera* and *Rhizophora* was markedly reduced at low oxygen concentrations; McMillan (1971) showed that *Avicennia germinans* would establish only in water depths of 5 cm or less and seedlings died at high temperatures. Persistent high waterlogging may kill young *Avicennia* plants (Clarke & Hannon, 1970).

The slow growth of many mangrove and salt marsh species may be a response to their difficult environment or an indication that they are poor competitors.

Zonation and its causal factors
The full zonation possible in a well developed mangrove stand was described by Macnae (1966) as containing six zones.

    1. The landward fringe, which has the greatest number of species and may contain any of the species listed below (the landward *Avicennia* zone is a finer subdivision of this zone).

2. *Ceriops* thickets, about 2 m high, which may have associated bare areas.
3. *Bruguiera* forest.
4. *Rhizophora* forest, which may also fringe creeks.
5. The seaward *Avicennia* zone, which is very narrow.
6. The *Sonneratia* zone, limited to the northeastern coast and the vicinity of Darwin.

Even in northeastern Australia, one or more of these zones is usually absent, and there is considerable local variation in zonation. More details of this variation were given by Chapman (1976), Lear & Turner (1977) and Saenger *et al.* (1977)

The fullest zonation noted in an Australian salt marsh is that given by Bridgewater (1975) for Victoria. Behind a frontal zone of *Avicennia*, nine complexes linked by structural and floristic similarities may occur.

1. Introduced *Spartina* (rare).
2. Extensive *Salicornia*-dominated zone (usual in New South Wales and Tasmania, and may occur in South and Western Australia).
3. Extensive *Arthrocnemum*-dominated zone (usual in South Australia, may occur in Western Australia and Tasmania, and absent in New South Wales).
4. *Suaeda* complex (sometimes noted in states other than Victoria).
5. *Puccinellia* complex (*Spartina* or other grasses usual in other states).
6. *Juncus* complex (well developed in New South Wales).
7. *Stipa* complex.
8. *Schoenus–Cotula* complex (the zone richest in species).
9. *Melaleuca* zone (*Casuarina* may replace this in New South Wales, South and Western Australia).

The zones may follow a sea–land sequence or occur as a mosaic related to minor topographical changes. Zones 7 and 8 are not well developed in other regions of Australia. *Distichlis distichophylla* is a common component of marshes in Victoria and Tasmania. In South Australia, *Atriplex* species commonly follow *Arthrocnemum* species in drier sites. Other comparisons between different regions were made by Chapman (1960) and Kratochvil *et al.* (1973).

Zonation patterns within local swamps and marshes have often been described as stages of succession, where the plants play a major role in modifying their environment. Autogenic factors are undoubtedly important, especially on prograding shorelines, but local zonation is governed by a variety of inter-related allogenic factors. Thom, Wright & Coleman (1975) considered that the patterns of mangroves they studied at Cambridge Gulf, Western Australia, are a 'result of geomorphic processes, which produce, modify and finally remove plant habitats'. The plants play a minor role in modifying the habitat. Clarke & Hannon (1967, 1969, 1970, 1971) found the governing factor explaining mangrove and salt marsh zonation in the Sydney district to be the tide–elevation–salinity influence in relation to the environmental requirements and tolerance ranges of the plant species. Diagrams of interactions between the factors which influence zonation were presented both by Clarke & Hannon (1969) and by Lear & Turner (1977). The main factors identified were physiography, tidal inundation, salinity, drainage and aeration, climate (especially rainfall and evaporation), nature of the soil and the biotic factor.

Geomorphic processes, as described by Thom *et al.* (1975), may determine which areas are suitable for colonisation at any given time. Within the local area the distribution and growth form of the plants is closely associated with the first four factors listed at the end of the previous paragraph. Very small differences in microtopography may play an important role in plant zonation (Clarke & Hannon, 1969), because of the relationship between physiography and the frequency and duration of tidal flooding. Salinity and waterlogging, which are closely related to physiography and flooding, also may play an important role in species distribution. Clarke & Hannon (1969, 1970) showed that the tolerances to salinity and to waterlogging of eight species when grown in culture correlated closely with their distribution in the field near Sydney. Regional climate is very influential in geomorphic processes and species availability, but climate influences local zonation patterns primarily through the modification of salinity. Several authors mention the importance of soil type to zonation and Walsh (1974) listed fine-grained alluvium as a requirement for mangroves. Clarke & Hannon (1967) gave the only detailed description of any soil in these Australian communities. They considered that the relatively coarse sandy soils of the Sydney district support mangroves without playing a major role in controlling plant distribution. Biotic factors (chiefly competition for light) were found (Clarke & Hannon, 1971) to influence the discreteness of zonation patterns. Gill (1975) reviewed the effect of several environmental factors (for instance, salt and flooding) on individual species.

Energy relationships

There is relatively little information available on the productivity of Australian mangrove and salt marsh communities, but estuarine swamps and marshes are generally considered to be highly productive. Data from Westernport Bay, Victoria, and three Sydney swamps are compared in Table 14.3 with data from Florida and Puerto Rico. The values estimated by Briggs (1977), based on complete annual leaf fall, may be too high, as Shapiro (1975) considered a leaf life of three years was indicated at Westernport Bay. The values from Westernport Bay are the lowest, probably because *Avicennia* is near the limits of its tolerance at this high latitude. The values for Sydney and those of Heald (1971) for Florida are fairly comparable, the difference perhaps being because of differences in latitude between the two sites. The higher values of Golley, Odum & Wilson (1962) and Miller (1972) are in the range expected, because these would include new growth.

Miller (1976) measured a high rate of total organic carbon production $(14-21 \text{ mg l}^{-1})$ exported from mangroves at Trinity Inlet ($16°58'S$; $145°47'E$), northern Queensland. The results of Shapiro (1975) show that roots constitute the major pool of nitrogen and phosphorus in the mangrove community at Westernport Bay, Victoria, with the major pathways for nutrient cycling occurring through turnover of leaves and roots. The mangrove community has a relatively small impact on nutrient cycling compared with the seagrasses. In Florida, however, Odum (1971) found that mangrove detritus played a major role in the food web.

Table 14.3. *Estimates of mangrove productivity at different locations*

Location	Principal species	Estimated productivity $(g \text{ m}^{-2} \text{ day}^{-1})$	Reference
1. Westernport Bay, Victoria	*Avicennia marina*	0.53 (leaf biomass)	Shapiro (1975)
2. Lane Cove, Sydney	*Avicennia marina*	3.3–4.1 (leaf/petiole biomass)	Briggs (1977)
3. Salt Pan Creek, Sydney	*Avicennia marina*	1.85 (litter fall)	Love (unpublished)
4. Botany Bay, Sydney	*Avicennia marina*	1.92 (litter fall)	Love (unpublished)
5. Florida, USA	*Rhizophora mangle*	2.41 (litter fall)	Heald (1971)
6. Florida, USA	*Rhizophora mangle*	3.51–5.55 (net photosynthesis)	Miller (1972)
7. Puerto Rico	*Rhizophora mangle*	3.40 (net photosynthesis)	Golley *et al.* (1962)

No Australian data are available for salt marsh productivity. In the United States, Barnes & Green (1972) cited production values ranging from 650 to 3700 g m^{-2} yr^{-1} for *Spartina* marshes.

The future of these communities

Mangroves have often been called 'land-builders'. From the results of studies of accretion at Westernport Bay, Bird (1971) indicated that mangroves cannot directly cause shoreline advance, but once established on areas of accretion they may influence the pattern of sedimentation. Bird considered that mangroves reduce the effects of offshore winds in moving sediment away and, also, they facilitate vertical accretion. Heinsohn & Spain (1974), after viewing the effects of cyclone Althea near Townsville, Queensland, considered that mangroves protect and stabilise coastal areas.

A variety of practical uses for the timber, leaves, fruit, nectar, bark, sap and even pneumatophores of mangroves have been found in Southeast Asia. Lear & Turner (1977) compiled an interesting list of uses for each species.

In recent years many swamps have been cleared, drained and filled to provide land for residential and industrial development near cities. Other swamps have been damaged by dredging, drifting sand, pollution, insecticides, mining, freshwater inflow and boat access. In view of the evidence that swamps and marshes have many useful roles – they are highly productive food sources in estuaries, mangroves stabilise the shoreline and have many potential uses, they provide a habitat for a variety of birds and other wildlife, and they have undoubted scientific interest – steps to ensure a better use of these resources are necessary. Attempts to regenerate selected mangrove stands have already started in Sydney and Perth, and it is hoped that this activity will gain momentum.

References

Atkinson, M.R., Findlay, G.P., Hope, A.B., Pitman, M.G., Saddler, H.D.W. & West, K.R. (1967). Salt regulation in the mangroves *Rhizophora mucronata* Lam. and *Aegialitis annulata* R.Br. *Australian Journal of Biological Sciences*, *20*, 589–99.

Barnes, R.S.K. & Green, J. (eds) (1972). *The Estuarine Environment*. Barking, Essex: Applied Science Publishers.

Beard, J.S. (1967*a*). An inland occurrence of mangroves. *Western Australian Naturalist*, *10*, 112–15.

Beard, J.S. (1967*b*). Some vegetation types of tropical Australia in relation to those of Africa and America. *Journal of Ecology*, *55*, 271–90.

Bird, E.C.F. (1971). Mangroves as land-builders. *Victorian Naturalist*, *88*, 189–97.

Bird, E.C.F. (1972). Mangroves on the Australian coast. *Australian Natural History, 17*, 167–71.

Boston, K.G. (1973). *Spartina* in Anderson's Inlet. *Victoria's Resources, 15*, 8–11.

Bridgewater, P.B. (1975). Peripheral vegetation of Westernport Bay. *Proceedings of the Royal Society of Victoria, 87*, 69–78.

Briggs, S.V. (1977). Estimates of biomass in a temperate mangrove community. *Australian Journal of Ecology, 2*, 369–73.

Brown, J.M.A., Outred, H.A. & Hill, C.F. (1969). Respiratory metabolism in mangrove seedlings. *Plant Physiology, 44*, 287–94.

Butler, A.J., Depers, A.M., McKillup, S.C. & Thomas, D.P. (1977). Distribution and sediments of mangrove forests in South Australia. *Transactions of the Royal Society of South Australia, 101*, 35–44.

Chapman, V.J. (1960). *Salt Marshes and Salt Deserts of the World*. London: Leonard Hill.

Chapman, V.J. (1976). *Mangrove Vegetation*. Lehre: Cramer.

Clarke, L.D. & Hannon, N.J. (1967). The mangrove swamp and salt marsh communities of the Sydney district. I. Vegetation, soils and climate. *Journal of Ecology, 55*, 753–71.

Clarke, L.D. & Hannon, N.J. (1969). The mangrove swamp and salt marsh communities of the Sydney district. II. The holocoenotic complex with particular reference to physiography. *Journal of Ecology, 57*, 213–34.

Clarke, L.D. & Hannon, N.J. (1970). The mangrove swamp and salt marsh communities of the Sydney district. III. Plant growth in relation to salinity and waterlogging. *Journal of Ecology, 58*, 351–69.

Clarke, L.D. & Hannon, N.J. (1971).The mangrove swamp and salt marsh communities of the Sydney district. IV. The significance of species interaction. *Journal of Ecology, 59*, 535–53.

Connor, D.J. (1969). Growth of grey mangrove (*Avicennia marina*) in nutrient culture. *Biotropica, 1*, 36–40.

Curtis, W.M. & Somerville, J. (1947). Boomer Marsh – a preliminary botanical and historical survey. *Proceedings of the Royal Society of Tasmania, 1947*, 151–7.

Field, C.D. (1976). Desalination and a mangrove. *Operculum, 5*, 53–5.

Gardner, C.A. (1923). Botanical notes. Kimberley Division of Western Australia. Western Australia Forests Department Bulletin No. 32.

Gill, A.M. (1975). Australia's mangrove enclaves: a coastal resource. *Proceedings of the Ecological Society of Australia, 8*, 129–46.

Golley, F., Odum, H.T. & Wilson, R.F. (1962). The structure and metabolism of a Puerto Rican red mangrove forest in May. *Ecology, 43*, 9–19.

Heald, E. (1971). *The production of organic detritus in a south Florida estuary*. University of Miami, Sea Grant Technical Bulletin No. 6.

Heinsohn, G.E. & Spain, A.V. (1974). Effects of a tropical cyclone on littoral and sub-littoral biotic communities and on a population of dugongs (*Dugong dugon* (Müller)). *Biological Conservation, 6*, 143–52.

Jennings, J.N. & Bird, E.C.F. (1967). Regional geomorphological characteristics of some Australian estuaries. In *Estuaries*, ed. G.H. Lauff, pp. 121–8. Washington: American Association for the Advancement of Science.

Jones, W.T. (1971). The field identification and distribution of mangroves in eastern Australia. *Queensland Naturalist, 20*, 35–51.

Kratochvil, M., Hannon, N.J. & Clarke, L.D. (1973). Mangrove swamp and salt marsh communities in southern Australia. *Proceedings of the Linnean Society of New South Wales, 97*, 262–74.

Lear, R. & Turner, T. (1977). *Mangroves of Australia*. Brisbane: University of Queensland Press.

Macnae, W. (1966). Mangroves in eastern and southern Australia. *Australian Journal of Botany*, *14*, 67–104.

McMillan, C. (1971). Environmental factors affecting seedling establishment of the black mangrove on the central Texas coast. *Ecology*, *52*, 927–30.

Miller, G.I. (1976). Export and production of organic detritus from north Queensland mangroves on a summer's day. *Operculum*, *5*, 56–60.

Miller, P.C. (1972). Bioclimate, leaf temperature and primary production in red mangrove canopies in south Florida. *Ecology*, *53*, 22–45.

Odum, W.E. (1971). *Pathways of energy flow in a south Florida estuary*. University of Miami, Sea Grant Technical Bulletin No. 7.

Osborn, T.G.B. & Wood, J.G. (1923). On the zonation of the vegetation in the Port Wakefield district, with special reference to the salinity of the soil. *Transactions of the Royal Society of South Australia*, *47*, 244–56.

Patton, R.T. (1942). Ecological studies in Victoria. VI. Salt marsh. *Proceedings of the Royal Society of Victoria*, *54*, 131–44.

Rains, D.W. & Epstein, E. (1967). Preferential absorption of potassium by leaf tissue in the mangrove *Avicennia marina*, an aspect of halophytic competence in coping with salt. *Australian Journal of Biological Sciences*, *20*, 847–57.

Saenger, P., Specht, M.M., Specht, R.L. & Chapman, V.J. (1977). Mangal and coastal salt-marsh communities in Australasia. In *Wet Coastal Ecosystems*, ed. V.J. Chapman, pp. 293–345. Amsterdam: Elsevier Scientific Publishing Company.

Sauer, J. (1965). Geographic reconnaisance of Western Australian seashore vegetation. *Australian Journal of Botany*, *13*, 39–69.

Scholander, P.F. (1968). How mangroves desalinate sea water. *Physiologia Plantarum*, *21*, 251–61.

Scholander, P.F., Hammel, H.T., Hemmingsen, E. & Carey, W. (1962). Salt balance in mangroves. *Plant Physiology*, *37*, 722–9.

Scholander, P.F., van Dam, L. & Scholander, S.L. (1955). Gas exchange in the roots of mangroves. *American Journal of Botany*, *42*, 92–8.

Shapiro, M.A. (1975). In *A Preliminary Report on the Westernport Bay Enironmental Study. Report for the Period 1973–1974*, pp. 366–79. Melbourne: Ministry for Conservation.

Troll, W. & Dragendorff, O. (1931). Über die luftwurzeln von *Sonneratia* Linn. f. und ihre biologische bedeutung. *Planta*, *13*, 311–473.

Thom, B.G., Wright, L.D. & Coleman, J.M. (1975). Mangrove ecology and deltaic-estuarine geomorphology: Cambridge Gulf–Ord River, Western Australia. *Journal of Ecology*, *63*, 203–32.

Van Steveninck, R.F.M., Armstrong, W.D., Peters, P.D. & Hall, T.A. (1976). Ultrastructural localisation of ions. III. Distribution of chloride in mesophyll cells of mangrove (*Aegiceras corniculatum* Blanco). *Australian Journal of Plant Physiology*, *3*, 367–76.

Waisel, Y. (1972). *Biology of Halophytes*. New York: Academic Press.

Walsh, G.E. (1974). Mangroves: a review. In *Ecology of Halophytes*, ed. T.J. Reimold & W.H. Queen, pp. 51–174. New York: Academic Press.

Wells, A.G. (1979). Distribution of mangrove species across northern Australia. In *Proceedings National Mangrove Workshop, April 1979*. Townsville: Australian Institute of Marine Science.

Wood, J.G. (1937). *Vegetation of South Australia*. Adelaide: Government Printer.

15

Freshwater wetlands
S.V. BRIGGS

Introduction

Wetland is an imprecise term that usually describes land permanently or temporarily covered by up to two metres of water. Wetland vegetation is influenced principally by climate, latitude, soil type, water depth and chemistry, fire and grazing regimes, and duration and frequency of flooding. Water depth, and seasonal and annual changes in water level and climate, affect vegetation within wetlands. As the divisions between terrestrial, wetland and aquatic communities are usually unclear, I have defined wetland vegetation in this chapter as vegetation in shallow, non-tidal water and on land subject to inundation. Terrestrial plants in dry swamps, lacustrine plants confined to water more than two metres deep and phytoplankton are excluded and saline, coastal wetland communities are considered in the previous chapter.

Australia's wetland vegetation is relatively poorly studied. The literature available to 1975 was summarised by Miles (1975), Smith (1975a, b, c, d) and Stanton (1975), and in Western Australia to 1977 by Chiffings (1977). Several regional surveys have been undertaken (e.g. Riggert, 1966; Goodrick, 1970; Corrick & Cowling, 1975, 1978; Jones, 1978), and Kats (1969) provided a general account (in Russian) of Australian wetlands. But there is no comprehensive description of the continent's wetland flora, and little is known of production, decomposition and nutrient relationships in Australian wetlands.

Many Australian aquatics derive from the tropics, Asia or Malaysia, but the majority of the tree, shrub and heath species of wetlands are endemic or confined to the southern hemisphere. The tropical influence is more obvious in northern Australia and endemism more so in the south (Sculthorpe, 1967; Kats, 1969). Representatives of the families Chenopodiaceae, Cyperaceae, Epacridaceae, Myrtaceae, Poaceae, Proteaceae, Restionaceae and aquatic species (Aston, 1973*a*) predominate in Australian wetlands (Kats, 1969). Wetland flora can be classified by structure (Specht, 1970), floristics or a combination of both. I have classified it initially by structure, and then by floristics; and have used the term 'swamp' to designate wetland communities, although the word usually has a more precise meaning (Tansley, 1939; Spence, 1967).

In this chapter I shall describe the distribution of the various Australian wetland types, illustrate some typical zonation patterns, and briefly discuss energy and nutrient relationships, and their wildlife and conservation value.

Distribution
Australia is a dry continent with low annual rainfall, high evaporation and high rainfall variability. Its wetlands are located principally on river floodplains, in internal drainage basins, in coastal sand country and in low-lying tableland areas. A few Australian wetlands are permanent, but most are ephemeral. Some dry out annually, whilst others carry water or remain waterlogged throughout the year. Some flood every year, others only once every fifty years. Water levels are rarely stable.

Wetlands are found throughout tropical and temperate Australia except for the Great Sandy Desert and the Nullarbor Plain. The largest areas in tropical Australia are on the floodplains of the rivers between the Kimberleys and Cape Arnhem, particularly on the Darwin sub-coastal plain, around the southern and eastern sides of the Gulf of Carpentaria, on Cape York Peninsula, and in the vicinity of Townsville and the Burdekin River (Stanton, 1975; Frith, 1977). In temperate Australia, coastal wetlands are located principally in northern New South Wales, eastern Victoria, especially the Gippsland Lakes area (38°00′S; 147°30′E), southeastern South Australia and southwestern Australia. Wetlands are found in mountain areas on the Atherton Tableland in Queensland, the New South Wales tablelands, the eastern highlands of Victoria, the Mt Lofty Ranges of South Australia and the central plateau of Tasmania. In the inland, wetlands are common along the Murray and Darling Rivers and their tributaries, in the channel country in southwestern Queensland, in central and northeastern South Australia, and in central and southern Western Australia.

Wetland types

Swamp forests

Mesophyll palm vine-forests. These seasonally-inundated forests are found on Cape York Peninsula, and between Cooktown and Ingham in Queensland. *Archontophoenix* spp. are usually dominant on fertile, basaltic soils, *Licuala* spp. are dominant on infertile soils derived from granite and schist, and *Calamus* spp. occur on both soil types (Webb, 1959, 1966, 1968; Stanton, 1975).

Paperbark swamp forests. Swamp forests of *Melaleuca* spp. occur on the coastal floodplains of northern Australia, in sand-dune country in coastal Queensland, on river floodplains and between dunes in coastal New South Wales, and in dune swales in southwestern Australia (Frith & Davies, 1961; Webb, 1966; Goodrick, 1970; Perry, 1970; Seddon, 1972; Stanton, 1975). Common species in *Melaleuca* paperbark forest in the Northern Territory are *Melaleuca viridiflora*, *M. cajuputi*, *M. leucaden-dron*, *Barringtonia gracilis* and *Pandanus* spp. (Story, 1969; Stocker, 1970; Williams, 1979). The relationships between these species have been described by Blake (1968*a*) and Williams (1979), and are shown in Figure 15.1. In Queensland, *Melaleuca leucadendron*, *M. viridiflora*, *M. nervosa* and *M. quinquenervia* predominate; and in New South Wales and south-

Figure 15.1. Generalised zonation in wetlands on the Darwin sub-coastal plain, Northern Territory (from Frith & Davies, 1961; Story, 1969; Stocker, 1970; Williams, 1979). F_1, *Melaleuca nervosa/M. viridiflora* forest; F_2, *M. leucadendron* forest; G_1, *Oryza fatua* grassland, G_2, *Pseudoraphis spinescens* grassland; Hfl, floating-leaved herbland; Hs, submerged herbland; S_1, *Eleocharis dulcis* sedgeland; S_2, *E. sphacelata* sedgeland.

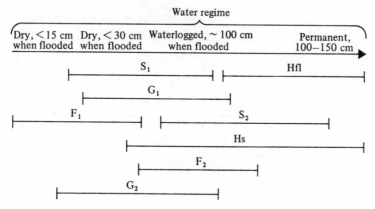

western Australia *M. quinquenervia* and *M. rhaphiophylla* respectively are common (Goodrick, 1970; Smith, 1973).

Paperbark swamp forests are usually inundated by up to a metre of water during the wet season and are dry or waterlogged at other times (Figures 15.1, 15.2). The wet season is in summer in northern Australia and in winter in the southwest. Sedges, and floating-leaved and submerged aquatics form a forest understorey in the north.

Swamp sclerophyll forests. Swamp sclerophyll forests are found on floodprone, clayey or sandy soils in coastal Queensland and New South Wales, eastern Victoria, southeastern South Australia and southwestern Australia, and along the Murray River and its tributaries in the inland. Typical species of these forests in coastal Queensland and northern New South Wales are *Eucalyptus robusta*, *Tristania suaveolens* and *Casuarina glauca* (White, 1930); in southern coastal New South Wales *Eucalyptus botryoides*, *E. robusta*, *Melaleuca* spp. and *C. glauca* are usual (Specht, Roe & Boughton, 1974). In eastern Victoria *E. botryoides* and in southeastern South Australia *E. ovata* are present, and in southwestern Australia *Eucalyptus rudis* and *Melaleuca* spp. are common (Figure 15.2) (Seddon, 1972; Specht *et al.*, 1974; Jones, 1978). Inland swamp sclero-

Figure 15.2. Generalised zonation in wetlands near Perth, Western Australia (from McComb & McComb, 1967; Seddon, 1972; Smith, 1973). F/W, *Eucalyptus rudis* forest/woodland; G, *Typha domingensis* grassland; He, emergent herbland; Hfl,s, floating-leaved and submerged herbland; S₁, *Scirpus validus* sedgeland; S₂, *Baumea juncea* sedgeland; S₃, *Gahnia trifida* sedgeland; S₄, *Lepidosperma gladiatum* sedgeland; S₅, *Baumea articulata* sedgeland; Sb, *Banksia littoralis/Leptospermum ellipticum* scrub; W₁, *Melaleuca parviflora* woodland; W₂, *M. rhaphiophylla* woodland.

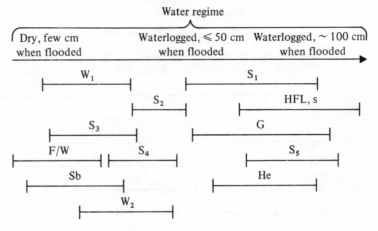

phyll forests consist of monospecific stands of *Eucalyptus camaldulensis* with an understorey of grasses, sedges, rushes and herbs, which flood for several weeks about every three out of five years (Figures 15.3, 15.5) (Dexter, 1967; Leigh & Noble, 1972).

Swamp woodlands
Paperbark swamp woodlands. These communities occur around inland salt lakes in Western Australia, and elsewhere in similar areas as paperbark forests (Story, 1969; Walter, 1971). *Melaleuca leucadendron, M. viridiflora, M. acacioides, M. nervosa* and shrubs predominate in northern Australia (Perry & Lazarides, 1964; Perry, 1970; Specht *et al.*, 1974; Kabay, George & Kenneally, 1977). *Melaleuca quinquenervia*, other species of *Melaleuca, Casuarina* spp. and shrubs are found in eastern Australia and *Melaleuca parviflora, M. rhaphiophylla, Eucalyptus rudis* and shrubs are common in southwestern Australia (Figure 15.2) (Seddon, 1972; Smith, 1973). *Melaleuca* spp., especially *M. thyoides*, and *Casuarina* spp. occur around salt lakes in inland Australia (Figure 15.4) (Walter, 1971; Beard, 1976*a*, *b*). Paperbark swamp woodlands in coastal Australia are seasonally inundated to several centimetres; elsewhere they are inundated less frequently.

Figure 15.3. Generalised zonation in freshwater swamps in northern Victoria (from Corrick & Cowling, 1975). F/W, *Eucalyptus camaldulensis* forest/ woodland; G$_1$, *Typha* sp. grassland; G$_2$, wet grassland; Hfl,f,e, floating-leaved, floating and emergent herbland; Hs, submerged herbland; S, *Eleocharis* spp. sedgeland; Sl, *Muehlenbeckia cunninghamii* shrubland; W, *E. largiflorens* woodland.

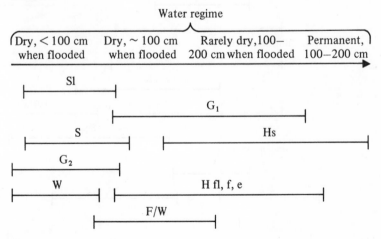

Swamp sclerophyll woodlands. These woodlands are distributed along watercourses, drainage depressions and floodplains mainly in inland Australia. *Eucalyptus microtheca, E. camaldulensis, E. terminalis, E. papuana* and *Bauhinia cunninghamii* are found as fringing forests on fine-textured cracking clays and alluvial soils in northern Australia (Pedley, 1967; Moore, Condon & Leigh, 1970). *Eucalyptus largiflorens* woodland occurs throughout New South Wales and *E. microtheca* woodland is common in northern New South Wales and southern Queensland along the Darling River and its tributaries. *Eucalyptus camaldulensis* woodland is widespread, but is particularly abundant in the western districts of Victoria, the lower southeast of South Australia and on the floodplains

Figure 15.4. Generalised zonation in and around claypan wetlands in inland Australia (from Perry & Christian, 1954; Specht, 1970; Walter, 1971; Beard & Webb, 1974, Paijmans, 1978). G, *Eragrostis australasica* grassland; Hfl/S, *Marsilea drummondii* herbland/*Eleocharis acuta* sedgeland; Sc$_1$, *Atriplex* spp./*Bassia* spp. shrubland; Sc$_2$, *Maireana pyrimidata*/*M. aphylla* shrubland: Sc$_3$, *Chenopodium auricomum*/*C. nitrariaceum* shrubland; Sc$_4$, *Atriplex nummularia* shrubland; Sc$_5$, *C. auricomum* shrubland; Sl, *Muehlenbeckia cunninghamii* shrubland; Ss$_1$, *Arthrocnemum* spp. shrubland; Ss$_2$, *Frankenia* spp. shrubland; Ss$_3$, *Pachycornia* spp. shrubland; We$_1$, *Eucalyptus largiflorens* woodland; We$_2$, *E. microtheca* woodland; Wm, *Melaleuca* spp. woodland.

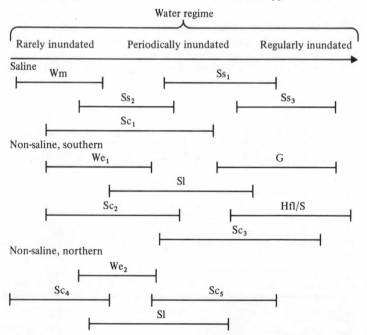

and watercourses of the Murray River and its tributaries (Moore *et al.*, 1970). *Eucalyptus camaldulensis* tends to occupy lower ground closer to watercourses than *E. largiflorens* and to be on more silty soils than *E. microtheca*. Shrubs, grasses, sedges and herbs usually form an understorey in swamp woodlands. Zonation in swamp woodland and nearby communities in the Macquarie Marshes, near Warren, in New South Wales is shown in Figure 15.5. Woodlands of *E. camaldulensis* are inundated less often and for shorter periods than forests, and *E. largiflorens* and *E. microtheca* woodlands are inundated less frequently than either.

Swamp scrubs and heaths
Swamp scrubs and heaths are described under the same heading although they differ structurally (Specht, 1970).

Swamp scrubs. These communities are typical of sandy or peaty soils in coastal areas. *Melaleuca nodosa* and *Leptospermum flavescens* are common in coastal Queensland; and *Melaleuca ericifolia, M. squarrosa* and *L. lanigerum* are common in coastal New South Wales, coastal and highland Victoria and northern and western Tasmania (Specht, 1970; Specht *et al.*, 1974). *Melaleuca halmaturorum* and *Leptospermum pubescens* occur in southeastern South Australia (L. Delroy, personal communication) and *Hypocalymma angustifolium, L. ellipticum* and *Banksia*

Figure 15.5. Generalised zonation in the Macquarie Marshes, New South Wales (from Paijmans, 1978). F/W, *Eucalyptus camaldulensis* forest/woodland; G_1, *Paspalum paspaloides* grassland; G_2, *Typha* sp. grassland; G_3, *Phragmites australis* grassland; Hfl,s, floating-leaved and submerged herbland; W, *E. largiflorens* woodland.

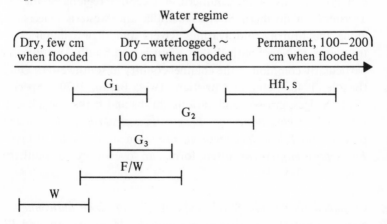

littoralis predominate in southwestern Australia (Seddon, 1972). Swamp scrubs are usually covered by a few centimetres of water for part of the year.

Swamp heaths. Swamp heaths are found in coastal and highland areas on peaty or sandy, acid, waterlogged soils. Standing water is typically present for part of the year. Coastal swamp heath species in southern Queensland and northern New South Wales include *Banksia serratifolia* (wallum), *B. robur*, *Leptospermum lanigerum*, *Xanthorrhoea* spp. and various members of the Epacridaceae, Fabaceae, Myrtaceae and Proteaceae (Bryan, 1970*a*; Specht *et al.*, 1974; Holliday & Watton, 1975). *Banksia*, *Bauera*, *Epacris*, *Hakea*, *Leptospermum*, *Xanthorrhoea* and *Sprengelia incarnata* are common in swamp heaths in southern coastal Australia east of the Murray River, and in northern and western Tasmania (Eardley, 1943; Jackson, 1965; Paton & Hosking, 1970; Frankenberg, 1971; Jones, 1978).

Baeckea, Banksia, Epacris, Kunzea, Leptospermum, Melaleuca and *Sprengelia incarnata* occupy waterlogged acid soils in tableland country in New South Wales, Victoria, Tasmania and southeastern South Australia (Pidgeon, 1938; Costin, 1954; Millington, 1954; Frankenberg, 1971; Specht *et al.*, 1974; Kirkpatrick, 1977). At higher elevations species of the genera *Callistemon*, *Epacris*, *Richea* and *Sphagnum* predominate (Costin, 1954; Jackson, 1965).

Swamp shrublands

Lignum shrublands. Shrublands dominated by *Muehlenbeckia cunninghamii* (lignum) occur on periodically-inundated, heavy-textured grey and brown soils in the southern and central regions of the Northern Territory, in northern South Australia and western Queensland, and along the Murray and Darling Rivers and their tributaries and distributaries in New South Wales, Victoria and South Australia. They are particularly common in the channel country in southwestern Queensland (Beadle, 1948; Perry & Christian, 1954; Bryan, 1970*b*; Specht *et al.*, 1974). Sedges, grasses, and various shrubs and herbs, including *Cyperus* spp., *Echinochloa turnerana*, *Eragrostis australasica*, *Chenopodium auricomum*, and *Trigonella suavissima* are frequent associates of lignum, and *Eucalyptus largiflorens* often forms an overstorey in southern areas (Figures 15.3, 15.4).

Chenopod shrublands. Shrublands of *Chenopodium auricomum*, *Chenopodium nitrariaceum*, *Atriplex nummularia*, *Maireana* spp. and *Rhagodia* spp. occupy periodically-flooded, heavy-textured soils in northern and central South Australia, western Queensland, the southern and central regions of the Northern Territory, and western New South Wales (Wood, 1937; Perry & Christian, 1954; Walter, 1971; Specht *et al.*, 1974 and Chapter 12 (Leigh, 1981)). A ground layer of *Eremophila* spp. and genera of the families Asteraceae, Fabaceae, Poaceae and other herbs, and an overstorey of *E. largiflorens* or *E. microtheca* (Figure 15.4)may be associated with the chenopods. Shrublands of *Atriplex* or *Bassia* commonly surround ephemeral salt lakes in inland Australia (Figure 15.4) (Wood, 1937; Walter, 1971). Chenopod swamps in northern areas flood every year, in southern areas less frequently (Perry & Christian, 1954).

Figure 15.6. *Muehlenbeckia cunninghamii* shrubland with an overstorey of *Eucalyptus microtheca* woodland in western New South Wales (Photo: E. Slater).

Samphire shrublands. Communities of samphires and associated plants colonise the edges and beds of periodically-inundated inland salt lakes in central and southern Western Australia, northern and central South Australia, southwestern and central Queensland, western New South Wales and northwestern Victoria (Pedley, 1967; Walter, 1971; Timms, 1972). *Arthrocnemum halocnemoides, Arthrocnemum* spp., *Pachycornia tenuis* and *Frankenia* spp., surrounded by *Nitraria schoberi* or *Melaleuca* spp. are the usual species in these shrublands (Figure 15.4) (Wood, 1937; Walter, 1971; Beard & Webb, 1974; Specht *et al.*, 1974).

Figure 15.7. *Eleocharis sphacelata* sedgeland surrounded by *Melaleuca quinquenervia* woodland in sand-dune country in coastal New South Wales (Photo: E. Slater).

Sedgelands

Eleocharis *sedgelands*. Three species of *Eleocharis*, namely *E. sphacelata*, *E. dulcis* and *E. equisetina*, are common in coastal Australia. *E. sphacelata* occupies permanent and semi-permanent wetlands. It is found in floodplain country in northern Australia, and on floodplains and in dune swales from Rockhampton to northern New South Wales, along the Victorian coast to the Murray River in South Australia, on the southeast coast of Tasmania, in the lagoons of the Murray River and its tributaries and on the tablelands of New South Wales and southern Queensland (Costin, 1954; Frith & Davies, 1961; Perry & Lazarides, 1964; Goodrick, 1970; Aston, 1973a; Briggs, 1976; McAlister, 1976; Paijmans, 1978). It may be associated with other species, particularly grasses (Figures 15.1, 15.7, 15.8).

Sedgelands dominated by *Eleocharis dulcis* are found in seasonally-inundated wetlands in northern Australia, particularly on the Darwin sub-coastal plain and near Townsville (Perry, 1960; Frith & Davies, 1961; Perry, 1970; Stanton, 1975; Paijmans, 1978). *Eleocharis dulcis* is often associated with other species (Figure 15.1), specifically *Oryza fatua* on the Darwin sub-coastal plain (Frith & Davies, 1961). Sedgelands of *E. equisetina* and associated floating-leaved, emergent and submerged aquatic herbs occupy seasonally-inundated soils on floodplains in northern New South Wales (Goodrick, 1970).

Two species of *Eleocharis* are common in inland Australia. *Eleocharis acuta* and *E. pallens* are found on regularly-inundated, heavy soils in southwestern Queensland, western New South Wales and along the

Figure 15.8. Generalised zonation in wetlands near Guyra on the northern tablelands of New South Wales (from Briggs, 1976). G_1, *Glyceria australis* grassland; G_2, *Paspalum paspaloides/Juncus articulatus* grassland, Hs, submerged herbland; Hfl,f,e, floating-leaved, floating and emergent herbland; S, *Eleocharis sphacelata* sedgeland.

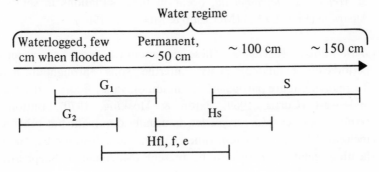

Murray River in South Australia and northern Victoria. These species commonly occur in *E. camaldulensis* woodlands or at the edge of ephemerally-flooded depressions. Associates include *Muehlenbeckia cunninghamii*, *Eragrostis australasica*, other grasses and the genera *Cyperus*, *Schoenus*, *Scirpus*, *Juncus* and *Marsilea* (Moore *et al.*, 1970).

Baumea *sedgelands*. These communities typically occur in coastal dune swales, and their species composition varies with latitude, water depth and permanency. *Baumea articulata*, *Cladium procerum* and *Lepironia articulata* are found in Queensland and northern New South Wales on soils which flood to a depth of one metre or more (Blake, 1968*b*; Goodrick, 1970). *Baumea articulata*, *Scirpus validus*, other Cyperaceae and *Juncus* spp. occupy similar sites in southeastern Australia, eastern South Australia and southwestern Australia (McComb & McComb, 1967; Land Conservation Council, 1974; Dodson & Wilson, 1975; Congdon & McComb, 1976; A.H. Corrick & F.I. Norman, personal communication).

Different species are found on soils seasonally-flooded by only a few centimetres of water. *Baumea rubiginosa*, *Gahnia sieberiana*, *Fimbristylis ferruginea*, *Schoenus* spp., *Cyperus* spp. and sometimes *Melaleuca* spp. are found at the tip of Cape York Peninsula, in the Cape Flattery–Cape Bedford district, between Hinchinbrook Island and Cairns and on sandy islands off the southern coast of Queensland (Webb, 1966; Blake, 1968*b*; Stanton, 1975). In New South Wales, eastern Victoria, southeastern South Australia and northwestern Tasmania similar sites are occupied by *Baumea rubiginosa*, *Baumea* spp. and members of the genera *Chorizandra*, *Gahnia*, *Lepidosperma*, *Leptocarpus*, *Restio* and *Sphagnum* (Eardley, 1943; Davies, 1964; Goodrick, 1970; Macphail & Shepherd, 1973; Specht *et al.*, 1974; Dodson & Wilson, 1975). *Baumea juncea*, *Gahnia trifida*, *Lepidosperma gladiatum*, sometimes with an overstorey of *Melaleuca rhaphiophylla*, occur in these situations in southwestern Australia (Figure 15.2) (McComb & McComb, 1967).

Button-grass sedgelands. These communities occur on waterlogged, periodically-inundated, peaty, infertile soils throughout much of Tasmania, but are particularly common on valley floors in the southwest and west (Curtis, 1969; Paton & Hosking, 1970). Button grass, *Gymnoschoenus sphaerocephalus*, is usually dominant but other species, including *Xyris operculata*, other Cyperaceae, Restionaceae, herbs and heath vegetation, may also be present (Macphail & Shepherd, 1973;

Jarman & Crowden, 1978). There is evidence that the spread of *G. sphaerocephalus* may be favoured by repeated burning (Curtis, 1969).

In low-lying depressions on the button-grass plains where the water table is permanently above the peat surface, *Leptocarpus tenax, Xyris operculata, Lepidosperma longitudinale, L. filiforme, Restio tetraphyllus*, other Cyperaceae, herbs such as *Utricularia, Drosera* and swamp-heath species may replace *G. sphaerocephalus* (Curtis & Somerville, 1947; MacPhail & Shepherd, 1973). *Gymnoschoenus sphaerocephalus, Lepidosperma, Restio, Xyris* and swamp heath species also occur on the tablelands of New South Wales (Pidgeon, 1938; Millington, 1954; Specht *et al.*, 1974).

Carex sedgelands. Sedgelands dominated by *Carex gaudichaudiana* occupy permanently-waterlogged, periodically-inundated sites in tableland New South Wales, Victoria and Tasmania (Costin, 1954; Specht *et al.*, 1974). *Carex gaudichaudiana* may be associated with *Eleocharis acuta*, other Cyperaceae, herbs and grasses to form fen, or with *Sphagnum* and other Cyperaceae on more acid soils to form bog (Costin, 1954 and Chapter 16, Costin, 1981).

Swamp grasslands
Swamp grasslands can be separated into five groups. The most widespread of these is classified as wet grassland, whilst tussock grassland is classified by its habit and the rest are classified by their dominant species.

Wet grasslands. These communities are most extensive in northern Australia (Moore, 1970; Stanton, 1975). *Pseudoraphis spinescens, Panicum paludosum, Paspalum paspaloides, Leersia hexandra, Echinochloa* and sedges, mainly *Cyperus, Eleocharis, Fimbristylis, Fuirena* and *Scirpus* colonise the seasonally-flooded, heavy soils in these areas. *Eriachne burkittii* and *Themeda australis* predominate on the less frequently flooded soils on higher ground (Perry, 1953, 1960; Frith & Davies, 1961; Story, 1969; Bryan, 1970*a*; Goodrick, 1970). Wet grasslands of *Paspalum paspaloides, Amphibromus neesii, Juncus, Carex appressa, Echinochloa crus-galli* and *Panicum obseptum* are present in Queensland and northern New South Wales, and *Paspalum paspaloides, Amphibromus neesii, Juncus, Carex appressa*, other Cyperaceae and *Polygonum* are present in southern New South Wales and eastern Victoria on coastal floodplains (Perry, 1953; Bryan, 1970*a*; Goodrick, 1970; A.H. Corrick & F.I. Norman, personal communication).

In tableland country, *Leersia hexandra* and Cyperaceae are found inland from Ingham and on the Atherton Tableland in Queensland (Kershaw, 1978; Paijmans, 1978), whilst *Glyceria australis, Paspalum paspaloides, Juncus articulatus* and floating-leaved and emergent aquatics are present on the northern tablelands of New South Wales (Figure 15.6) (Briggs, 1976). *Pseudoraphis spinescens, Amphibromus neesii, Paspalidium, Poa labillardieri, Juncus* and Cyperaceae commonly form on understorey in *Eucalyptus camaldulensis* forests and woodlands in southern inland Australia (Leigh & Noble, 1972).

Phragmites *grasslands*. There are two types of *Phragmites* grassland in Australia. *Phragmites karka* grasslands are confined to the tropics. They occur in stands interspersed with sedges, particularly *Eleocharis sphacelata*, and floating-leaved aquatic herbs on the Darwin sub-coastal plain, and in swamps on river floodplains and between sand dunes in coastal Queensland (Frith & Davies, 1961; Aston, 1973a; Paijmans, 1978).

Phragmites australis grasslands occur in temperate areas. Major occurrences are in eastern Queensland south of Rockhampton, coastal and inland New South Wales, particularly in the Macquarie Marshes, coastal and inland Victoria, particularly around the Gippsland Lakes, southeastern South Australia including the mouth of the Murray River, and southeastern Tasmania, mainly in the estuaries of the Huon and Derwent Rivers (Goodrick, 1970; Cowling, 1972; Aston, 1973a; Corrick & Cowling, 1975; L. Delroy, personal communication). Grasslands of *P. australis* occur on sandy soils in coastal areas and on a variety of soil types in the inland, and are usually inundated for part of the year by up to one metre of water.

Typha *grasslands*. Communities of *Typha* spp. are widespread throughout coastal and southern inland Australia. *Typha domingensis* is present in northern coastal, southeastern inland Australia and in sand-dune lakes in southwestern Australia. It is particularly common in depressions fed by irrigation water in the Riverina area of New South Wales and in northern Victoria (Frith, 1959; Aston, 1973a; Smith, 1973). In coastal eastern Australia *T. domingensis* is usually replaced by *Typha orientalis* (Aston, 1973a). *Typha* communities are not soil specific, and they exist under a variety of water regimes from permanent inundation by up to a metre and a half of water to shallower inundation with seasonal drying.

Tussock grasslands. These communities occur on waterlogged, acid soils in highland regions of southeastern Australia. Two types were identified by Costin (1954). Wet tussock grassland inhabits slightly acid, alluvial or colluvial soils below sub-alpine and alpine altitudes in the southern tablelands of New South Wales. Sod tussock grassland inhabits distinctly acid, usually humic, soils in sub-alpine and alpine areas in the southern tablelands and Barrington Tops region of New South Wales, and in the Australian Capital Territory, Victoria and Tasmania. Wet tussock grassland consists principally of *Poa caespitosa*, *Themeda australis*, *Festuca muelleri*, *Juncus* and *Carex tereticaulis*, and the principal species in sod tussock grassland are *P. caespitosa*, *T. australis*, *Danthonia nudiflora* and *Calorophus lateriflorus*. Either community may adjoin *Carex gaudichaudiana* sedgeland or swamp heath. There is evidence that heath invasion of sod tussock grasslands is favoured by anthropogenic disturbance (Costin, 1954).

Canegrass grasslands. Swamps dominated by canegrass, *Eragrostis australasica*, are common in low-lying flood country in southwestern Queensland, northeastern South Australia, western New South Wales and northwestern Victoria on a variety of soil types. Canegrass is often found on more clayey soils and in areas subject to more prolonged but shallower flooding than *Muehlenbeckia cunninghamii*. *Eucalyptus largiflorens* and *M. cunninghamii* sometimes form an overstorey, and ground layer plants include *Eleocharis* spp., *Carex* spp., *Scirpus* spp. and aquatic herbs (Figure 15.4) (Frith, 1959).

Swamp herblands
Swamp herblands include floating, floating-leaved, emergent and submerged aquatic herb communities. They are found on a variety of soils in low-lying country on the floodplains of most coastal rivers, in swales between coastal sand dunes, in depressions in tableland country and in ephemeral, semi-permanent and permanent wetlands in inland Australia. Location, principally latitude, and water regime influence their species composition. The distribution of aquatic herbs in Australia is documented by Aston (1973*a*).

Floating and floating-leaved herblands. *Nelumbo nucifera*, *Nymphaea capensis*, *N. gigantea*, *Nymphoides indica*, *Ottelia ovalifolia* and *Eichhornia crassipes* are typical northern Australian floating and floating-leaved species (Perry, 1953; Frith & Davies 1961; Frith, 1977). *Nymphaea capensis*, *N. gigantea*, *Potamogeton tricarinatus*, *Ottelia ovalifolia*, *Nymphoides indica*, *Lemna* sp., *Azolla filiculoides* and *Eichhornia cras-*

sipes are found in permanent and semi-permanent floodplain and sand-dune wetlands in coastal New South Wales (Goodrick, 1970), and *Potamogeton tricarinatus, Azolla pinnata, A. filiculoides, Ottelia ovalifolia, Lemna trisulca* and *L. minor* are usually present in similar country in Victoria (Aston, 1973a; A.H. Corrick & F.I. Norman, personal communication). *Azolla filiculoides, L. minor* and *P. tricarinatus* are found in southeastern South Australia and *L. minor, A. filiculoides, Spirodela oligorrhiza, P. tricarinatus* and *Ottelia ovalifolia* are present in southwestern Australia (Aston, 1973a; Smith, 1973).

In inland Australia, floating and floating-leaved communities are principally associated with the Murray and Darling Rivers and their tributaries. Major species include *Azolla filiculoides, A. pinnata, Potamogeton tricarinatus, Lemna minor, Spirodela oligorrhiza* and *Marsilea* spp. (Beadle, 1948; Frith, 1959; Knight & Smith, 1961; Corrick & Cowling, 1975). The water level in floating-leaved communities usually remains between one and two metres (Figures 15.1, 15.2, 15.3, 15.5, 15.8), but may fall to zero during dry periods, especially in clay pans.

Floating-leaved and floating herbland communities are frequently found in *Melaleuca leucadendron* swamp forests in tropical Australia and in *M. ericifolia* swamp scrub in Victoria (Frith & Davies, 1961; A.H. Corrick & F.I. Norman, personal communication). *Eleocharis sphacelata, Phragmites karka, P. australis, Typha* spp., and submerged and emergent vegetation are also common associates.

Submerged and emergent herblands. These communities occupy a range of sites from those covered by less water and inundated less frequently than floating and floating-leaved herblands, to those inundated by up to two metres of water. They are often associated with floating and floating-leaved plants (Figures 15.1, 15.2, 15.3, 15.5, 15.8). Submerged and emergent herbs in northern Australia include *Triglochin procera, Caldesia oligococca, Limnophila indica, Ludwigia adscendens, Ceratophyllum demersum, Monochoria cyanea, Utricularia* spp., *Myriophyllum* spp., *Eriocaulon* spp., *Vallisneria spiralis, Chara* sp. and filamentous algae (Frith & Davies, 1961; Story, 1969; Aston, 1973a; Kabay *et al.*, 1977; Paijmans, 1978; Williams, 1979). In coastal New South Wales and Victoria *Ludwigia peploides, Najus marina, V. spiralis, T. procera, Myriophyllum propinquum, Utricularia* spp., *Potamogeton crispus, P. ochreatus, Nitella* sp. and filamentous algae are common (Goodrick, 1970; A.H. Corrick & F.I. Norman, personal communication).

Triglochin procera, Lepilaena preissii, Nasturtium officinale, Potamoge-

ton pectinatus, *Myriophyllum propinquum*, *Polygonum serrulatum*, *Ranunculus inundatus*, *Chara*, *Nitella* and filamentous algae are present in southeastern Australia (Wood, 1937; Eardley, 1943; Braithwaite & Norman, 1974; Specht *et al.*, 1974; Dodson & Wilson, 1975); and *T. procera*, *L. preissii*, *Najus marina*, *Ruppia maritima*, *P. pectinatus*, *P. ochreatus*, *Nitella*, *Chara baueri*, *Villarsia albiflora*, *Polygonum serrulatum* and filamentous algae occur in coastal sand dune swamps in southwestern Australia (Smith, 1973; Congdon & McComb, 1976; Atkins, Congdon, Finlayson & Gordon, 1977).

Typical emergent and submerged species in coastal Tasmania are *Myriophyllum* spp., *Hydrocotyle* spp., *Limosella lineata* and *Liparophyllum* (Specht *et al.*, 1974). In higher areas of Tasmania, particularly on the central plateau, *T. procera*, *Myriophyllum pedunculatum*, *Liparophyllum*, *Potamogeton* spp., *Claytonia australasica* and *Centrolepis monogyna* predominate (Specht *et al.*, 1974; Cheng & Tyler, 1976). Submerged and emergent communities in tableland southeastern Australia include *Myriophyllum propinquum*, *Potamogeton* spp., *Chara* sp., *Hydrocotyle peduncularis*, *Ranunculus inundatus*, *Lilaeopsis australica*, *T. procera* and *Utricularia* spp. (Costin, 1954; Briggs, 1976; McAlister, 1976). In inland southeastern Australia *Myriophyllum propinquum*, *M. elatinoides*, *M. verrucosum*, *Ludwigia peploides*, *R. inundatus* and *Vallisneria spiralis* usually occur (Corrick & Cowling, 1975). *Lepilaena australis*, *Damasonium minus*, *Callitriche* sp., *Crassula helmsii*, *Isoetes drummondii* and *Chara australis* are present in claypans in eastern, inland Western Australia (Knight & Smith, 1961).

Zonation
Wetland plant communities are usually distinctly zoned. With decreasing water depth, submerged and floating-leaved plants are commonly succeeded by emergents, and then by terrestrial vegetation (Spence, 1964; Sculthorpe, 1967; Mitchell, 1978). Tansley (1939) described a typical sequence as being from open water to swamp, to fen and then to woodland. Not all Australian wetlands exhibit classical zonation patterns. Some that do are described by Eardley (1943), McComb & McComb (1967), Briggs (1976) and Kershaw (1978), and typical community changes with water depth are shown in Figures 15.1–15.6. Zonation with water depth is usually confined to perennial species, but in many Australian wetlands ephemerals predominate.

On the floodplains of the Murray River and its tributaries terrestrial grasses and herbs are rapidly replaced by sedges, swamp grasses and aquatics as water levels rise. The process occurs even more quickly following local rain in claypans and dry lakes. Wetland vegetation in northern Australia is also affected by changes in water regime. The boundaries of the grass, sedge and aquatic communities alter considerably between wet and dry seasons.

Most of temperate Australia has erratic rainfall and high evaporation rates and the tropical north has marked wet and dry seasons. Under such conditions water permanency and stability often exert a greater effect on wetland vegetation than water depth. Grasses, sedges and aquatics colonise wherever water lies and only develop distinct zonation patterns in areas where water is present for some time. Perennial wetland vegetation is usually zoned with depth, but in tropical and particularly in temperate Australia, ephemeral and annual wetland species are generally indistinctly zoned as a result of highly variable water levels and frequent drying.

Energy and nutrient relationships

Production
Production and decomposition rates vary considerably between wetlands. Productivity is largely governed by nutrient supply, water availability, temperature and sunlight. Australian wetland productivity has received little attention but Mitchell (1978) considers that productivities of Australian aquatic plants probably do not differ greatly from those elsewhere. Some overseas wetland communities, particularly those dominated by *Scirpus*, *Typha* and *Phragmites*, produce over 2000 and up to 4600 g dry matter m^{-2} year^{-1}, whilst others, notably *Sphagnum* bogs, produce less than 400 g dry matter m^{-2} year^{-1} (Moore & Bellamy, 1974). The reason for this difference is nutrient supply. The productive wetland communities receive an adequate supply of nutrients in run-off water, whereas the unproductive *Sphagnum* bogs rely on local rainfall and are severely nutrient-limited (Moore & Bellamy, 1974).

Nutrient supply, water availability and temperature principally influence production rates in Australian wetlands. Communities such as button-grass sedgelands on the central plateau of Tasmania, *Sphagnum* bogs on the southern tablelands of New South Wales and swamp heaths in coastal sandy areas occur on infertile soils, have low productivities and are probably nutrient-limited. Low temperatures may also limit production in mountain regions. Elsewhere water is often limiting. Sedge and aquatic herb communities in paperbark forests in northern

Australia are only productive during the wet season, and are limited by lack of water during the dry. Their annual productivity is probably relatively high because of rapid growth in the wet season due to relatively fertile soils and high temperatures. Sedge and aquatic herb communities in lignum shrublands and *Eucalyptus camaldulensis* forests lie dormant as seeds or rhizomes during dry periods but grow rapidly after flooding.

Phytoplankton have not been considered in this chapter but they often contribute considerably to wetland production. Their contribution in relation to macrophytes depends mainly on water turbidity and depth, soil and water chemistry, shading effects, and may be partly governed by duration and season of flooding in wetlands. Terrestrial producers also provide an organic input to wetlands, especially as particulate and dissolved matter in run-off water and wind-blown leaves.

Decomposition

Only a small proportion of the organic material produced by wetland plants is directly used by animals. Most of it is degraded instead by microbial decomposers, and sometimes indirectly provides food for invertebrate detritivores (Wetzel, 1975). Rates of decomposition in wetland plants govern nutrient availability, and accumulation of undecomposed plant remains or peat. Decomposition is influenced by constancy of water supply, temperature, pH, plant fibre content, nutrient conditions and plant morphology (Moore & Bellamy, 1974; Gallagher, 1978).

Decomposition rates are slow and peat accumulates under permanently waterlogged, cold, acid and low-nutrient conditions. Conversely, warm conditions, readily available nutrients and alternate wetting and drying induce rapid decomposition of dead plant material. Peat accumulates in a few Australian wetlands, notably in button-grass and *Carex* sedgelands in southeastern tableland country, in swamp scrubs and heaths in coastal and tableland regions and in some coastal *Baumea* and *Lepironia* sedgelands.

But usually in Australia, conditions for peat accumulation do not occur. Most of the continent's wetlands are ephemeral as a result of the highly variable and generally low rainfall. Warm conditions prevail and the floodplain soils are relatively fertile. Under these conditions, decomposition rates in wetlands could be expected to be high and peat accumulation to be minimal. Peat formation is therefore not a general feature of Australian wetlands.

Nutrient cycling

Wetland plants have the capacity to absorb and release large quantities of nutrients. Nutrient uptake and release vary with seasonal environmental factors, plant morphology and phenology, soil and water chemistry, water regime and plant species. Differences between submerged, floating-leaved and emergent herb species are usually small compared with differences between wetland types. Data for Australian wetland species are lacking but some generalisations can be made from the scanty information available and results obtained elsewhere. Sclerophyll shrub species in swamp heaths and shrubs invariably occur on soils low in nitrogen and phosphorus (Groves, 1968), and *Sphagnum* bogs are exceedingly deficient in most plant nutrients (Moore & Bellamy, 1974). Conversely, floating-leaved, submerged and emergent aquatic herb and some grassland communities usually occur in relatively nutrient-rich situations, such as on floodplains.

Wetland vegetation which is permanently flooded or waterlogged, especially on acid soils, tends to decompose and to return nutrients to the environment slowly. Australian communities of this type include *Sphagnum* bogs, some *Baumea* sedgelands, button-grass sedgelands, and most swamp heaths and scrubs. Conditions in these communities are anaerobic and the dead organic matter usually forms peat instead of decaying. Large quantities of nutrients are thereby locked up.

The reverse situation occurs in wetland communities which dry out regularly. Here, aerobic conditions stimulate bacterial action and accelerate breakdown of organic matter and nutrient release. Nutrient uptake and release is probably therefore extremely rapid in some Australian wetlands. In this category are sedge and aquatic herb communities on floodplains in northern Australia, which are inundated annually, and similar communities in ephemeral wetlands on the floodplains of the Murray and Darling Rivers and their tributaries. Kadlec (1969) suggested that the rapid growth of plants in freshly flooded wetlands was in response to increased nutrient availability, which followed accelerated decomposition during the dry period.

Wildlife and conservation

Wildlife

Wetlands provide habitat for a wide variety of fauna including invertebrates, fish, amphibians, reptiles, birds and mammals. The present discussion will be confined to what are perhaps the most obvious of these, the waterbirds. Waterbirds do not occur on all wetland types. They are commonly found in paperbark and inland swamp sclerophyll forests and woodlands, lignum and samphire shrublands, *Eleocharis* sedgelands, *Phragmites*, *Typha* and canegrass grasslands, and swamp herblands. They usually only inhabit these communities when water is present and will move extensively in search of wetlands with water (Frith, 1977). Flooded *Eucalyptus camaldulensis* forests and lignum shrublands, with their sedge and aquatic herb understoreys, support large numbers of waterbirds, particularly ducks. During the wet season, numerous waterbirds are found in sedge, grass and aquatic herb communities on the extensive floodplains of northern coastal Australia. The birds in these areas move to more permanent sites, usually wetlands with aquatic herbs and sedges, as water levels recede during the dry season (Frith & Davies, 1961; Lavery, 1970).

Waterbirds are typically absent from swamp heaths and scrubs, and only small numbers are found in *Baumea*, button grass and *Carex* sedgelands. In some instances birds are discouraged from these communities by dense plant growth (Aston, 1973*b*) but there is probably a more fundamental reason for their low use. Permanently waterlogged, peat-accumulating wetlands on infertile sites, where productivity, rates of decomposition and nutrient release are generally low, support few waterbirds. Conversely, flooded wetlands on fertile sites, especially those which dry out periodically, usually support large numbers of waterbirds. Their higher initial fertility and increased nutrient turnover results in higher plant and invertebrate production, and larger numbers of birds. Management of overseas wetlands for waterbirds aims to maintain high productivity and nutrient availability by regular draining, drying and reflooding (Moore & Bellamy, 1974).

Conservation

Wetlands were once considered as useful for little more than rubbish dumping. In recent years this attitude has changed and their value as flood storage areas, for removing nutrients from polluted waters, as wildlife habitat and for recreation is now being realised. As we better understand wetlands our ability to manage them is increasing. It is to be hoped that knowledge of, and our ability to conserve and manage Australian wetlands, outpaces their destruction.

The assistance of many people with the preparation of this chapter is gratefully acknowledged. Drs L.W. Braithwaite, E.O. Campbell, P.J. Myerscough, F.I. Norman, A.R. Williams and Mr A.H. Corrick allowed me access to unpublished data, and Drs J. Burrell, R.K. Crowden, J.B. Kirkpatrick, P.S. Lake, A.J. McComb, and Messrs R.D. Barker, P.G. Bayliss, S.J. Cowling, F.H.J. Crome, L. Delroy, M.T. Maher and J.G. Tracey supplied information. Drs J. Burrell, J. Caughley, R.K. Crowden, A.J. McComb, P.J. Myerscough, F.I. Norman, L.J. Webb and Messrs A.H. Corrick, S.J. Cowling, L. Delroy and G.C. Stocker made helpful suggestions on an early draft. Thanks are due to the National Parks and Wildlife Foundation for the provision of funds, and to the Director of the National Parks and Wildlife Service (NSW) and the Chief of the Division of Wildlife Research, CSIRO for the use of facilities.

References

Aston, H.I. (1973*a*). *Aquatic Plants of Australia*. Melbourne: University Press.

Aston, H.I. (1973*b*). Bird habitat in wetlands. *Victoria's Resources, 15*, 16–19.

Atkins, R.P., Congdon, R.A., Finlayson, C.M. & Gordon, D.M. (1977). Lake Leschenaultia – an oligotrophic artificial lake in Western Australia. *Journal of the Royal Society of Western Australia, 59*, 65–70.

Beadle, N.C.W. (1948). *The Vegetation and Pastures of Western New South Wales*. Sydney: Government Printer.

Beard, J.S. (1976*a*). *The Vegetation of the Dongara Area, Western Australia*. Perth: Vegmap Publications.

Beard, J.S. (1976*b*). *Vegetation of Newdegate and Bremer Bay Areas*, 2nd edn. Perth: Vegmap Publications.

Beard, J.S. & Webb, M.J. (1974). *Vegetation Survey of Western Australia – Great Sandy Desert*. Perth: University of Western Australia Press.

Blake, S.T. (1968*a*). A revision of *Melaleuca leucadendron* and its allies (Myrtaceae). *Contributions from the Queensland Herbarium* No. 1.

Blake, S.T. (1968*b*). The plants and plant communities of Fraser, Moreton and Stradbroke Islands. *Queensland Naturalist, 19*, 23–30.

Braithwaite, L.W. & Norman, F.I. (1974). *The 1972 open season on waterfowl in south-western Australia*. CSIRO, Australia, Division of Wildlife Research Technical Paper No. 29.

Briggs, S.V. (1976). Comparative ecology of four New England wetlands. M. Nat. Res. thesis, University of New England.

Bryan, W.W. (1970*a*). Tropical and sub-tropical forests and heaths. In *Australian Grasslands*, ed. R.M. Moore, pp. 101–11. Canberra: Australian National University Press.

Bryan, W.W. (1970*b*). Tropical pastures. In *The Australian Environment*, 4th ed. (rev.), ed. G.W. Leeper, pp. 83–93. Melbourne: CSIRO, Australia & Melbourne University Press.

Cheng, D.M.H. & Tyler, P.A. (1976). Nutrient economies and trophic status of Lakes Sorrell and Crescent, Tasmania. *Australian Journal of Marine and Freshwater Research, 27*, 151–64.

Chiffings, A.W. (1977). *An Inventory of Research and Available Information on Wetlands in Western Australia*. Perth: Department of Conservation and Environment.

Congdon, R.A. & McComb, A.J. (1976). The nutrients and plants of Lake Joondalup, a mildly eutrophic lake experiencing large seasonal changes in volume. *Journal of the Royal Society of Western Australia, 59*, 14–23.

Corrick, A.H. & Cowling, S.J. (1975). A survey of the wetlands of Kerang, Victoria. *Fisheries and Wildlife, Victoria, Paper* No. 5.

Corrick, A.H. & Cowling, S.J. (1978). A survey of wetlands in the Lake Cooper area, Victoria. *Fisheries and Wildlife, Victoria, Paper* No. 17.

Costin, A.B. (1954). *A Study of the Ecosystems of the Monaro Region of New South Wales*. Sydney: Government Printer.

Costin, A.B. (1981). Alpine and sub-alpine vegetation. In *Australian Vegetation*, ed. R.H. Groves, pp. 361–376. Cambridge University Press.

Cowling, S.J. (1972). Wetland habitats. In *Birds of Victoria – Inland Waters*, pp. 7–10. Melbourne: Gould League of Victoria.

Curtis, W.M. (1969). The vegetation of Tasmania. In *Tasmanian Year Book* No. 3, pp. 55–9. Hobart: Government Printer.

Curtis, W.M. & Somerville, J. (1947). Boomer Marsh – a preliminary botanical and historical survey. *Proceedings of the Royal Society of Tasmania, 1947*, 151–7.

Davies, J.L. (1964). A vegetation map of Tasmania. *Geographical Review, 54*, 249–53.

Dexter, B.D. (1967). Flooding and regeneration of river red gum, *Eucalyptus camaldulensis*, Dehn. *Forests Commission of Victoria Bulletin* No. 20.

Dodson, J.R. & Wilson, I.B. (1975). Past and present vegetation of Marshes Swamp in south-eastern South Australia. *Australian Journal of Botany, 23*, 123–50.

Eardley, C.M. (1943). An ecological study of the vegetation of Eight Mile Creek Swamp, a natural South Australian fen formation. *Transactions of the Royal Society of South Australia, 67*, 200–23.

Frankenberg, J. (1971). *Nature Conservation in Victoria*, ed. J.S. Turner. Melbourne: Victorian National Parks Association.

Frith, H.J. (1959). The ecology of wild ducks in inland New South Wales. 1. Waterfowl habitats. *CSIRO Wildlife Research, 4*, 1–97.

Frith, H.J. (1977). *Waterfowl in Australia*, 2nd edn. Sydney: A.H. & A.W. Reed.

Frith, H.J. & Davies, S.J.J.F. (1961). Ecology of the magpie goose, *Anseranas semipalmata* Latham (Anatidae). *CSIRO Wildlife Research, 6*, 1–141.

Gallagher, J.L. (1978). Decomposition processes: summary and recommendations. In *Freshwater Wetlands – Ecological Processes and Management Potential*, ed. R.E. Good, D.E. Whigham & R.L. Simpson, pp. 145–54. New York: Academic Press.

Goodrick, G.N. (1970). *A survey of wetlands of coastal New South Wales*. CSIRO Australia, Division of Wildlife Research Technical Memorandum No. 5.

Groves, R.H. (1968). Physiology of sclerophyllous shrubs in south-eastern Australia. *Proceedings of the Ecological Society of Australia, 3*, 40–5.

Holliday, I. & Watton, G. (1975). *A Field Guide to Banksias*. Adelaide: Rigby.

Jackson, W.D. (1965). Vegetation. In *Atlas of Tasmania*, ed. J.L. Davies, pp. 31–4. Hobart: Lands & Survey Department.

Jarman, S.J. & Crowden, R.K. (1978). *A survey of vegetation from the lower Gordon River and associated catchments*. South West Tasmania Resources Survey Paper No. 12.

Jones, W. (1978). *The Wetlands of the South-east of South Australia.* Adelaide: Nature Conservation Society of South Australia.

Kabay, E.D., George, A.S. & Kenneally, K.F. (1977). Environment of the Drysdale River National Park, North Kimberley, Western Australia. *Wildlife Research, Western Australia, Bulletin, 6,* 13–30.

Kadlec, A.F. (1969). Techniques of wetland management. *Wisconsin Department of Natural Resources Research Report* No. 45.

Kats, N.Y. (1969). Swamps of Australia and Tasmania. *Bjull. Mosk Obsc. Ispyt. Prir (Otd. Biol.), 74,* 106–16. (Translated from Russian).

Kershaw, A.P. (1978). The analysis of aquatic vegetation on the Atherton Tableland, north-east Queensland. *Australian Journal of Ecology, 3,* 23–42.

Kirkpatrick, J.B. (1977). Native vegetation of the west coast region of Tasmania. In *Landscape and Man,* ed. M.R. Banks and J.B. Kirkpatrick, pp. 55–80. Launceston: Royal Society of Tasmania.

Knight, J. & Smith, G.G. (1961). Aquatic plants from Mingenew. *Western Australian Naturalist, 7,* 205.

Land Conservation Council (1974). *Report on the East Gippsland Study Area.* Melbourne: Land Conservation Council.

Lavery, H.J. (1970). Studies of waterfowl (Anatidae) in north Queensland. 4. Movements. *Queensland Journal of Agricultural and Animal Sciences, 27,* 411–24.

Leigh, J.H. (1981). Chenopod shrublands. In *Australian Vegetation,* ed. R.H. Groves, pp. 276–292. Cambridge University Press.

Leigh, J.H. & Noble, J.C. (1972). *Riverine Plain of New South Wales – Its Pastoral and Irrigation Development.* Canberra: CSIRO, Australia, Division of Plant Industry.

Macphail, M. & Shepherd, R.R. (1973). Plant communities at Lake Edgar, south-west Tasmania. *Tasmanian Naturalist, 34,* 1–23.

McAlister, E.J. (1976). Land use and changes in fresh water lagoons of the New England tableland, Australia. *Agriculture & Environment, 3,* 77–81.

McComb, J.A. & McComb, A.J. (1967). A preliminary account of the vegetation of Loch McNess, a swamp and fen formation in Western Australia. *Journal of the Royal Society of Western Australia, 50,* 105–12.

Miles, J.T. (1975). *A review of literature and other information on wetlands in New South Wales.* CSIRO, Australia, Division of Land Use Research Technical Memorandum No. 75/6.

Millington, R.J. (1954). *Sphagnum* bogs of the New England plateau, New South Wales. *Journal of Ecology, 42,* 328–44.

Mitchell, D.S. (1978). *Aquatic Weeds in Australian Inland Waters.* Canberra: Australian Government Publishing Service.

Moore, P.D. & Bellamy, D.J. (1974). *Peatlands.* London: Elek Science.

Moore, R.M. (ed.) (1970). *Australian Grasslands.* Canberra: Australian National University Press.

Moore, R.M., Condon, R.W. & Leigh, J.H. (1970). Semi-arid woodlands. In *Australian Grasslands,* ed. R.M. Moore, pp. 228–45. Canberra: Australian National University Press.

Paijmans, K. (1978). *A reconnaissance of four wetland pilot study areas.* CSIRO, Australia, Division of Land Use Research Technical Memorandum No. 78/3.

Paton, D.F. & Hosking, W.J. (1970). Wet temperate forests and heaths. In *Australian Grasslands,* ed. R.M. Moore, pp. 141–58. Canberra: Australian National University Press.

Pedley, L. (1967). Vegetation of the Nogoa-Belyando area. *CSIRO Australia, Land Research Series, 18,* 138–69.

Perry, R.A. (1953). The vegetation communities of the Townsville-Bowen regions. *CSIRO Australia, Land Research Series, 2,* 44–54.

Perry, R.A. (1960). Pasture lands of the Northern Territory, Australia. *CSIRO Australia, Land Research Series* No. 5.

Perry, R.A. (1970) Vegetation of the Ord-Victoria area. *CSIRO Australia, Land Research Series, 28,* 104–19.

Perry, R.A. & Christian, C.S. (1954). Vegetation of the Barkly region. *CSIRO Australia, Land Research Series, 3,* 78–112.

Perry, R.A. & Lazarides, M. (1964). Vegetation of the Leichhardt-Gilbert area. *CSIRO Australia, Land Research Series, 11,* 152–91.

Pidgeon, I.M. (1938). The ecology of the central coastal area of New South Wales. II. Plant succession on the Hawkesbury sandstone. *Proceedings of the Linnean Society of New South Wales, 63,* 1–26.

Riggert, T.L. (1966). *Wetlands of Western Australia – a Study of the Swan Coastal Plain.* Perth: Department of Fisheries and Fauna.

Sculthorpe, C.D. (1967). *The Biology of Aquatic Vascular Plants.* London: Edward Arnold.

Seddon, G. (1972). *Sense of Place.* Perth: University of Western Australia Press.

Smith, A.J. (1975*a*). *A review of literature and other information on Victorian wetlands.* CSIRO, Australia, Division of Land Use Research Technical Memorandum No. 75/5.

Smith, A.J. (1975*b*). *A review of literature and other information on wetlands in South Australia.* CSIRO, Australia, Division of Land Use Research Technical Memorandum No. 75/7.

Smith, A.J. (1975*c*). *A review of literature and other information on wetlands in Western Australia.* CSIRO, Australia, Division of Land Use Research Technical Memorandum No. 75/8.

Smith, A.J. (1975*d*). *A review of literature and other information on Tasmanian wetlands.* CSIRO, Australia, Division of Land Use Research Technical Memorandum No. 75/9.

Smith, G.G. (1973). A guide to the coastal flora of south-western Australia. *Western Australian Naturalists Club Handbook,* No. 10.

Specht, R.L. (1970). Vegetation. In *The Australian Environment,* 4th edn. (rev.), ed. G.W. Leeper, pp. 44–67. Melbourne: CSIRO, Australia & Melbourne University Press.

Specht, R.L., Roe, E.M. & Boughton, V.H. (1974). Conservation of major plant communities in Australia and Papua New Guinea. *Australian Journal of Botany, Supplementary Series,* No. 7.

Spence, D.H.N. (1964). The macrophytic vegetation of lochs, swamps and associated fens. In *The Vegetation of Scotland,* ed. J.H. Burnett, pp. 306–425. Edinburgh: Oliver & Boyd.

Spence, D.H.N. (1967). Factors controlling the distribution of fresh-water macrophytes with particular reference to the lochs of Scotland. *Journal of Ecology, 55,* 147–69.

Stanton, J.P. (1975). *A preliminary assessment of wetlands in Queensland.* CSIRO, Australia, Division of Land Use Research Technical Memorandum No. 75/10.

Stocker, G.C. (1970). The effects of buffaloes on paperbark forests in the Northern Territory. *Australian Forest Research, 5,* 29–34.

Story, R. (1969). Vegetation of the Adelaide-Alligator River area. *CSIRO Australia, Land Research Series, 25,* 114–30.

Tansley, A.G. (1939). *The British Islands and their Vegetation.* Cambridge University Press.

Timms, B.V. (1972). Inland salt lakes as bird habitats. In *Birds of Victoria – Inland Waters*, p. 11. Melbourne: Gould League of Victoria.

Walter, H. (1971). *Ecology of Tropical and Sub-tropical Vegetation*. Edinburgh: Oliver & Boyd.

Webb. L.J. (1959). A physiognomic classification of Australian rainforests. *Journal of Ecology*, 47, 551–70.

Webb, L.J. (1966). The identification and conservation of habitat types in the wet tropical lowlands of north Queensland. *Proceedings of the Royal Society of Queensland*, 78, 59–86.

Webb, L.J. (1968). Environmental relationships of the structural types of Australian rainforest vegetation. *Ecology*, 49, 296–311.

Wetzel, R.G. (1975). *Limnology*. Philadelphia: Saunders.

White, C.T. (1930). Queensland vegetation. In *Australian Association for the Advancement of Science Handbook*, pp. 53–62. Brisbane: Australian Association for the Advancement of Science.

Williams, A.R. (1979). Vegetation and stream pattern as indicators of water movement on the Magela floodplain, Northern Territory. *Australian Journal of Ecology*, 4, 239–48.

Wood, J.G. (1937). *The Vegetation of South Australia*. Adelaide: Government Printer.

16

Alpine and sub-alpine vegetation
A.B. COSTIN

The distribution of high mountain country in Australia is related primarily to summer temperatures, as in other alpine, arctic and sub-antarctic regions of the world (Daubenmire, 1954; Tranquillini, 1979). Tree growth is limited to areas where the mean temperature of the warmest month is 10°C or greater; in areas with summer temperatures below this level trees do not survive. Such is the case for the alpine, treeless areas both in Tasmania and mainland Australia (Costin, 1967). The high mountain country comprises this treeless alpine zone and also the sparsely timbered sub-alpine zone which abuts the alpine region at a slightly lower altitude. This chapter describes the vegetation in both these regions of Australia. Factors other than summer temperature also contribute to the characteristic identities of these two regions, for example, frequent frosts, surficial solifluction, low winter temperatures, persistent snow cover (of at least 1 month, but never lying permanently in Australia), and restricted biological production (Costin, 1973).

Using the above definition of tree-line or alpine conditions, the areas mapped in Figure 16.1 are situated above an altitude of about 1370 to 1525 m on the mainland and above about 915 m in Tasmania. The apparently unusually low sub-alpine limit in Tasmania is related to its cool oceanic summer climate. The lower altitudinal limits may be further reduced on part of the Tasmanian Central Plateau, and also on the cold air plains of the mainland, by unfavourable site factors, such as cold air drainage (Moore & Williams, 1976), wet soils, exposure to strong winds or periodic fires.

The areas mapped in Figure 16.1 comprise about 6480 km² of Tasmania (about 10% of that island state) and about 5180 km² of mainland Australia. The latter area occurs only in New South Wales,

Figure 16.1. Distribution of high mountain areas in southeastern Australia (from Costin, 1980).

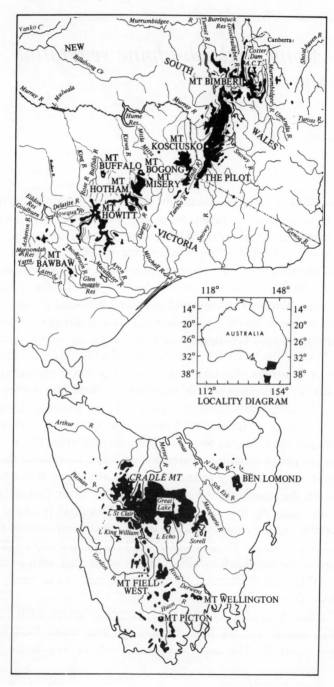

Victoria and the Australian Capital Territory and comprises about 0.07% of the total mainland area. Together, they comprise approximately 0.15% of the Australian land surface. Despite this apparently small overall area, the alpine and sub-alpine regions are of great significance to the use of land in Australia in terms of water yield, recreation and nature conservation, as will be discussed subsequently.

In this chapter I wish to describe briefly the complex of types of vegetation in the alpine and sub-alpine regions of Australia and comment on the effects of man on these different vegetation types. But first it is necessary to summarise the environment in which this vegetation complex occurs.

Alpine and sub-alpine environments

The two most extensive areas of alpine and sub-alpine vegetation are the Central Plateau of Tasmania and the Snowy Mountains of New South Wales (including Mt Kosciusko, 2228 m in elevation, the highest peak). These regions have been described in Banks (1973) and Costin (1954) respectively and the interested reader is referred to these two publications for detail. The high mountain areas of Victoria, of which the Bogong High Plains is the largest, have a less continuous distribution and are described in Carr & Turner (1959) and Costin (1957a, 1962).

Physiographically, most of the Australian alpine and sub-alpine regions occur as plateaux, often defined by relatively abrupt scarps (Costin, 1957b; Davies, 1965). In the Pleistocene these plateaux were characterised by glacial and/or periglacial conditions, the effects of which contributed to the uniqueness of the high mountain environments, compared with most of the remainder of Australia. Glacial activity was widespread in Tasmania (Davies, 1965) and led to formation of the glacial lakes, cirques and moraines so apparent in the Tasmanian mountain landscape. On the mainland, glacial activity was more localised (see, e.g., Galloway, 1962). Associated periglacial activity was widespread in both areas and produced extensive slope deposits. The limited evidence available (Costin, 1971, 1972) suggests that on the mainland, the last period of periglacial and glacial conditions extended from about 32 000 to 15 000 B.P. and locally, perhaps until about 9000 B.P. A less severe cold interval was experienced about 3000 to 1500 years ago. The sequence of late Quaternary climates in Tasmania appears to have been broadly similar (Colhoun, 1978).

The present climate is characterised by relatively low temperatures by Australian standards and relatively heavy precipitations. The latter occur because of the situation of the plateaux in the belt of westerly winds and the associated low pressure influences arising in the sub-antarctic region. The level of precipitation varies from about 760 mm at some of the relatively dry sub-alpine sites to more than 2540 mm at leeward alpine sites. In summer, temperatures are mild to cool and thunderstorms are common. Frosts may occur in any month of the year and snowfalls are common in the alpine areas from about June to early October.

The glacial and periglacial conditions referred to above have affected soil development considerably in these areas. Alpine humus soils are well developed on periglacial and glacial deposits, especially on the mainland (Costin, 1954); on areas of glacial erosion, lithosols are well developed on well drained sites, as in Tasmania, and peats in locally wet situations.

The diversity of present and past climates and microclimates and the range of soil types (Costin, 1954) existing in the alpine and sub-alpine environments in large measure correlates well with the diverse pattern of vegetation types. Costin, Gray, Totterdell & Wimbush (1979: p. 29) summarise this complex ecological situation, in referring to the Kosciusko alpine area, thus:

> ... we now see in the Kosciusko alpine area the remains of an ancient peneplain surface, with its two main rock types of granite and altered sediments. This surface has been uplifted and variously dissected by different forms of erosion. On the steep western slopes vigorous river action has cut back deep valleys almost to the Great Dividing Range itself, whereas on the more gently sloping eastern side river erosion has been slower and the headwaters of some of the streams (notably the upper Snowy River) are still in their ancient pre-uplift condition. Glacial and periglacial erosion and deposition have diversified the landscape further. This landscape now interacts with the climate to produce still greater variation in the alpine environment. In particular, the movement of the moisture-bearing subantarctic weather systems more or less at right angles across the long north-south axis of the Main Range produces large differences in the amount and persistence of precipitation on different aspects – from relatively light and non-persistent snow cover on the wind- and sun-exposed west-facing slopes to deep semi-permanent snow patches in sheltered leeward sites. Therefore, we now find a very diverse group of environments, or habitats, for plant growth in the Kosciusko alpine area. A correspondingly wide range of distinctive plant communities (or vegetation types) has developed in response to this diversity ...

Table 16.1. *Major plant communities of Australian high mountain areas (after Costin, 1962, 1980, Costin et al., 1979)*

Vegetation type	Main dominants	Distribution
Sod tussock grassland	*Poa* spp. *Danthonia nudiflora* *Empodisma minus* *Themeda australis* (locally)	Widespread along valleys and in basins of cold air drainage, especially in sub-alpine areas. Alpine humus soils.
Tall alpine herbfield	*Celmisia* spp. *Poa* spp. *Helipterum albicans* ssp. *alpinum* *Chionochloa frigida* (locally) *Craspedia* spp. *Euphrasia* spp.	The main alpine community above tree line on mainland; more restricted in Tasmania. Mostly on freely drained, relatively deep alpine humus soils.
Short alpine herbfield	*Plantago* spp. *Neopaxia australasica* *Caltha introloba* *Brachycome stolonifera* *Ranunculus niphophilus*	Local occurrence beneath alpine snow patches with persistent (> 8 months) snow cover, especially on mainland. Snow patch meadow soils, acid fen peats.
Fen	*Carex gaudichaudiana* *Danthonia nudiflora* *Festuca muelleri* *Eleocharis acuta* *Poa* spp.	Widespread locally in wet, acid, almost level alpine and sub-alpine situations influenced by mineral soil. Acid fen peats, acid marsh soils.

Table 16.1 (*continued*)

Vegetation type	Main dominants	Distribution
Bog	*Carex gaudichaudiana* *Sphagnum cristatum* *Epacris* spp. *Callistemon sieberi* *Richea continentis* *Restio australis* *Carpha* spp. *Astelia* spp.	Widespread locally in wet, acid valley situations and around hillside springs, both in alpine and sub-alpine areas; relatively little influence of mineral soil. Bog peats.
Fjaeldmark	*Coprosma* spp. *Colobanthus* spp. *Epacris petrophila* *Epacris microphylla* *Chionohebe densifolia* *Ewartia nubigena* *Drapetes tasmanicus* *Helipterum albicans* ssp. *alpinum*	Local alpine occurrences above persistent snow patches. Local very wind-exposed alpine sites.
Cushion heath	*Abrotanella forsteroides* *Donatia novaezelandiae* *Phyllachne colensoi* *Pterygopappus lawrencii* *Astelia* spp. *Oreobolus* spp.	Wet alpine situations in Tasmania.
Wet heath	*Epacris* spp. *Kunzea muelleri*	Locally common in damp situations marginal to bog.

Heath

Oxylobium ellipticum
Podocarpus lawrencei
Leucopogon montanus
Phebalium ovatifolium
Orites lancifolia
Prostanthera cuneata
Acacia alpina
Hovea purpurea var. *montana*
Tasmannia xerophila
Bossiaea foliosa
Kunzea peduncularis
Baeckea gunniana
Callistemon sieberi
Pherosphaera hookeriana ⎫
Microcachrys tetragona ⎬ Tasmania
Diselma archeri ⎭

Widespread in rocky situations, especially in Tasmania. Lithosols, alpine humus soils.

Sub-alpine woodland

Eucalyptus pauciflora ssp. *niphophila*
E. stellulata
E. pauciflora pendulous form (locally)
Eucalyptus coccifera
E. gunnii
Athrotaxis spp.
Nothofagus gunnii

Widespread sub-alpine community on mainland. Relatively deep alpine humus soils.

Widespread sub-alpine community in Tasmania.

Vegetation

The major vegetation types in the alpine and sub-alpine regions have been described for Tasmania by Jackson (1965, 1973) and for the mainland by Costin (1954, 1957*b*); they are summarised in Table 16.1.

A number of so-called alliances have been recognised which are common to both areas; but there are also differences. For example, shrub-dominated communities, including the distinctive cushion heaths and conifer heaths, are more important in Tasmania. Beyond the tree-line (on the mainland above about 1830 m), alpine herbfields dominated by species of *Celmisia* and *Poa* occur on soils relatively free of stones and predominantly of the alpine humus type. Heaths dominated by *Oxylobium* and *Podocarpus* species occur on rockier sites and they are especially extensive in Tasmania. Sod tussock grassland occurs in broad sub-alpine valleys and in the basins subject to cold-air drainage referred to earlier. Wet areas, both in the sub-alpine and alpine belts, have *Sphagnum* bogs, *Carex* fens and cushion heaths (the latter only in Tasmania), which are usually ringed by wet heaths (*Epacris* spp.) at their

Figure 16.2. *Eucalyptus pauciflora* in Kosciusko National Park, New South Wales. Photo: C.J. Totterdell.

margins. On sites with more prolonged snow cover there are two main communities: short alpine herbfield dominated by species of *Plantago* and *Neopaxia* in the moist snowpatch sites below a fjaeldmark community occurring in higher drier snowpatch areas and dominated by species of *Coprosma* and *Colobanthus*. A fjaeldmark dominated by *Epacris* spp. and *Chionohebe densifolia* is restricted to exposed, very windy, relatively snow-free, alpine sites.

The sub-alpine woodlands occur between about 915 and 1200 m altitude in Tasmania and between about 1400 and 1830 m on the mainland; in Tasmania they are dominated by *Eucalyptus coccifera* and *E. gunnii* and on the mainland by *E. pauciflora* ssp. *niphophila* (Table 16.1; Figure 16.2). In Tasmania, as well as these *Eucalyptus*-dominated sub-alpine woodlands, distinctive rainforest thickets of deciduous beech (*Nothofagus gunnii*) and stands of *Athrotaxis cupressoides* and *A. selaginoides* occur. Shrubby understoreys are more common in Tasmania than on the mainland where grasses and forbs predominate.

With increasing elevation the proportions of the different phytogeographic elements change (Table 16.2). For example, the so-called 'southern' or Antarctic element in the flora increases proportionally from tableland to alpine communities and the proportions of both the autochthonous Australian and the 'tropical' elements decrease. A high proportion of the southern element, showing similarities with elements of the New Zealand, South American, Antarctic and South African floras, has been considered since the time of Hooker (1860) to be evidence for earlier land connections between these countries and this claim seems

Table 16.2. *Geographical elements in floras of Monaro Region of New South Wales, according to elevation (from Costin, 1980)*

Environment	Geographical element (%)			
	Southern (Antarctic)	Cosmopolitan	Australian	Tropical ('Malesian')
Tableland (610–915 m)	5	36	50	9
Montane (915–1525 m)	6	33	53	7
Sub-alpine (1525–1830 m)	14	48	36	2
Alpine (> 1830 m)	22	46	31	1

even more plausible now with our increased understanding of plate tectonics (Raven & Axelrod, 1972; Campbell, 1975).

The flora is also characterised by a high degree of endemism (Costin, 1980), especially in the genera *Abrotanella*, *Athrotaxis*, *Chionochloa*, *Colobanthus*, *Craspedia*, *Diselma*, *Donatia*, *Microcachrys*, *Microstrobos*, *Plantago* and *Ranunculus*. There is also a considerable degree of regional endemism, as around Mt Kosciusko, where at least 10% of the alpine plants are endemic to that area (Costin *et al.*, 1979). Some of the characteristic insects and crustaceans are also endemic and show southern affinities (Costin, 1980), possibly indicating a co-evolution of some sections of the fauna and flora.

The various plant communities have undergone considerable change since they were recognised as being useful to man and the next section describes some of these changes.

Man's use of the alpine/sub-alpine vegetation

Use by the Aborigines
Man's association with the alpine regions begins with the Aborigines, although it is probable that the cold, inhospitable glacial and periglacial conditions prevented them from using the high country at least until the last few thousand years. They seem to have migrated annually to the mountains from lower areas, for example, to feast on Bogong moths (*Agrotis infusa*). On the mainland, these moths swarm in late spring in large numbers to the mountains where they rest during the summer among rocks and in crevices. The Aborigines collected and cooked them each summer, before both the moths and their Aboriginal hunters migrated to lower altitudes with the onset of autumn. This appears to have been the only use of the land prior to the time of European settlement. The low intensity of use by Aboriginal man and the virtual absence of large indigenous mammals meant that the pastoral practices of European man had the potential greatly to modify the alpine/sub-alpine vegetation, ill-adapted as it was to utilisation by domestic herbivores.

Effects of European settlement
Grazing. By the late nineteenth century the practice of summer grazing of sheep, cattle and occasionally horses for a period of more than 50 years, combined with the associated practice of 'burning off' the high mountain pastures to improve their palatability, had already changed the condition

of much of the vegetation. Plant cover was reduced, soil erosion promoted, and some plant species, such as *Aciphylla glacialis*, *Chionochloa frigida* and *Ranunculus anemoneus*, became rare and disappeared locally under this combined regime of grazing, trampling and burning. Yet it was not until 1944 that a small part of the mainland alpine area – that around Mt Kosciusko itself – was withdrawn from grazing. Subsequently, all areas in Kosciusko National Park above about 1370 m were similarly protected. By this time, however, various feral animals had become established, such as the hare, rabbit, brumby (wild horse), fox and feral cat and they are thought to exert a subtle and persistent influence on some of the vegetation as well as the more obvious one on the native fauna. Relatively few alien plants have become naturalised, but their number is increasing, for example from six to almost thirty in the Mt Kosciusko area during the last 30 years (Costin *et al.*, 1979).

In some of the other alpine and sub-alpine regions in Victoria and Tasmania grazing by domestic herbivores continues, despite the evidence from experience and research in New South Wales that, with selective grazing, the condition of much of the vegetation will not improve or will deteriorate further. However, such grazing is on a decreasing scale, as additional areas are recommended for parks and reserves (e.g. Land Conservation Council of Victoria, 1979).

Over a period of the last 20 or more years, since sheep and cattle grazing and associated burning off ceased at Kosciusko, there have been many significant changes in vegetation condition and trend in both the alpine and sub-alpine regions. These have been monitored regularly and are the subject of three recent publications (Wimbush & Costin, 1979*a*, *b*, *c*), some of the main findings of which are summarised below.

> Established snowgrass tussocks (of *Poa* spp.) and/or shrubs have spread on to adjacent bare ground.
> Formerly bare ground and intertussock spaces have also been colonised by various minor and major herbs.
> There has been an increase in the number and abundance of species of major forbs, as co- and sub-dominants in the widespread snowgrass communities.
> Snowgum has regenerated around existing trees, but not on deforested areas.
> Partial regeneration of *Sphagnum* bogs has occurred, except where the bogs are incised by gullies and drainage lines.

These trends will undoubtedly continue with time until most of the different vegetation types are relatively stable with fluctuations occurring mainly in response to variations in climate. Some knowledge of the grazing pressures exerted by native and feral animals will help our understanding of when such stability has been attained, an aspect which is the subject of current research (J.H. Leigh & D.J. Wimbush, personal communication).

Water. On mainland Australia the high mountain catchments contribute about 25% of the average annual yield of the largest Australian water catchment, the River Murray system. In a country with such limited water resources it is not surprising that as early as the 1920s concern for the state of the vegetative cover and soils of these alpine and sub-alpine regions was expressed. Water from the western slopes of Kosciusko was first collected by the Hume Dam, situated on the Murray River and built at this time, whence it is reticulated for subsequent use in the irrigation areas in New South Wales, Victoria and South Australia. This purpose was also part of the concept of the Snowy Mountains Scheme, begun in 1948 and completed in 1972, as well as the purpose of generating peak-load hydro-electric power. In Victoria and Tasmania the catchments are used mainly for hydro-electric power generation, for the development of which there is still further scope.

The results of studies of several aspects of catchment hydrology in the Australian Alps are summarised in Costin (1966). No single type of plant cover (i.e., herbs, shrubs or trees) was found to be ideal for all catchment requirements. The most effective control of surface run-off and soil loss was provided by a virtually continuous herbaceous cover. But this cover was found to be a relatively inefficient collector of wind-blown snow, rain, fog and cloud, and had limited value in delaying snowmelt. These aspects of water collection were best provided by trees, preferably as fairly open stands, or denser groups of trees with small clearings or strips between them. The extra water collected by the trees was not offset by greater evapotranspiration which was found to be much the same in all non-ground-water vegetation. Quality and continuity of yield were further improved by bogs and similar ground-water communities, but evapotranspiration from these areas is probably greater.

In summary, an open-forest or woodland, underlain by dense herbaceous cover, with ground-water communities along seepages and drainage lines gives water of optimum quality and quantity. Long-term protection of the vegetation from both livestock grazing and fires helps achieve this condition.

Recreation and tourism. The alpine and sub-alpine regions are all close to the main centres of population in southeastern Australia. The use of these regions for tourism has increased exponentially in the last 30 years and shows no sign of abating. The building of roads and hotels for tourist purposes commenced in the early 1900s and continues to the present day. In places, such as the ski resort complexes at Thredbo and Perisher, New South Wales, tourism has had and will continue to have a major effect on land use, in the area itself and in adjoining areas, especially as it conflicts with nature conservation.

The network of roads initially built in conjunction with the construction of hydro-electric power schemes, has opened up large areas of alpine and sub-alpine country for scenic driving, camping sites, tourist villages and ski resorts. These multifarious uses usually involve some destruction and disturbance of the vegetation, both directly and from the accelerated run-off of rain water and melted snow, as does the use of four-wheel drive and over-snow vehicles. Even foot traffic along walking tracks is causing local damage to some of the most restricted plant communities (Wimbush & Costin, 1973).

National parks and nature conservation. To a large extent, preservation of the scientific values of the high mountain areas is compatible with their use for such activities as wilderness-based recreation and for water harvesting. These uses are incompatible, however, with grazing and sophisticated, highly commercial tourism, such as ski resorts.

Considering Australia as a whole, the high mountain ecosystems appear to be more adequately reserved than any other (Newsome, 1973, and Chapter 18, Specht, 1981). In Tasmania and New South Wales most of the high mountain country is reserved in national parks. The situation is less satisfactory in Victoria, but is improving (Land Conservation Council of Victoria, 1979).

Whilst preservation of a large and representative sample of alpine and sub-alpine vegetation in Australia seems assured, there is no room for complacency when it comes to managing these communities for fauna and flora conservation (Wimbush & Costin, 1973). Herein lies the challenge for nature conservation and its practitioners in the immediate future.

Other uses. Commercial forestry has no significance in the alpine and sub-alpine regions of Australia as yet. Intermittent mining, such as for gold at Kiandra, New South Wales, predominantly in the 1860s, has had only local effects on the vegetation and is of minor significance overall at present.

Conclusions

Much of this chapter has dwelt on the uniqueness of the alpine and sub-alpine environments and their vegetation in relation to those of some other, more widespread vegetation types. Their distribution is restricted and yet it is much studied and a seemingly-adequate sample of it has been preserved for the future. Conflicts in the use of this land have been and are acute, largely because its uses for water harvesting, recreation and for nature conservation are mostly incompatible with grazing, commercial tourism and some hydro-electric developments. At the national level, these conflicts have sometimes been resolved in State parliaments and increasingly are being subjected to closer and wider scrutiny by an articulate and concerned public.

Even in situations where land use is nominally in the hands of a single authority, such as the National Parks and Wildlife Service of New South Wales which manages Kosciusko National Park, conflicts in land use still arise. The current plan of management (National Parks and Wildlife Service of New South Wales, 1974) is an attempt to minimise such conflicts and maximise the benefits to the different categories of users through zoning the land, with appropriate management for each zone. In so far as public comment is invited in the formulation and revision of such plans, the public is now more directly involved in decision-making on land use.

References

Banks, M.R. (ed.) (1973). *The Lake Country of Tasmania.* Hobart: Royal Society of Tasmania.

Campbell, K.S.W. (ed.) (1975). *Gondwana Geology.* Canberra: Australian National University Press.

Carr, S.G.M. & Turner, J.S. (1959). The ecology of the Bogong High Plains. I. The environmental factors and the grassland communities. *Australian Journal of Botany*, 7, 12–33.

Colhoun, E.A. (1978). The Late Quaternary environment in Tasmania as backdrop to man's occupance. *Records of the Queen Victoria Museum, Launceston*, Number 61.

Costin, A.B. (1954). *A Study of the Ecosystems of the Monaro Region of New South Wales.* Sydney: Government Printer.

Costin, A.B. (1957a). *High Mountain Catchments in Victoria in Relation to Land Use.* Melbourne: Soil Conservation Authority.

Costin, A.B. (1957b). The high mountain vegetation of Australia. *Australian Journal of Botany*, 5, 173–89.

Costin, A.B. (1962). Ecology of the High Plains. I. *Proceedings of the Royal Society of Victoria*, 75, 327–37.

Costin, A.B. (1966). Management opportunities in Australian high mountain catchments. In *Forest Hydrology*, ed. W.E. Sopper & H.W. Lull, pp. 565–77. Oxford and New York: Pergamon Press.

Costin, A.B. (1967). Alpine ecosystem of the Australian region. In *Arctic and Alpine Environments*, ed. H.E. Wright & W.H. Osburn, pp. 57–87. Bloomington: Indiana University Press.

Costin, A.B. (1971). Vegetation, soils, and climate in Late Quaternary southeastern Australia. In *Aboriginal Man and Environment in Australia*, ed. D.J. Mulvaney & J. Golson, pp. 26–37. Canberra: Australian National University Press.

Costin, A.B. (1972). Carbon-14 dates from the Snowy Mountains area, southeastern Australia, and their interpretations. *Quaternary Research*, 2, 579–590.

Costin, A.B. (1973). Characteristics and use of Australian high country. In *The Lake Country of Tasmania*, ed. M.R. Banks, pp. 1–23. Hobart: Royal Society of Tasmania.

Costin, A.B. (1980). Vegetation of high mountains in Australia. In *Ecological Biogeography in Australia*, ed. A. Keast, pp. 717–32. The Hague: Junk.

Costin, A.B., Gray, M., Totterdell, C.J. & Wimbush, D.J. (1979). *Kosciusko Alpine Flora*. Melbourne: CSIRO, Australia & Collins.

Daubenmire, R. (1954). Alpine timberlines in the Americas and their interpretation. *Butler University (Indianopolis, USA) Botanical Studies*, 11, 119–36.

Davies, J.L. (1965). Landforms. In *Atlas of Tasmania*, ed. J.L. Davies, pp. 19–22. Hobart: Department of Lands & Survey.

Galloway, R. (1962). Glaciation in the Snowy Mountains: a re-appraisal. *Proceedings of the Linnean Society of New South Wales*, 88, 180–98.

Hooker, J.D. (1860). *The Botany (of) the Antarctic Voyage*, Part III. *Flora Tasmaniae*, vol. 1. London: Lovell Reeve.

Jackson, W.D. (1965). Vegetation. In *Atlas of Tasmania*, ed. J.L. Davies, pp. 30–5. Hobart: Department of Lands & Survey.

Jackson, W.D. (1973). Vegetation of the Central Plateau. In *The Lake Country of Tasmania*, ed. M.R. Banks, pp. 61–86. Hobart: Royal Society of Tasmania.

Land Conservation Council of Victoria (1979). *Final Recommendations Alpine Area*. Melbourne: Land Conservation Council.

Moore, R.M. & Williams, J.D. (1976). A study of a sub-alpine woodland-grassland boundary. *Australian Journal of Ecology*, 1, 145–53.

National Parks and Wildlife Service of New South Wales (1974). *Kosciusko National Park Plan of Management*. Sydney: National Parks and Wildlife Service.

Newsome, A.E. (1973). The adequacy and limitations of flora conservation for fauna conservation in Australia and New Zealand. In *Nature Conservation in the Pacific*, ed. A.B. Costin & R.H. Groves, pp. 93–110. Canberra: Australian National University Press.

Raven, P.H. & Axelrod, D.I. (1972). Plate tectonics and Australasian paleobiogeography. *Science*, 176, 1379–86.

Specht, R.L. (1981). Conservation of vegetation types. In *Australian Vegetation*, ed. R.H. Groves, pp. 393–410. Cambridge University Press.

Tranquillini, W. (1979). *Physiological Ecology of the Alpine Timberline*. Berlin: Springer-Verlag.

Wimbush, D.J. & Costin, A.B. (1973). *Vegetation mapping in relation to ecological interpretation and management in the Kosciusko alpine area.* CSIRO, Australia, Division of Plant Industry Technical Paper Number 32.

Wimbush, D.J. & Costin, A.B. (1979*a*). Trends in vegetation at Kosciusko. I. Sub-alpine grazing trials 1957–1971. *Australian Journal of Botany, 27,* 741–87.

Wimbush, D.J. & Costin, A.B. (1979*b*). Trends in vegetation at Kosciusko. II. Sub-alpine range transects 1959–1978. *Australian Journal of Botany, 27,* 789–831.

Wimbush, D.J. & Costin, A.B. (1979*c*). Trends in vegetation at Kosciusko. III. Alpine range transects 1959–1978. *Australian Journal of Botany, 27,* 833–71.

17

Desert vegetation

G.M. CUNNINGHAM

Deserts are difficult to define and in the world literature there appears to be no common agreement as to what constitutes a desert. The situation is no different on the Australian scene with numerous authors attempting, by a variety of criteria and indices, to define the arid zone. Delineation of deserts on the basis of changes in vegetation or soils is also difficult since many vegetation communities and soil types found in desert areas often extend into relatively more humid environments.

Probably the most important factor influencing the description of an area as a desert is climate. Although McGinnies (1968) noted that establishment of desert boundaries on the basis of climatic data is difficult, the most satisfactory delineations of deserts in Australia to date are based on climatic information.

In various publications (see e.g. Slatyer & Perry, 1969), the arid lands of Australia have been divided into semi-arid and arid sections. The arid lands occupy the central core of the continent, generally between latitudes 19° and 31°S, and are surrounded by the semi-arid areas. It is these arid lands which are the subject of this chapter.

McGinnies, Goldman & Paylore (1968) presented a map of the boundaries of the Australian arid lands based on the classification developed by Meigs (1960). This is not greatly different from that proposed subsequently by Moore (1973). Austin & Nix (1978) attempted to define moisture index and bioclimatic regions within the Australian rangelands. Their central bioclimatic region approximates the area shown by McGinnies *et al.* (1968) and Moore (1973) as arid land, or desert.

Features of the Australian desert

Climate

If we accept, without attempting to be too precise in defining boundaries, that the desert of Australia is the arid land forming the central core of the continent, it is important to realise that the climate varies dramatically throughout the area. Average annual rainfall varies from 165 mm at Forrest (30°51′S; 128°06′E) in the south to 360 mm at Tennant Creek (19°38′S; 134°11′E) in the north; from 210 mm at Tibooburra (29°26′S; 142°01′E) and 291 mm at Windorah (25°26′S; 142°39′E) in the east to 264 mm at Onslow (21°40′S; 115°07′E) in the west. Seasonal incidence varies from almost equal summer/winter rainfall occurrence in the south to a major predominance of summer rainfall in the north.

Physiography

No treatment of the vegetation of deserts in Australia would be complete without a brief account of the physiography of the area, since physiography can exert a major influence on vegetation. Mabbutt (1969) recognised six types of desert based on physiographic features.

> *Mountain and piedmont deserts*, which include the major ranges and rugged uplands in the arid zone.
> *Riverine deserts*, which include the alluvial floodplains of the major rivers.
> *Shield deserts*, consisting of broadly undulating plateaux with low ridges, breakaways and salt lakes.
> *Desert clay plains*, comprised mainly of dark grey cracking clay soils with some interspersion of sandy- and stony-surfaced soils.
> *Stony deserts*, with their characteristic mesas and stony table-lands with a silcrete gibber mantle.
> *Sand deserts*, featuring level to undulating sandplains, parallel longitudinal sand dunes, linked crescentic dunes and reticulate dunes with the longitudinal dunes having by far the most extensive occurrence.

Overall, in arid Australia, Mabbutt (1969) estimated that sand deserts are the most extensive (32%) followed by shield desert (23%), desert clay plains (13%) and stony desert (12.5%), with riverine deserts occupying only 4.5% of the area. Whilst these figures apply to the whole arid and semi-arid zone, it could be expected that the various physiographic types exist in similar proportions in the desert lands being discussed here.

Soils

Soils of the Australian desert are quite variable and range in texture from sands to self-mulching clays, but by far the most common are the red siliceous sands and red earthy sands (Stace *et al.*, 1968). Other major soil types recorded from the desert lands are desert loams; solonised brown soils; grey, brown and red calcareous soils; calcareous red earths; red earths; lithosols; grey, brown and red cracking clay soils with self-mulching surfaces; grey cracking clays with massive surfaces; red saline loams (solonchaks); red brown earths; calcareous sands; alluvial soils; red and brown hardpan soils; and ironstone gravels.

Vegetation

General

The earliest observers of the desert vegetation were the explorers who traversed these areas in the nineteenth century (e.g. Sturt, 1849; Stuart, 1865; Gosse, 1874; Forrest, 1875). Their accounts are rather sketchy, although they do provide a generalised picture of land forms and vegetation. Previous to this time the only inhabitants of the desert were Aborigines who roamed nomadically through the region and left no documentation of the vegetation.

Since then, the vegetation of the desert areas has been studied to differing extents in different parts of the continent with the intensity of study usually varying from State to State. Much of the published material refers to the desert fringes, or semi-arid zones, with a lesser coverage of the driest areas.

In New South Wales, where there is only a relatively small area of desert, the vegetation of the whole area was described by Beadle (1948), and Collins (1923, 1924), Burrell (1973) and Milthorpe (1972*a*, *b*, 1973) provided more detailed accounts of selected parts of the area.

Queensland desert lands have been studied to some extent. Details of the vegetation in the areas covered by these studies can be found in Blake (1938), Christian *et al.* (1954), Boyland (1970), and in the *Western Arid Region Land Use Survey* (1974).

For South Australia, Specht (1972) treated the arid area relatively briefly but recognised five broad vegetation types within the arid lands of that State. Within each of these broad vegetation types, a further number of distinct communities were delineated. A further overview of the vegetation communities of the South Australian arid lands may be had from the map, *Vegetation and Range Site Plan of the Arid Zone,* produced by the South Australian Pastoral Board. The main references relating to the vegetation of the South Australian desert are listed by Specht (1972).

Vegetation studies in the Northern Territory have provided a good account of the whole of the desert area in terms of broad community descriptions. The works of Perry (1960) and Perry & Lazarides (1962) are major contributions to the literature of this field. More detailed accounts of the vegetation of particular desert areas are provided by Crocker (1946) in the Simpson Desert and Chippendale (1963) for the 'western desert' area of the Northern Territory. Crocker's study area extends into Queensland and is complemented by that of Boyland (1970).

The vast areas of desert in Western Australia are now well represented in the published literature. With the availability of the various vegetation survey sheets covering the Great Sandy Desert (Beard & Webb, 1974), Great Victoria Desert (Beard, 1974), Nullarbor (Beard, 1975a), Pilbara (Beard, 1975b) and Murchison (Beard, 1976) regions. In addition, Wilcox & Speck (1963) described the vegetation and pastures of the Wiluna–Meekatharra areas, and a detailed account of the vegetation of the Gascoyne River catchment is contained in Wilcox & McKinnon (1972). The other major vegetation studies of the Western Australian arid zone are those of Speck *et al.* (1960) in the North Kimberley, Speck *et al.* (1964) in the West Kimberley and Stewart *et al.* (1970) in the Ord–Victoria regions of Western Australia and the the Northern Territory. The areas discussed in these papers are really peripheral to the true desert lands.

Much of the available information on the vegetation of desert lands has been utilised by Carnahan (1976) in preparing his map of the natural vegetation of Australia. Although the map covers the whole continent, the vegetation types which occur in the desert lands are easily discernible and can be placed into six broad categories. These are the communities dominated by low trees (< 10 m high), tall shrubs (> 2 m high), low shrubs (< 2 m high), hummock grasses, tussock grasses, forbs and the salt lakes. Within each of these broad classifications, a number of different communities can be identified.

Low trees. Included in this group are the *Eucalyptus* woodlands with a low shrub understorey or a ground cover dominated by spinifex (*Triodia* spp.) or various tussock grasses; the *Acacia* woodlands with an understorey of chenopod or other low and tall shrubs and a spinifex or tussock grass ground cover; the *Casuarina*-dominated communities with an

Acacia and other low and tall shrub understorey or a spinifex grass ground-layer; communities dominated by *Owenia* spp. or various other low tree species with a spinifex or tussock grass cover. Within this vegetation type, Carnahan separated communities into two different foliage-cover classes, those with less than 10% and those with 10–30% foliage cover.

Vegetation dominated by low trees occurs scattered through the desert lands but the communities with the most dense foliage-cover are concentrated in the east and south.

Tall shrubs. Tall shrub communities include those dominated by *Eucalyptus* with spinifex grass cover; *Acacia* shrubs with lower stratum dominated by chenopod shrubs, spinifex, tussock grasses or genera of the Asteraceae; *Hakea* shrubs with a tussock grass understorey; as well as those co-dominated by *Eucalyptus* and *Acacia* species or various other tall shrubs. As with the low trees, two foliage-cover classes (less than 10% and 10–30%) were recognised by Carnahan within the desert lands.

Low shrubs. The well-known Chenopod (*Atriplex, Maireana, Chenopodium*) shrublands (see Chapter 12, Leigh, 1981) occur within this vegetation type together with the areas dominated by low *Acacia* shrubs with a spinifex or tussock grass ground-cover. Lesser areas are dominated by various other low shrub species. The Chenopod shrublands within the Australian desert generally occur in the southern part, whilst the low *Acacia* shrublands are found in small areas in the northwest of the desert. Foliage cover of the tallest species in these low shrublands can be differentiated into the less than 10% and 10–30% classes.

Hummock grasslands. The areas of spinifex-dominated grassland without tree or shrub cover are small within the total desert area and confined to the far west of the desert. They are described briefly in Chapter 13 by Groves & Williams (1981).

Tussock grasslands. Mitchell grasses (*Astrebla* spp.) are the most prominent tussock grasses dominating this vegetation type (see also Chapter 13). Its most concentrated occurrence is in the eastern section of the region with isolated areas being mapped in the northwest. Other tussock grasses dominate much smaller areas of these grasslands. The foliage cover of *Astrebla* stands varies from 10 to 70% over the desert area; other tussock grasslands within the area show similar variations in foliage cover. The most dense stands occur in small areas in the extreme northwest of the desert.

Forblands. Significant areas in the east of the desert are dominated by chenopodiaceous forbs, such as *Atriplex* and *Bassia* species, with various members of the Asteraceae being present. These communities are found in southwestern Queensland and northeastern South Australia, often on duplex soils with a dense gibber (stone) mantle.

Salt lakes. These are scattered throughout the desert area but particularly in the central and western section. Vegetation may be absent on these surfaces or may be dominated by salt-tolerant species such as samphires (*Arthrocnemum* spp.).

Detail

There is no study of the Australian desert which provides a quantitative picture of the proportions of the desert lands covered by the various vegetation communities. From the map of Perry (1960), however, it is possible to obtain such an assessment for the desert country in the Northern Territory south of the 19°S parallel of latitude (Table 17.1). Over two-thirds of the area can be classified as spinifex country on which hard- and soft-spinifex (*Triodia basedowii* and *T. pungens* respectively) are the dominant species in the ground stratum. The next most extensive vegetation type is the woodlands with short grass-forb pastures, which accounts for over 19% of the area. Pastures of barley mitchell grass (*Astrebla pectinata*) occupy about 5% of the area, and bluebush-saltbush lands cover about 1%. The remainder is comprised of rugged mountainous land (6%) and major salt lakes (0.3%).

It is of interest to examine the composition of desert vegetation to ascertain which families and species are represented. Perry & Lazarides (1962) recorded more than 1200 species from 336 genera and 94 families in their study of the vegetation of the Alice Springs area in the Northern Territory. In the desert country of far northwestern New South Wales (between 141° and 144°E; and 29° and 30°S), G.M. Cunningham, P.L. Milthorpe & W.E. Mulham (unpublished observations) recorded provisionally 592 species representing 247 genera and 66 families. Boyland (1970), in a limited examination of the Simpson Desert National Park in southwestern Queensland, recorded 98 species from 64 genera and 27 families.

All three species-collections revealed that the four most prominent families in the area are Asteraceae, Chenopodiaceae, Leguminosae (comprised of Caesalpiniaceae, Fabaceae, Mimosaceae) and Poaceae. These differ somewhat from the Crocker collections during the 1939 Simpson Desert Expedition where the most prolifically represented families were the Leguminosae, Chenopodiaceae, Myoporaceae and Amaranthaceae (Eardley, 1946, quoted by Boyland, 1970).

In the study of Perry & Lazarides (1962) these four families account for about 40% of the total species present. In the areas examined by

Table 17.1. *Proportion of desert country in the Northern Territory occupied by various vegetation types (after Perry, 1960)*

Vegetation type		Area occupied (%)
Spinifex Country		
Soft spinifex (*Triodia pungens*) plains		18.03
Hard spinifex (*Trioida basedowii*) plains		18.52
Spinifex dunefields		31.54
	Sub-Total	68.09
Mitchell Grass Country		
Barley mitchell grass (*Astrebla pectinata*) pastures		4.72
Barley mitchell grass plains interspersed with soft		
spinifex plains		0.09
	Sub-Total	4.81
Chenopod Country		
Northern bluebush (*Chenopodium auricomum*)		
shrublands		0.12
Bladder saltbush (*Atriplex vesicaria*) and southern		
bluebush (*Maireana astrotricha*) shrublands		1.27
	Sub-Total	1.39
Short Grass-Forb Country		
Low woodlands with short grasses and forbs		10.76
Sparse woodlands with short grasses and forbs on		
alluvial plains		2.90
Hilly country with lowlands supporting short grass-		
forb pastures		5.50
	Sub-Total	19.16
Rugged Country		
Rugged inaccessible country		6.21
	Sub-Total	6.21
Major Salt lakes		0.33
	Sub-Total	0.33

Cunningham *et al.* (unpublished observations) and Boyland (1970), these same families accounted for 54% and 53%, respectively, of the species recorded.

Chippendale (1963) in his traverses of the 'western desert' of the Northern Territory (between 18° and 26°S and 129° and 132°E) described the vegetation in terms of the recognisable habitats in the area. He noted twelve broad habitat-types, including burnt country, the most important being the sandplains; sand dunes and swales; river and creek beds; river and creek fringes and floodplains; ranges and adjacent valleys; and salt lakes. Within these broader habitats, Chippendale noted many different ecological niches.

Burnt areas are a feature of the desert environment and may be extensive at times, particularly after good seasons. For example, 93.6×10^6 ha, or 12.2% of the continent, including much desert land, was burnt by wildfires in the 1974–5 fire season (Luke & McArthur, 1978). The communities establishing on these areas frequently exhibit a different structure and floristic composition from the mature but dynamic communities which eventually develop.

Crocker (1946), after his east–west traverse of the Simpson Desert, grouped the vegetation into twelve associations, each with an identifiable habitat. He concluded that soil factors, including the moisture regime, appear to be the most important environmental factors governing the distribution of the vegetation in and about the Simpson Desert. In the desert proper, Crocker noted the consistency in the vegetation of the dune crests and the inter-dune corridors throughout, despite the natural variation in the individual sandhill communities. He added that 'the relative abundance of species varies a great deal, but the most important ones are the same practically throughout'.

Chippendale (1963), in comparing his 'western desert' with the Simpson Desert, noted that the latter is probably the more arid of the two and that although there are similarities between the vegetation in both areas, one significant difference from the pastoral viewpoint is the fewer edible trees and shrubs in the 'western desert'.

The impact of man on desert vegetation

Australian desert lands have been utilised by Aborigines in their nomadic wanderings for thousands of years. Their use of these areas was strictly dependent, however, on the availability of water and they were often forced to retreat from the desert in times of severe drought. The Aborigines utilised desert vegetation as food but the impact of food-gathering activities would have been limited in extent. A more striking impact resulted from intentional and wild fires associated with aboriginal use. Although substantial areas of the desert are set aside at present as aboriginal reserves, much of the area is not now used in the truly traditional manner by these people.

The advent of European man, his settlements, his mines and his flocks produced more definite changes to the desert vegetation over much of continent. It should be emphasised, however, that significant areas of Western Australia, South Australia and the Northern Territory are ungrazed by domestic stock even at the present time.

Newman & Condon (1969) attempted to assess the extent of deterioration of the grazed arid lands of Australia since the time of first pastoral settlement. These authors postulated that 70% of the grasslands, 90% of the shrublands, 60% of the low woodlands and floodplains and alluvial plains, 30% of the spinifex grasslands and 50% of the mountain and hill vegetation have undergone minor to severe degeneration in relation to the expected pristine condition.

On a regional scale, the grazed and 'watered' lands of the Alice Springs district in Central Australia were examined by Condon, Newman & Cunningham (1969) to determine the extent of deterioration of the various vegetation types as a result of grazing and other use. The

Table 17.2. *Proportions (%) of the grazed area of each vegetation type in the Alice Springs area in the various condition classes identified by Condon, Newman & Cunningham (1969)*

Vegetation Type	Very Poor	Poor	Mediocre	Fair	Good	Very Good
1. Mitchell grass	—	—	—	11.9	31.1	57.0
2. Shortgrass-forbs on young alluvium	—	13.6	20.4	25.3	34.7	6.0
3. Shortgrass-forbs on flat or undulating country	0.5	5.8	17.2	32.8	42.7	1.0
4. Saltbush-bluebush country	7.4	39.5	42.0	11.0	—	—
5. Alternating hills and lowlands	1.1	13.0	25.7	53.0	7.2	—
6. Sandplains and dunefields with spinifex	—	—	4.5	15.6	79.1	0.8
7. Salt lakes	—	—	4.6	81.5	13.9	—
8. Mountains and hills	—	20.7	1.4	71.1	6.7	0.1
Mean	1.13	11.58	14.48	37.75	26.93	8.11

proportions of each vegetation type deemed to equate with each of the six classes chosen by the authors are contained in Table 17.2. These data show the degree of change which has occurred in this area since pastoral settlement, since only about one third of the total grazed and 'watered' area now remains in the good and very good classes.

Wilcox & McKinnon (1972) in their survey of the 63147 km² Gascoyne catchment in the desert area of Western Australia noted that 9412 km² (or 14.9%) are badly eroded and will become irreversibly degraded unless they are removed from grazing, and that 33160 km² (52.5%) are degraded and have some erosion.

Degradation of the utilised sections of the Australian deserts has come from over-use of these lands by domestic livestock as well as by uncontrolled grazing of feral animals, such as rabbits, horses, donkeys, camels and goats, which may often cause very serious degradation.

Despite these illustrations of how man has influenced the Australian desert vegetation it should be emphasised, in all fairness, that there are extremely large areas of desert land which are not utilised and which are in a virtually pristine state.

Other current uses of the Australian desert range from weapons research to tourism. Tourism is of particular concern in desert areas since vegetation is relatively easily destroyed and nutrient and hydrological regimes readily upset in these fragile ecosystems. Already, changes have occurred from tourist use of the Ayers Rock area where trampling damage is evident on the dune used for viewing the Rock at sunset and access roads have intercepted run-off with consequent disastrous effects on areas of vegetation. Fortunately, these and other problems associated with tourist use of these areas are recognised and studies such as those at Ayers Rock (Hooper *et al.*, 1973; Lacey, 1973) have been commissioned to enable formulation of policies on proper use of these areas.

Overview

The Australian desert vegetation is varied in structure and floristic composition and particular communities are often restricted to definite habitats. It is only in comparatively recent times that the full range of the desert communities has been documented and attempts made to map the extent of these communities.

A relatively large section of the desert is utilised for grazing by sheep and cattle. This and other activities have resulted in major changes in the vegetation on the utilised desert land. Despite these changes on the used areas there is an even larger section of the desert which is not used in any way except on rare occasions by Aborigines. This land is virtually untouched except for use by feral animals.

Other uses of the desert include tourism, weapons research and access for various public utilities. All of these have some impact on the vegetation but these effects can be minimised in the future by proper planning based on a fuller understanding of the desert vegetation and its closely coupled interactions with the environment.

References

Austin, M.P. & Nix, H.A. (1978). Regional classification of climate and its relation to Australian rangeland. In *Studies of the Australian Arid Zone. III. Water in Rangelands*, ed. K.M.W. Howes, pp. 9–17. Melbourne: CSIRO, Australia, Division of Land Resources Management.

Beadle, N.C.W. (1948). *The Vegetation and Pastures of Western New South Wales*. Sydney: Government Printer.

Beard, J.S. (1974). *Vegetation Survey of Western Australia – Great Victoria Desert*. Perth: University of Western Australia Press.

Beard, J.S. (1975a). *Vegetation Survey of Western Australia – Nullarbor*. Perth: University of Western Australia Press.

Beard, J.S. (1975b). *Vegetation Survey of Western Australia – Pilbara*. Perth: University of Western Australia Press.

Beard, J.S. (1976). *Vegetation Survey of Western Australia – Murchison*. Perth: University of Western Australia Press.

Beard, J.S. & Webb, M.J. (1974). *Vegetation Survey of Western Australia – Great Sandy Desert*. Perth: University of Western Australia Press.

Blake, S.T. (1938). The plant communities of western Queensland and their relationships, with special reference to the grazing industry. *Proceedings of the Royal Society of Queensland*, 49, 156–204.

Boyland, D. (1970). Ecological and floristic studies in the Simpson Desert National Park, south western Queensland. *Proceedings of the Royal Society of Queensland*, 82, 1–16.

Burrell, J.P. (1973). Vegetation of Fowlers Gap Station. In *Lands of Fowlers Gap Station, New South Wales*, ed. J.A. Mabbutt. University of New South Wales Research Series No. 3, pp. 175–95.

Carnahan, J.A. (1976). Natural vegetation. In *Atlas of Australian Resources*, Second Series. Canberra: Department of National Resources.

Chippendale, G.M. (1963). Ecological notes on the 'Western Desert' area of Northern Territory. *Proceedings of the Linnean Society of New South Wales*, 88, 54–66.

Christian, C.S., Noakes, L.C., Perry, R.A., Slatyer, R.O., Stewart, G.A. & Traves, D.M. (1954). *Survey of the Barkly Region, Northern Territory and Queensland, 1947–48*. CSIRO, Australia, Land Research Series No. 3.

Collins, M.I. (1923). Studies in the vegetation of arid and semi-arid New South Wales. I. The plant ecology of the Barrier district. *Proceedings of the Linnean Society of New South Wales*, 48, 229–66.

Collins, M.I. (1924). Studies in the vegetation of arid and semi-arid New South Wales. II. The botanical features of the Grey Range and its neighbourhood. *Proceedings of the Linnean Society of New South Wales*, 49, 1–18.

Condon, R.W., Newman, J.C. & Cunningham, G.M. (1969). Soil erosion and pasture degeneration in central Australia. IV. *Journal of the Soil Conservation Service of New South Wales*, *25*, 295–321.

Crocker, R.L. (1946). The Simpson Desert Expedition, 1939 – Scientific Report No. 8 – The soils and vegetation of the Simpson Desert and its borders. *Transactions of the Royal Society of South Australia*, *70*, 235–58.

Eardley, C.M. (1946). The Simpson Desert Expedition, 1939 – Scientific Report No. 7 – Botany. I. Catalogue of plants. *Transactions of the Royal Society of South Australia*, *70*, 145–74.

Forrest, J. (1875). *Explorations in Australia*. London: Sampson Low, Marston, Low and Searle (Reprinted (1969), New York: Greenwood Press).

Gosse, W.C. (1874). *Report and Diary of Mr W.C. Gosse's Central and Western Exploring Expedition, 1873*. Adelaide: Government Printer (Reprinted (1973), as Australiana Facsimile Editions No. 71, Adelaide: Libraries Board of South Australia).

Groves, R.H. & Williams, O.B. (1981). Natural grasslands. In *Australian Vegetation*, ed. R.H. Groves, pp. 293–316. Cambridge University Press.

Hooper, P.T., Sallaway, M.M., Latz, P.K., Maconochie, J.R., Hyde, K.W. & Corbett, L.K. (1973). *Ayers Rock – Mt Olga National Park Environmental Study, 1972*. Canberra: Australian Government Publishing Service.

Lacey, J.A. (1973). *The Proposed Master Plan for the Conservation of the Environment of the Ayers Rock – Mt Olga National Park*. Alice Springs: Northern Territory Reserves Board.

Leigh, R.H. (1981): Chenopod shrublands. In *Australian Vegetation*, ed. R.H. Groves, pp. 276–292. Cambridge University Press.

Luke, R.H. & McArthur, A.G. (1978). *Bushfires in Australia*. Canberra: Australian Government Publishing Service.

Mabbutt, J.A. (1969). Landforms of arid Australia. In *Arid Lands of Australia*, ed. R.O. Slatyer & R.A. Perry, pp. 11–32. Canberra: Australian National Unviersity Press.

Mabbutt, J.A., Litchfield, W.H., Speck, N.H., Sofoulis, J., Wilcox, D.G., Arnold, J.M., Brookfield, M. & Wright, R.L. (1963). *General Report on Lands of the Wiluna-Meekatharra area, Western Australia, 1958*. CSIRO, Australia, Land Research Series No. 7.

McGinnies, W.G. (1968). Vegetation of desert environments. In *Deserts of the World – An Appraisal of Research into their Physical and Biological Environments*, ed. W.G. McGinnies, B.J. Goldman & P. Paylore, pp. 375–566. Tucson: University of Arizona Press.

McGinnies, W.G., Goldman, B.J. & Paylore, P. (eds) (1968). *Deserts of the World – An Appraisal of Research into their Physical and Biological Environments*. Tucson: University of Arizona Press.

Meigs, P. (1960). *Distribution of Arid Homoclimates – Eastern Hemisphere, Western Hemisphere*. United Nations Maps 392 & 393 (Revision 1). Paris: UNESCO.

Milthorpe, P.L. (1972*a*). Vegetation of the Fowlers Gap–Calindary area. In *Lands of the Fowlers Gap–Calindary Area, New South Wales*, ed. J.A. Mabbutt. University of New South Wales Research Series No. 4, pp. 119–34.

Milthorpe, P.L. (1972*b*). Pasture and pasture lands of the Fowlers Gap–Calindary area. In *Lands of the Fowlers Gap–Calindary Area, New South Wales*, ed. J.A. Mabbutt. University of New South Wales Research Series No. 4., pp. 135–51.

Milthorpe, P.L. (1973). Pasture lands of Fowlers Gap station. In *Lands of Fowlers Gap Station, New South Wales*, ed. J.A. Mabbutt. University of New South Wales Research Series No. 3, pp. 197–216.

Moore, R.M. (1973). Australian arid shrublands. In *Arid Shrublands*, ed. D.N. Hyder, pp. 6–11. Denver: Society for Range Management.

Newman, J.C. & Condon, R.W. (1969). Land use and present condition. In *Arid Lands of Australia*, ed. R.O. Slatyer & R.A. Perry, pp. 105–32. Canberra: Australian National University Press.

Perry, R.A. (1960). *Pasture lands of the Northern Territory, Australia*. CSIRO, Australia, Land Research Series No. 5.

Perry, R.A. & Lazarides, M. (1962). Vegetation of the Alice Springs area. CSIRO, Australia, Land Research Series No. 6, pp. 208–36.

Rainfall Statistics, Australia. (1966). Melbourne: Commonwealth Bureau of Meteorology.

Slatyer, R.O. & Perry, R.A. (eds) (1969). *Arid Lands of Australia*. Canberra: Australian National University Press.

Specht, R.L. (1972). *The Vegetation of South Australia*, 2nd edn. Adelaide: Government Printer.

Speck, N.H. (1963). Vegetation of the Wiluna – Meekatharra Area. In *Lands of The Wiluna – Meekatharra Area, Western Australia, 1958*, ed. J.A. Mabbutt. CSIRO, Australia, Land Research Series No. 7. pp. 143–61.

Speck, N.H., Bradley, J., Lazarides, M., Patterson, R.A., Slatyer, R.O., Stewart, G.A. & Twidale, C.R. (1960). *The lands and pastoral resources of the North Kimberley area, W.A.* CSIRO, Australia, Land Research Series No. 4.

Speck, N.H., Wright, R.L., Rutherford, G.R., Fitzgerald, K., Thomas, F., Arnold, J.M., Basinski, J.J., Fitzpatrick, E.A., Lazarides, M. & Perry, R.A. (1964). *General report on lands of the West Kimberley area, W.A.* CSIRO, Australia, Land Research Series No. 9.

Stace, H.C.T., Hubble, G.D., Brewer, R., Northcote, K.H., Sleeman, J.R., Mulcahy, M.J. & Hallsworth, E.G. (1968). *A Handbook of Australian Soils*. Glenside, South Australia: Rellim Technical Publications.

Stewart, G.A., Perry, R.A., Paterson, S.J., Traves, D.M., Slatyer, R.O., Dunn, P.R., Jones, P.J. & Sleeman, J.R. (1970). *Lands of the Ord–Victoria area, Western Australia and Northern Territory*. CSIRO, Australia, Land Research Series No. 28.

Stuart, J. McD. (1865). *The Journals of John McDouall Stuart During the Years 1858, 1859, 1860, 1861 and 1862*, 2nd edition, ed. W. Hardman. London: Oxley & Co. (reprinted (1975), as Australiana Facsimile Edition No. 198, Adelaide: Libraries Board of South Australia).

Sturt, C. (1849). *Narrative of an Expedition into Central Australia Performed Under the Authority of Her Majesty's Government, During the Years 1844, 5 and 6*, 2 vols. London: T. & W. Boone (Reprinted (1969), New York: Greenwood Press).

Western Arid Region Land Use Study. Part 1 (1974). Queensland Department of Primary Industries, Division Land Utilisation, Technical Bulletin No. 12.

Wilcox, D.G. & McKinnon, E.A. (1972). *A Report on the Condition of the Gascoyne Catchment*. Perth: Departments of Agriculture and Lands & Surveys.

Wilcox, D.G. & Speck, N.H. (1963). Pastures and pasture lands of the Wiluna–Meekatharra area. In *Lands of The Wiluna–Meekatharra Area, Western Australia, 1958*, ed. J.A. Mabbutt. CSIRO, Australia, Land Research Series No. 7, pp. 162–77.

CONSERVATION

18

Conservation of vegetation types

R.L. SPECHT

The Australian flora

The plant species collected in Australia by the early explorers created great interest and amazement when examined by European botanists of the eighteenth and nineteenth centuries. Much attention was focussed on the floras of heathlands and dry sclerophyll forest of the Hawkesbury Sandstone around Sydney, New South Wales, (see Chapter 7, Gill, 1981), and on the flora of the sandplain/laterite soils centred on Perth, Western Australia. The great diversity of colourful and unusual flowering species in these floras made them particularly exciting. Although many regarded the flora as typically Australian, it was soon realised that it was closely related to the heathland ('fynbos') flora of southwestern Cape Province on the other side of the Indian Ocean.

Similarly, the *Nothofagus* flora of southeastern Australia was seen to be closely related to the cool temperate, closed-forest floras of New Zealand and of southern Chile, and later was found in New Caledonia and New Guinea.

The relict pockets of tropical/sub-tropical closed-forest around the coast of northern and eastern Australia were clearly related to the rainforest floras of Malesia and in India (Chapter 4, Webb & Tracey, 1981); but also had close relatives at the family and generic level in Africa and South America.

The grass genus *Themeda*, widespread across Australia, possesses species almost identical with *Themeda triandra* in Africa. Tropical/sub-tropical Australian grass genera show the same close intercontinental affiliations.

The evidence strongly points to the conclusion that the Australian flora is but a part of the angiosperm flora which developed and became widespread across the super-continent of Gondwanaland before it broke up into the present-day continental plates. As has been discussed already in Chapter 2 (Walker & Singh, 1981), New Guinea, New Caledonia and New Zealand were part of the Australasian tectonic plate and show close floristic relationships. However, the flora on the Australasian plate became isolated from the other continents early in the Tertiary (Specht, 1980c). It was not until the mid-Miocene that the New Guinea section of the Australasian tectonic plate came in contact with the island archipelago extending from southeast Asia. Migration then occurred northward from the Australasian plate at least up into the Malay Peninsula but apparently little further into Asia; at the same time, a number of Malesian plants invaded New Guinea and northern Australia. Contact between the two land masses was relatively brief, however, and occurred at a time when the climate of the Australian continent was becoming more arid and grasses, chenopods and acacias began to appear in the fossil records of central and eastern Australia.

The continent of Australia can thus be considered as a 'floating museum' of Gondwanan flora, containing many genera and species which retain extremely primitive structures in both floral and leaf morphology (Specht, Roe & Boughton, 1974; Melville, 1975). Examples of these primitive plants are scattered throughout the whole continent; plants showing primitive floral characteristics are particularly prominent in the closed-forests and related vegetation of northeastern Australia (Figure 18.1a); plants with primitive leaf morphology are characteristic of the sclerophyll (heath) flora, with greatest concentrations in southwest Western Australia and near Sydney (Figure 18.1b).

Until mid-Miocene, the climate of Australia was 5 to 10°C warmer and more humid than at present. Seasonal fluctuations in rainfall led to the development of extensive areas of lateritic soils across the continent. At least in southern Australia where fossil evidence is available, the floral elements were widespread across the continent. It was only in the mid-Miocene that some evidence of aridity appeared in the geological records and this dryness has continued until the present day with periods of extreme intensity. Aridity induced widespread destruction; the formerly continuous flora survived in disjunct areas often separated by thousands of kilometres (Figures 18.1c, d).

Today, remnants of the primitive Gondwanan flora, after considerable speciation in certain areas, survive in relict pockets of diverse vegetation across the continent; restricted and rare species abound.

In a recent survey made by a working group established by the Australian Council of Nature Conservation Ministers (CONCOM), 2053 species of *'plants at risk'* (those species considered to be endangered, vulnerable or not represented in conservation areas; these include species which are (1) known only from the type collection or type locality, (2) restricted endemics, (3) rare species, (4) species of geographical importance) were listed for the continent (Hartley & Leigh, 1979). Results on the number of species at risk in regional subdivisions of Australia, are reproduced from this publication (Figure 18.2).

Unfortunately, only one third of Australia lies in the humid/sub-humid zones (the rest lies in the arid zone) and this part of the continent must support the ever-increasing pressures of urbanisation, tourism,

Figure 18.1*a*. Distribution of 126 genera (592 species) of seed plants retaining primitive floral characteristics (after Specht *et al.*, 1974). *b*. Distribution of 381 species of angiosperms retaining primitive morphological characters (after Specht *et al.*, 1974). *c*. Distribution of *Eucryphia* (Eucryphiaceae) and *Calycopeplus* (Euphorbiaceae) in Australia. *d*. Distribution of *Borya* (Xanthorrhoeaceae) and *Strangea* (Proteaceae) in Australia.

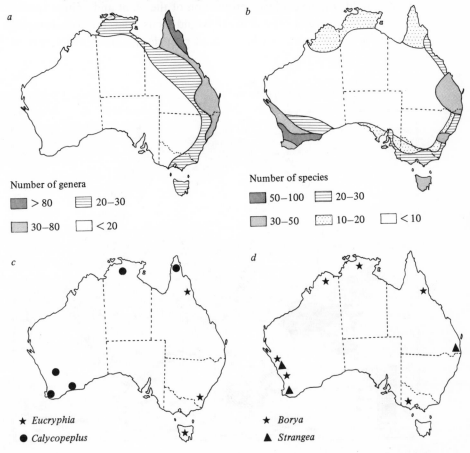

recreation, farming, forestry, water conservation, etc. The greatest number of 'plants at risk' are located in these areas. Before long, even the most common plants may be included in the list of 'plants at risk'.

Plant formations and alliances

There are many logistic problems in acquiring and maintaining satisfactory conservation areas to ensure the survival of the 2053 'plants at risk' (Figure 18.2), plus the countless numbers of plants which may become endangered over the next century.

Australian ecologists, working co-operatively during the International Biological Programme (IBP) (1964–74), recommended the establishment of a *National System of Ecological Reserves* (Specht *et al.*, 1974; Fenner, 1975) which would conserve representative examples of the major Australian ecosystems. A network of reserves, based on major ecosystems, would conserve a large proportion of the plant and animal species recorded in Australia. The network would focus long-term conservation objectives on those areas which would provide the greatest diversity of

Figure 18.2. Number of 'species at risk' in regional sub-divisions of Australia (from Hartley & Leigh, 1979).

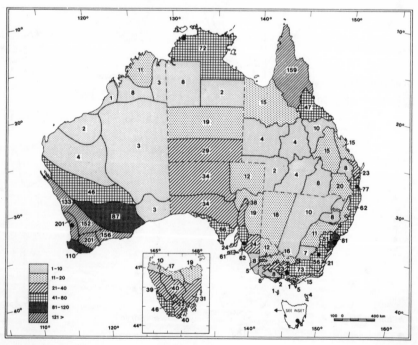

Table 18.1. *Structural formations in Australia, their percentage area, and number of major alliances (in parenthesis) included in each formation*

Life form and height of tallest stratum	Foliage projective cover of tallest stratum			
	100–70%	70–30%	30–10%	<10%
Trees[a] >30 m	Tall closed-forest	Tall open-forest 1.64% (41)	(Tall woodland)	(Tall open-woodland)
Trees 10–30 m	Closed-forest	Open-forest	Woodland 17.84% (102)	Open-woodland
Trees <10 m	Low closed-forest 0.92% (79)	Low open-forest 3.61% (153)	Low woodland 4.81% (113)	Low open-woodland 5.80% (102)
Shrubs[a] >2 m	Closed-scrub	Open-scrub 7.69% (77)	Tall shrubland 9.89% (56)	Tall open-shrubland 0.22% (24)
Shrubs 25 cm – 2 m[b]	Closed-heath (86)	Open-heath 1.47% (75)	Low shrubland 4.00% (58)	Low open-shrubland 14.39% (17)
Shrubs <25 cm[b]	—	—	Fellfield	Open-fellfield
Hummock grasses <2 m	—	—	Hummock grassland 3.95% (23)	Open-hummock grassland 14.71%
Graminoids, herbs, ferns, etc.	Closed-grassland	Grassland	Open-grassland 6.69%	Ephemeral herbland
	1.13% (121)			

Littoral complex 0.37%; Alpine complex 0.30%; Salt lakes 0.57%.
[a] A *tree* is defined as a woody plant usually with a single stem; a *shrub* is a woody plant with many stems arising at or near the base.
[b] These categories may be subdivided into (1) sclerophyllous and (2) semi-succulent/succulent.

ecosystems, a wide range of plant and animal species, as well as a wide variety of heterogeneous habitats in which evolution could proceed far into the future. In effect, this approach would channel limited manpower and resources into those conservation areas likely to achieve the greatest long-term benefit, rather than dissipate these energies on the emotional issues of conserving a single 'rare and endangered' species.

As a basis for the achievement of the aim to establish a National System of Ecological Reserves, the major vegetation formations and alliances were defined for each State of Australia (Specht, 1970; Specht *et al.*, 1974; Specht, 1981*a*) and a map of the natural vegetation of Australia was prepared (Carnahan, 1976). The major formations were defined on a two-way table relating Foliage Projective Cover and height/life form of the tallest stratum in the plant community (Table 18.1). The area of Australia covered by each formation is shown as a percentage on this Table together with the number of major alliances which have been recognised.

Conservation reserves
On 1 March 1872, the Congress of the United States of America formally proclaimed the marvels of the Yellowstone area as a National Park. The area was set aside as 'pleasuring-ground for the benefit and enjoyment of the people' and the act provided for 'the preservation, from injury or spoilation, of all timber, mineral deposits, natural curiosities or wonders . . . and their retention in their natural condition'.

The conservation movement, begun in the United States, soon swept the world. Australia did not lag behind and, in fact, established the world's second national park, Royal National Park, south of Sydney, in 1879, only seven years after Yellowstone National Park was proclaimed. National Park, Belair, near Adelaide, followed in 1891, and Tower Hill National Park in Victoria in 1892. Western Australia had a Parks and Reserves Act in 1895.

In Australia, responsibility for conservation policies is shared between State and Federal authorities. In general, regional policies are the responsibility of the States; national or international policies are developed by the Federal Authorities in consultation with State Authorities. Considerable variation existed between State Acts dealing with conservation, although, in recent years, some integration of policies has been achieved by regular meetings of the Council of Nature Conservation Ministers (CONCOM).

In general, four major conservation categories have been adopted by the States (see J.G. Mosley, in Specht *et al.*, 1974).

1. *National Parks* protect and preserve indigenous plants and animals, and features of scenic, scientific or historical interest; maintain the existing environment of the area; and provide for the education and enjoyment of visitors to the area and encourage and control such visitors. National Parks can only be revoked or altered by an Act of Parliament.
2. *Wildlife Reserves, Fauna Sanctuaries, Nature Reserves, Conservation Parks*, etc. are for the protection and care of wildlife, the propagation of wildlife and the promotion and study of wildlife. Compared with National Parks, Wildlife Reserves may be used as scientific reference areas, for the development of habitat management techniques, maintenance of habitat for rare species and for formal education. In most States, these reserves are not as secure as National Parks, being subject to change by Governor-in-Council or by gazettal.
3. *Flora Reserves, Forest Reserves* have been established under the Forestry Act in New South Wales and Victoria where sample areas representative of the main forest communities have been dedicated. Revocation needs an Act of Parliament in New South Wales, but is not defined in Victoria.
4. *Marine Parks, Aquatic Parks, Fisheries Habitat Reserves* are being established by governmental marine authorities to conserve samples of marine ecosystems. At present, the definition and administration of these reserves is in a state of flux. Some reserves still preserve the right of the public to fish, gather worms, oysters and other molluscs.

A fifth category of reserve termed a *Scientific Reference Area* or a *Nature Reference Site*, is being erected in Victoria (J.S. Turner & T.C. Chambers, personal communication) and in Queensland a *Nature Reference Site* is being erected (H. Lavery and C.S. Sandercoe, personal communication). This reserve is an area of significant ecological interest on which facilities will be provided to allow for the intensive, long-term monitoring of the total ecosystem, or some part of it, to be undertaken.

Conservation legislation and reserves in Australia, although initiated a century ago, have had a checkered history, progress in the declaration of conservation reserves having depended largely on the enthusiasm of public conservation organisations and the pressure-campaigns which these organisations initiated.

In most cases the actions were not seen in a state-wide or national perspective. The action of the Australian Academy of Science in initiating in 1959 a survey of national parks and reserves throughout Australia enabled planners to see the State and national conservation problems in perspective (Specht & Cleland, 1961, 1964; Day, 1968; Frankenberg, 1971; Specht *et al.*, 1974; Fenner, 1975; Specht, 1975*a*, 1978). As a nation, we accepted our '*evolutionary responsibility*' (Frankel, 1970) and are now in the position to plan for a *National System of Ecological Reserves* (Specht *et al.*, 1974; Fenner, 1975).

Table 18.2 shows the proportion of each Australian State proclaimed as Conservation Reserves by June 1971 (when the IBP conservation survey of Australia was completed) against June 1978 (or 1979 for some States). The conservation movement is proceeding apace, though still somewhat unplanned. Ecological surveys of the States of South Australia (Laut *et al.*, 1977), Western Australia (J.S. Beard), Victoria (Land Conservation Council), northwestern New South Wales (J. Pickard) and southwestern Queensland (D.E. Boyland *et al.*) should form the basis for a rational system of conservation areas.

In the field of conservation, the distribution of plants and animals and the migration of birds, fish, etc. are not limited by international boundaries. UNESCO, through its 'Man and the Biosphere' (MAB) Programme, initiated in 1974 a campaign for the establishment of an international system of Biosphere Reserves (*vide* the plan for a *National System of Ecological Reserves* in Australia).

The primary objectives of Biosphere Reserves are listed below.

(a). To conserve for present and future use the diversity and integrity of biotic communities of plants and animals within natural ecosystems, and to safeguard the genetic diversity of species on which their continuing evolution depends.

(b). To provide areas for ecological and environmental research including, particularly, base-line studies, both within and adjacent to such reserves, such research to be consistent with objective (a) above.

(c). To provide facilities for education and training.

The ideal Biosphere Reserve should be representative of a widespread and important ecosystem and should be large, internally diverse and, as far as possible, self-contained.

It should include the following areas.

(a). An area of the ecosystem in an undisturbed state (the 'core' area) which should be kept undisturbed.

(b). An area or areas under various forms of economic use which should continue, but be controlled (e.g. sustained forestry or grazing).
(c). An area under stable forms of traditional use which should continue under control (e.g. shifting cultivation).
(d). Various stages of the modification or degradation of the original ecosystem.

Although the concept of Biosphere Reserves is still novel, eleven Australian conservation reserves, which conform to a number of the above criteria, have been accepted (as at 30 June 1979) as Biosphere Reserves by the MAB Bureau. They are as follows.

1. Uluru (Ayers Rock–Mount Olga) National Park, Northern Territory – 132 500 ha.
2. Danggali Conservation Park, South Australia (33°40′S; 140°40′E) – 253 230 ha.
3. Kosciusko National Park, New South Wales – 627 218 ha.
4. Yathong Nature Reserve, New South Wales – 107 241 ha.
5. The Unnamed Conservation Park of South Australia (28°30′S; 130°00′E) – 2 132 600 ha.
6. Croajingalong National Park, Victoria (37°40′S; 149°30′E) – 86 000 ha.
7. Murray Valley, Victoria – 125 000 ha.
8. Southwest National Park, Tasmania – 403 240 ha.
9. Macquarie Island Nature Reserve, Tasmania – 12 353 ha.
10. Prince Regent River Nature Reserve, Western Australia – 633 825 ha.
11. Fitzgerald River National Park, Western Australia – 242 727 ha.

Asbestos Range National Park (4281 ha in Tasmania, 41°07′S; 146°41′E) is under consideration.

A national system of ecological reserves
Following the conservation survey conducted by Australian ecologists as part of the International Biological Programme, the Australian Academy of Science sponsored a symposium entitled 'A National System of Ecological Reserves in Australia' (Fenner, 1975).

The basic premises proposed during the symposium are listed below.

(a). Samples of all major ecosystems in Australia should be conserved in reserves which could be termed 'ecological reserves'.

(b). As most animals depend on particular plant communities for food, shelter, and nest-sites and since plant communities can be relatively easily studied and should be conserved for their own sake, one can readily substitute major plant community for major ecosystem in (a).

(c). Replicate samples are needed to cover the variation observed throughout the geographical range of major ecosystems.

(d). The replicate samples of major ecosystems should provide a satisfactory spatial system of habitats for migratory animals.

(e). The 'ecological reserve' should be of sufficient size (Slatyer, 1975; Kitchener *et al.*, 1980) so that:

   (i) there is sufficient space for the perpetuation of larger animal and plant species;

   (ii) man's unplanned environmental impact on the ecosystems within the 'ecological reserve' is minimal;

   (iii) diverse environments (necessary for mobile animals and for the continued conservation and evolution of both plants and animals in response to environmental stress) are included in the reserve.

The various conservation reserves already proclaimed in the States of Australia would provide the framework for this proposed network of 'ecological reserves'. It was necessary, therefore, to determine what ecosystems were included in the existing reserves, how well they were conserved and how well the current network of reserves covered all the major ecosystems in Australia.

Almost 1200 major alliances have been recognised throughout the continent of Australia. The conservation status of about half of these alliances could be considered to be reasonable, though not necessarily ideal, by mid-1971 (Specht *et al.*, 1974; Table 18.3). Small States such as Tasmania had a good conservation record (based both on area and on diversity of ecosystems conserved). The large State of Queensland, with almost a quarter of the area (9767 km^2) of the total area (42 890 km^2) of National Parks declared for the whole of Australia, lagged behind other States in the diversity of ecosystems conserved.

Since the 1950s, the various State Authorities have become increasingly aware of their responsibilities in the field of conservation. Determined efforts have been, and are being made, to improve the deficiencies in total area, in diversity of ecosystems conserved and in the protection of these areas for the future. The progress made between 1971 and 1978–9 is summarised in Tables 18.2 and 18.3.

In summary, many Australian plant communities (formations and alliances) are now reasonably well conserved in the network of conservation reserves proclaimed throughout Australia. Undoubtedly, efforts need to be made in all regions of Australia to overcome deficiencies in the National System of Ecological Reserves. However, the following major Australian plant formations are virtually absent from or poorly conserved in the network.

1. Tropical/sub-tropical tussock grasslands (*Astrebla, Dichanthium*) in the coastal and semi-arid zones and in southeastern Queensland (see Everist & Webb, 1975; IUCN, 1980).
2. *Acacia aneura* (mulga) and related *Acacia* tall shrubland communities of the semi-arid zone of Queensland, New South Wales, Northern Territory and South Australia. (Western Australia has several reserves containing mulga but these are located at the edge of the main distribution of the mulga.)
3. Low shrubland (shrub steppe) communities dominated by *Atriplex* spp. and *Maireana* (*Kochia*) spp. in semi-arid southern Australia.
4. Temperate tussock grasslands (*Themeda, Danthonia, Lomandra*) of western Victoria and South Australia.
5. *Acacia harpophylla* (brigalow) open-forests in central and southeastern Queensland.
6. Savannah woodland communities (dominated by many *Eucalyptus* spp.) in the wheat belt of southeastern Queensland, New South Wales, Victoria, Tasmania, South Australia and Western Australia.
7. The mallee open-scrub, *Eucalyptus socialis* alliance, in the wheat belt of northwestern Victoria and the adjacent Murray Lands of South Australia.

The first three formations listed above form a vital part of the pastoral industry in the semi-arid zone of Australia; large, but degraded, conservation reserves could still be acquired. The last four formations lie within the wheat belt/improved pasture zone of Australia. Only small 'islands', plus roadside corridors, still remain. Research programs aimed at assessing the conservation potential of the remnants of the original ecosystems and developing a conservation plan are essential (see Kitchener *et al.*, 1980 for a discussion of the wheat belt of southwest Western Australia).

Table 18.2. Percentage of the States of Australia proclaimed as National Parks and other Conservation Reserves by June 1971 (Specht et al., 1974) and by June 1978 or 1979 (National Parks Authorities of each State, Goldstein, 1979)

State	Area of State (km²)	Conservation category[a]	Conservation Reserves Area (km²)		Conservation Reserves (% of State)	
			1971	1978 or 1979	1971	1978 or 1979
Australian Capital Territory	2 400	National Parks	—	—	—	—
		Other Reserves	47	98	1.96	4.08
New South Wales	801 600	National Parks	9 427	17 346	1.18	2.16
		Other Reserves	1 920	3 449	0.24	0.43
Northern Territory	1 346 200	National Parks	2 258	12 003[b]	0.17	0.89[b]
		Other Reserves	45 131	45 516[b]	3.35	3.38[b]
Queensland	1 727 200	National Parks	9 767	21 927[b]	0.57	1.27[b]
		Other Reserves	nil + 60[c]	639[b] + 769[c]	—	0.04[b]
South Australia	984 000	National Parks	1 655	4 650[b]	0.17	0.47[b]
		Other Reserves	33 341	36 949[b]	3.39	3.75[b]

Tasmania	67 800	National Parks	3 775	6 253	5.57	9.22
		Other Reserves	5 222 + 124[d]	725 + 124[d]	7.70	1.07
Victoria	227 600	National Parks	2 051	4 281[b]	0.90	1.88[b]
		Other Reserves	861	4 283[b]	0.38	1.88[b]
Western Australia	2 525 500	National Parks	13 957	44 631	0.55	1.77
		Other Reserves	91 436	75 208	3.62	2.98
Total	7 682 300	National Parks	42 890	111 091	0.56	1.45
		Other Reserves	177 958 + 184	166 867 + 893	2.32	2.17

[a] Categories of conservation reserves recognized by State Authorities: New South Wales, National Park, Nature Reserve, Game Reserve, Flora Reserve (in State Forest); Northern Territory, National Park, Conservation Reserve, Nature Park, Wildlife Sanctuary; Queensland, National Park, Fauna Reserve, Environmental Park, Fisheries Habitat Reserve; South Australia, National Park, Conservation Park, Game Reserve, Recreation Park; Tasmania, National Park, Nature Reserve, Conservation Area, State Reserve; Victoria, National Park, Faunal Reserve, Game Reserve, Game Refuge, State Park, Coastal Park, Forest Park, etc. (under Forests Commission); Western Australia, National Park, Nature Reserve.
[b] June 1979.
[c] Fisheries Habitat Reserves.
[d] Macquarie Island.

Table 18.3. *Numbers of major plant alliances per structural formation which were reasonably well conserved on existing reserves in each State of Australia by June 1971 (Specht et al., 1974) and by June 1979 (National Parks Authorities of each State)*

Structural formation	Estimated area in Australia (km²)	New South Wales			Northern Territory			Queensland			South Australia			Tasmania			Victoria			Western Australia		
		Total	1971	1979[a]	Total	1971	1979	Total	1971	1979	Total	1971	1979	Total	1971	1979	Total	1971	1979	Total	1971	1979[a]
Closed-forest(s)	71 377	14	8		2	2	2	24	12	21	—	—	—	26	21	21	6	3	5	8	3	
Tall open-forest/woodland(s)	126 514	5	4	Not available	—	—	—	2	2	2	—	—	—	18	16	16	11	6	8	5	3	Not available
Open-forest/low open-forest	278 182	32	18		3	1	2	26	6	14	17	9	10	35	21	24	24	12	23	20	12	
Woodland	1 374 354	23	13		7	5	6	21	1	12	8	6	6	12	8	8	11	5	8	22	11	
Low woodland	370 175	26	8		17	8	11	14	1	7	13	9	9	19	6	9	6	4	4	18	11	
Open-woodland(s)	446 709	2	0		8	4	4	15	1	6	10	6	7	17	7	8	8	3	3	42	11	
Closed-scrub/closed-heathland	small area	12	5		—	—	—	9	5	9	2	1	1	39	34	36	10	6	8	15	8	
Open-scrub	591 829	16	10		2	0	0	5	1	4	19	12	12	14	6	8	9	5	7	12	7	
Open-heathland	113 157	17	12		—	—	—	8	1	4	7	6	7	26	22	24	7	3	4	10	7	
Tall shrubland	762 193	10	4		13	6	7	4	0	1	3	3	3	5	4	4	4	0	2	17	9	
Low shrubland	308 298	7	4		6	0	0	7	0	0	18	11	11	11	9	9	4	0	3	5	3	
Tall open-shrubland	16 817	—	—		3	1	1	3	3	3	3	3	3	—	—	—	2	1	2	13	9	
Low open-shrubland	66 962	4	4		—	—	—	4	2	2	—	—	—	3	2	3	—	—	—	6	2	
Hummock grassland	2 471 453	—	—		7	7	7	3	1	1	3	2	3	—	—	—	1	1	1	9	4	
Grassland (alpine)	23 260	9	9		—	—	—	—	—	—	—	—	—	48	47	47	9	0	3	—	—	
Grassland (lowland)	602 473	25	13		5	1	1	13	2	5	24	8	17	32	25	28	13	4	7	6	3	
Total[b]	7 682 300[b]	202	112		73[b]	35	41	158	38	91	127	76	89	305	228	245	125	53	88	208[b]	103	

[a] Insufficient survey data available to estimate the 1979 status of conservation in the States of New South Wales and Western Australia. The adequacy of conservation of alliances in existing reserves in all States (1979 data) has not been evaluated by survey.

[b] Freshwater, mangrove, and coastal dune complexes and salt lakes, totalling 59 547 hectares are not included in the list of the str...

Conservation management

Once a satisfactory system of 'ecological reserves' has been established, the greatest difficulty faced by conservation authorities is the maintenance of the conserved ecosystems for posterity. In the creation of conservation reserves, we have accepted *'evolutionary responsibility'* as indicated by Frankel (1970). Ecosystems are dynamic systems, which change continually with time. In order to maintain one ecosystem in a 'conservation island' surrounded by a miscellany of alternative land-use systems needs far-sighted long-range planning. For example, the mound-building mallee-fowl (*Leipoa ocellata*) needs a good supply of *Acacia* seed as food for much of the year (Frith, 1962); as seed supply is poor in young stands of mallee regenerating after fire, and also in senescing stands, careful management to promote a series of age-classes is necessary. This will ensure a reasonable supply of *Acacia* seed for the mallee-fowl in the small 'island' reserves which will eventually be all that will be left of the formerly widespread mallee communities.

Some of the management problems faced by 'island' reserves may be classed as external, such as hydrology, atmospheric and water pollution, soil erosion, wind damage, weeds, vermin, pathogens, fire, etc. On a continent as dry as Australia, the hydrological relations of the total landscape determine the nature of the various ecosystems found in the region. Water movement from adjacent areas will affect water tables which influence the development of closed-forests, coastal dune vegetation, mangrove vegetation, wet-heathlands, etc. (Specht, 1972; Specht, Salt & Reynolds, 1977). Changes in adjacent land-use, outside the control of the Conservation Authority, may affect the water supply to the conservation area. Unless controlled by an overall-planning authority, little co-operation between adjacent land owners can be expected in dealing with the wide range of external environmental impacts on the conservation area.

In the case of internal impacts, the conservation authorities have a little more control. Many of the serious environmental impacts appear to be solved by controlling man and his domestic animals. Problems of compaction, erosion, litter, weeds, pathogens, vermin, fire, etc. are in this category. However, there is always someone who flaunts the controls. it is thus necessary to initiate research programs which will solve the problems created by the minority. For example, small amounts of litter strewn along pathways through sclerophyll (heath) vegetation usually lead to an increase in nutrient level along the paths and the invasion of weeds and herbs which out-compete the native flora (Specht, 1963;

Heddle & Specht, 1975). In order to maintain the conservation objectives of these reserves, it is necessary to reduce the nutrient level, thus changing the balance back in favour of the sclerophyllous plants. It is believed that this can be achieved by the application of kaolinitic clays along the tracks (Specht, 1975*b*, 1980*d*).

Considerable ecological research is necessary to provide a background for wise long-range management programs. Part of this background can be developed by intensive studies of a single ecosystem over a long period of time (see Specht, Rayson & Jackman, 1958, and Specht, 1980*a*, for heathland vegetation; Hall, Specht & Eardley, 1964, for arid-zone vegetation; Ashton, 1976, for tall open-forest; Groves, 1979, for various grassland communities; Specht & Morgan, 1981, for examples from the humid to the arid zone).

Many of the changes which may occur within a conservation reserve due to man's impact are likely to be solved, however, by the knowledge not only of the long-term internal dynamics of the ecosystem, but also of the ecophysiological principles which maintain the equilibrium interface between that ecosystem and its neighbours (Specht, 1980*c*, 1981*b*). Until recently, there were many areas of natural or semi-natural vegetation where ecological studies could be undertaken. Much of this landscape has now disappeared, or has been incorporated into National Parks where research, in many States, is not permitted. It is thus essential that a series of research reserves such as Biosphere Reserves (UNESCO, 1974) or the Queensland Nature Reference Sites (H. Lavery & C.S. Sandercoe, personal communication) be created so that an adequate research background, on which wise management programs are based, can be developed.

This chapter was prepared with the assistance of State and Federal National Parks and Wildlife Authorities.

References

Ashton, D.H. (1976). The development of even-aged stands of *Eucalyptus regnans* F. Muell. in central Victoria. *Australian Journal of Botany*, *24*, 397–414.

Carnahan, J.A. (1976). Natural vegetation. In *Atlas of Australian Resources*, Second Series. Canberra: Department of National Resources.

Day, M.F. (1968). *National Parks and Reserves in Australia*. Australian Academy of Science Report No. 9.

Everist, S.L. & Webb, L.J. (1975). Two communities of urgent concern in Queensland: Mitchell grass and tropical closed-forests. In *A National System of Ecological Reserves in Australia*, ed. F. Fenner, pp. 39–52. Australian Academy of Science Report No. 19.

Fenner, F. (ed.) (1975). *A National System of Ecological Reserves in Australia.* Australian Academy of Science Report No. 19.

Frankel, O.H. (1970). Variation – the essence of life. *Proceedings of the Linnaean Society of New South Wales*, *95*, 158–69.

Frankenberg, J. (1971). *Nature Conservation in Victoria*, ed. J.S. Turner. Melbourne: The Victorian National Parks Association.

Frith, H.J. (1962). *The Mallee Fowl: The Bird That Builds an Incubator*. Sydney: Angus & Robertson.

Gill, A.M. (1981). Patterns and processes in open-forests of *Eucalyptus* in southern Australia. In *Australian Vegetation*, ed. R.H. Groves, pp. 152–176. Cambridge University Press.

Goldstein, W. (ed.) (1979). Australia's 100 years of National Parks. *Parks and Wildlife*, *2(3–4)*, 1–160. Sydney: National Parks and Wildlife Service.

Groves, R.H. (1979). The status and future of Australian grasslands. *New Zealand Journal of Ecology*, *2*, 76–81.

Hall, E.A.A., Specht, R.L. & Eardley, C.M. (1964). Regeneration of the vegetation on Koonamore Vegetation Reserve, 1926–1962. *Australian Journal of Botany*, *12*, 205–64.

Hartley, W. & Leigh, J. (1979). Plants at risk in Australia. *Australian National Parks and Wildlife Service Occasional Paper* No. 3.

Heddle, E.M. & Specht, R.L. (1975). Dark Island heath (Ninety-Mile Plain, South Australia). VIII. The effect of fertilizers on composition and growth, 1950–1972. *Australian Journal of Botany*, 23, 151–64.

IUCN (International Union for Conservation of Nature and Natural Resources) (1980). *World Conservation Strategy*, 1196 pp. Gland, Switzerland: IUCN.

Kitchener, D.J., Chapman, A., Dell, J., Muir, B.G. & Palmer, M. (1980). Lizard assemblage and reserve size and structure in the Western Australian wheat belt – some implications for conservation. *Biological Conservation*, *17*, 25–62.

Laut, P., Heyligers, P.C., Keig, G., Löffler, E., Margules, C., Scott, R.M. & Sullivan, M.E. (1977). *Environments of South Australia, Provinces 1–8 & Handbook*. Canberra: CSIRO, Australia, Division of Land Use Research.

Melville, R. (1975). The distribution of Australian relict plants and its bearing on angiosperm evolution. *Botanical Journal of the Linnaean Society*, *71*, 67–88.

Slatyer, R.O. (1975). Ecological reserves: size, structure and management. In *A National System of Ecological Reserves in Australia*, ed. F. Fenner, pp. 22–38. Australian Academy of Science Report No. 19.

Specht, R.L. (1963). Dark Island heath (Ninety-Mile Plain, South Australia). VII. The effect of fertilizers on composition and growth, 1950–1960. *Australian Journal of Botany*, *11*, 67–94.

Specht, R.L. (1970). Vegetation. In *The Australian Environment*, ed. G.W. Leeper, 4th edn (rev.), pp. 44–67. Melbourne: CSIRO, Australia & Melbourne University Press.

Specht, R.L. (1972). Water use by perennial, evergreen plant communities in Australia and Papua New Guinea. *Australian Journal of Botany*, *20*, 273–99.

Specht, R.L. (1975*a*). The report and its recommendations. In *A National System of Ecological Reserves in Australia*, ed. F. Fenner pp. 11–21. Australian Academy of Science Report No. 19.

Specht, R.L. (1975*b*). A heritage inverted: our flora endangered. *Search*, *6*, 472–7.

Specht, R.L. (1978). *In Wildness is the Preservation of the World*. Sixth Romeo Watkins Lahey Memorial Lecture. Brisbane: National Parks Association of Queensland.

Specht, R.L. (1980*a*). Responses of selected ecosystems: heathlands and related shrublands. In *Fire and the Australian Biota*, ed. A.M. Gill, R.H. Groves & I.R. Noble, pp. 395–415. Canberra: Australian Academy of Science.

Specht, R.L. (1980*b*). Ecophysiological principles determining the biogeography of major vegetation formations in Australia. In *Ecological Biogeography in Australia*, ed. A. Keast, pp. 299–332. The Hague: Junk.

Specht, R.L. (1980*c*). Evolution of the Australian flora: some generalisations. In *Ecological Biogeography in Australia*, ed. A. Keast, pp. 783–806. The Hague: Junk.

Specht, R.L. (1980*d*). Conservation: Australian heathlands. In *Ecosystems of the World*. Vol. 9B, *Heathlands and Related Shrublands*. ed. R.L. Specht, pp. 235–40. Amsterdam: Elsevier Scientific Publishing Company.

Specht, R.L. (1981*a*). Structural attributes – foliage projective cover and standing biomass. In *Vegetation Classification in the Australian Region*, ed. A.N. Gillison & D.J. Anderson, (in press). Canberra: CSIRO, Australia & Australian National University Press.

Specht, R.L. (1981*b*). Developments in terrestrial ecology in Australia. In *Handbook of Contemporary Developments in World Ecology*, ed. E.J. Kormondy & J.F. McCormick, (in press). Westport, Conn. USA: greenword press.

Specht, R.L. (1981*c*). Heathlands. In *Australian Vegetation*, ed. R.H. Groves, pp. 253–275. Cambridge University Press.

Specht, R.L. & Cleland, J.B. (1961). Flora conservation in South Australia. I. The preservation of plant formations and associations recorded in South Australia. *Transactions of the Royal Society of South Australia*, 85, 177–96.

Specht, R.L. & Cleland, J.B. (1964). Flora conservation in South Australia. II. The preservation of species recorded in South Australia. *Transactions of the Royal Society of South Australia*, 87, 63–92.

Specht, R.L. & Morgan, D.G. (1981). The balance between the foliage projective covers of overstorey and understorey strata in Australian vegetation. *Australian Journal of Ecology*, 6 (in press).

Specht, R.L., Rayson, P. & Jackman, M.E. (1958). Dark Island heath (Ninety-Mile Plain, South Australia). VI. Pyric succession: changes in composition, coverage, dry weight, and mineral nutrient status. *Australian Journal of Botany*, 6, 59–88.

Specht, R.L., Roe, E.M. & Boughton, V.H. (ed.) (1974). Conservation of major plant communities in Australia and Papua New Guinea. *Australian Journal of Botany Supplement* No. 7.

Specht, R.L., Salt, R.B. & Reynolds, S. (1977). Vegetation in the vicinity of Weipa, north Queensland. *Proceedings of the Royal Society of Queensland*, 88, 17–38.

UNESCO (1974). Criteria and guidelines for the choice and establishment of biosphere reserves. Final report. Paris: UNESCO Programme on Man and the Biosphere (MAB) No. 22.

Walker, D. & Singh, G. (1981). *Vegetation history. In Australian Vegetation*, ed. R.H. Groves, pp. 26–43. Cambridge University Press.

Webb, L.J. & Tracey, J.G. (1981). The rainforests of northern Australia. In *Australian Vegetation*, ed. R.H. Groves, pp. 67–101. Cambridge University Press.

AUTHOR INDEX

Page numbers in italic type are bibliographical references

INDEX OF SCIENTIFIC NAMES

SUBJECT INDEX

Aborigines, *see* Man, Aboriginal
Acacia
  adaptations to moisture stress,
    185, 199–200
  communities: distribution, 198, 200–2,
    211, 213–19, 221–2; utilisation, 200,
    210–11, 213, 223, 224
  low woodlands: of central and Western
    Australia, 213–19; of mulga, 213–19
  open-forests: conservation status, 403,
    406; of bendee, 203–4; of blackwood,
    209; of brigalow, 205–7 (interspersed
    with rainforest) 73, 205; of gidgee,
    207–9 (interspersed with *Astrebla*
    grassland) 209; of lancewood, 202–3;
    of northeastern Australia, 200–2,
    203–11; (their utilisation) 210–11; of
    southern Australia, 211–13 (their utili-
    sation) 213; on deep fine-textured al-
    kaline soils, 204–10; on shallow
    coarse-textured acid soils, 202–4
  shrublands: conservation status, 403,
    406; of *A. ancistrocarpa*, 222; of bas-
    tard mulga, 219; of central and
    Western Australia, 213–24; of kanji,
    222; of mulga, 213–19 (interspersed
    with Chenopod shrubland) 219 (inter-
    spersed with eucalypt woodland) 216
    (interspersed with mallee eucalypts)
    218, 222–3, 228; of sandhill mulgas,
    220–1; of snakewood, 221; of turpen-
    tine mulga, 219–20; of witchetty bush,
    220; utilisation, 223–4; with hummock
    grasses, 221–3; with tussock grasses,
    213–21

  woodlands: of boree, 210; of georgina
    gidgee, 210; of lancewood, 202–3; of
    myall, 212–13; of northeastern Austra-
    lia, 204–10 (their utilisation) 210–11; of
    southern Australia, 211–13 (their utilisa-
    tion) 213; of western myall, 212; on deep
    fine-textured alkaline soils, 204–10
Alien ferns, 47, 48
Alien flora:
  classification, 50–1;
  deficiencies in knowledge, 59–60;
  extent, 47, 49–50
Alien plants:
  definition, 44–5
  distribution in Australia, 46, 50, 56–8, 60
    (in relation to areas of origin) 56–8
    (in tall open-forests) 142
  origins, history and spread, 52–6; of
    Echinochloa spp., 54; of *Hordeum* spp.,
    54; of thistles, 54; of *Xanthium* spp.,
    53–4
  rates of spread, 56
  variation in species, 58–9
Alpine and sub-alpine vegetation
  Aborigines in, *see* Man, Aboriginal, in
    alpine and sub-alpine vegetation
  definition, 361
  distribution, 361–3, 365–7
  environments, 363–4
  floristic alliances, 365–7, 368
  grazing effects, *see* Grazing, in alpine and
    sub-alpine vegetation
  hydrology, 372
  national parks and nature conservation,
    373